Structured Matrices and Polynomials

Unified Superfast Algorithms

To My Parents

Victor Y. Pan

Structured Matrices and Polynomials

Unified Superfast Algorithms

Springer Science+Business Media, LLC

Victor Y. Pan
Department of Mathematics
and Computer Science
Lehman College, CUNY
Bronx, NY 10468-1590

Library of Congress Cataloging-in-Publication Data

Pan, Victor.
    Structured matrices and polynomials : unified superfast algorithms / Victor Y. Pan.
        p. cm.
    Includes bibliographical references and index.
    ISBN 978-1-4612-6625-9   ISBN 978-1-4612-0129-8 (eBook)
    DOI 10.1007/978-1-4612-0129-8
        1. Matrices–Data processing. 2. Polynomials–Data processing. I. Title.

    QA188.P364 2001
    512.9'434–dc21
                                                          2001035928
                                                          CIP

AMS Subject Classifications: Primary: 12Y05, 68Q40, 65F05, 65F10, 93B40, 47B35
                             Secondary: 65Y20, 68Q25,15A09, 47A57, 47A40, 93B25, 93B28

Printed on acid-free paper.
©2001 Springer Science+Business Media New York
Originally published by Birkhäuser Boston in 2001
Softcover reprint of the hardcover 1st edition 2001

ISBN 978-1-4612-6625-9   SPIN 10835910

Typeset by the author
Cover design by Mary Burgess, Cambridge, MA

9 8 7 6 5 4 3 2 1

Contents

Preface   ix

Glossary   xix

List of Tables   xxiii

List of Figures   xxv

1   **Computations with Structured Matrices. Introduction**   1
    1.1   Application areas, our subjects and objectives   1
    1.2   Four features of structured matrices   3
    1.3   Displacements (some history, definitions and examples)   4
    1.4   COMPRESS, OPERATE, DECOMPRESS   7
    1.5   Basic operations with compressed structured matrices   10
    1.6   Two classes of superfast algorithms. The case of Toeplitz-like matrices   12
    1.7   Unified algorithms and algorithmic transformations   13
    1.8   Numerical and algebraic versions of structured matrix algorithms.
          Flops and ops   15
          1.8.1   Numerical implementation   15
          1.8.2   Algebraic implementation   17
          1.8.3   Unification and separation of numerical and algebraic versions
                  of algorithms   17
          1.8.4   Computational complexity estimates   17
    1.9   Limitations of the unification   18
    1.10 Organization of the book   19
    1.11 Notes   19
    1.12 Exercises   21

2   **Toeplitz/Hankel Matrix Structure and Polynomial Computations**   23
    2.1   Definitions and preliminaries   23
    2.2   Fast Fourier transform (polynomial version)   25
    2.3   Fast Fourier transform (matrix version)   26
    2.4   Polynomial multiplication (with extensions)   27
    2.5   Polynomial division and triangular Toeplitz matrix inversion   30
    2.6   The algebra of f-circulant matrices. Diagonalizations by DFT's . . .   33

2.7   Algebra of polynomials modulo a polynomial and the Frobenius matrix
      algebra   36
2.8   Polynomial gcd and the (Extended) Euclidean Algorithm   40
2.9   Applications and extensions of the EEA: modular division, rational
      function reconstruction, Padé approximation, and linear recurrence span   46
2.10 Matrix algorithms for rational interpolation and the EEA   50
2.11 Matrix approach to Padé approximation   52
2.12 Conclusion   55
2.13 Notes   56
2.14 Exercises   57
2.15 Appendix. Pseudocodes   61

3   **Matrix Structures of Vandermonde and Cauchy Types and Polynomial and
    Rational Computations**   **73**
    3.1   Multipoint polynomial evaluation   73
    3.2   Modular reduction and other extensions   76
    3.3   Lagrange polynomial interpolation   78
    3.4   Polynomial interpolation (matrix method), transposed Vandermonde
          matrices, and composition of polynomials   80
    3.5   Chinese remainder algorithm, polynomial and rational Hermite
          interpolation, and decomposition into partial fractions   82
    3.6   Cauchy matrix computations. Polynomial and rational interpolation
          and multipoint evaluation   88
    3.7   Loss (erasure)-resilient encoding/decoding and structured matrices . .   93
    3.8   Nevanlinna–Pick interpolation problems   96
    3.9   Matrix Nehari Problem 100
    3.10 Sparse multivariate polynomial interpolation 101
    3.11 Diagonalization of Matrix Algebras. Polynomial Vandermonde
          Matrices and Discrete Sine and Cosine transforms 103
    3.12 Conclusion 108
    3.13 Notes 109
    3.14 Exercises 113

4   **Structured Matrices and Displacement Operators**   **117**
    4.1   Some definitions and basic results 118
    4.2   Displacements of basic structured matrices 120
    4.3   Inversion of the displacement operators 122
    4.4   Compressed bilinear expressions for structured matrices 125
    4.5   Partly regular displacement operators 134
    4.6   Compression of a generator 137
          4.6.1   SVD-based compression of a generator in numerical
                  computations with finite precision 137
          4.6.2   Elimination based compression of a generator in computations
                  in an abstract field 139
    4.7   Structured matrix multiplication 141

4.8   Algorithm design based on multiplicative transformation of operator
matrix pairs 144
4.9   Algorithm design based on similarity transformations of operator
matrices 147
4.10 Conclusion 148
4.11 Notes 149
4.12 Exercises 150

5   Unified Divide-and-Conquer Algorithm   **155**
5.1   Introduction. Our subject and results 155
5.2   Complete recursive triangular factorization (CRTF) of general matrices 156
5.3   Compressed computation of the CRTF of structured matrices 159
5.4   Simplified compression of Schur complements 161
5.5   Regularization via symmetrization 162
5.6   Regularization via multiplicative transformation with randomization . 164
5.7   Randomization for structured matrices 167
5.8   Applications to computer algebra, algebraic decoding, and numerical
rational computations 168
5.9   Conclusion 170
5.10 Notes 170
5.11 Exercises 172

6   Newton-Structured Numerical Iteration   **177**
6.1   Some definitions and preliminaries 178
6.2   Newton's iteration for root-finding and matrix inversion 179
6.3   Newton-Structured Iteration 181
6.4   Compression of the displacements by the truncation of their smallest
singular values 182
6.5   Compression of displacement generators of approximate inverses
by substitution 184
6.6   Bounds on the norms of the inverse operator 186
    6.6.1   Introductory comments 186
    6.6.2   Bounds via powering the operator matrices 187
    6.6.3   The spectral approach 187
    6.6.4   The bilinear expression approach 188
    6.6.5   Estimation using the Frobenius norm 188
6.7   Initialization, analysis, and modifications of Newton's iteration
for a general matrix 189
6.8   How much should we compress the displacements of the computed
approximations? 193
6.9   Homotopic Newton's iteration 195
    6.9.1   An outline of the algorithms 195
    6.9.2   Symmetrization of the input matrix 196
    6.9.3   Initialization of a homotopic process for the inversion of
general and structured matrices 196

6.9.4   The choice of step sizes for a homotopic process.
Reduction to the choice of tolerance values 198
6.9.5   The choice of tolerance values 199
6.9.6   Variations and parallel implementation of
the Newton-Structured Iteration 199
6.10 Newton's iteration for a singular input matrix 201
6.11 Numerical experiments with Toeplitz matrices 202
6.12 Conclusion 204
6.13 Notes 214
6.14 Exercises 215

7   Newton Algebraic Iteration and Newton-Structured Algebraic Iteration   219
7.1   Some introductory remarks 219
7.2   Newton's algebraic iteration: generic algorithm 220
7.3   Specific examples of Newton's algebraic iteration 223
7.4   Newton's algebraic iteration for the inversion of general matrices . . . 226
7.5   Newton and Newton-Structured Algebraic Iterations for characteristic
polynomials and the Krylov spaces 227
7.6   Extensions of Krylov space computation 229
7.7   Inversion of general integer matrices and solution of general integer
linear systems 230
7.8   Sequences of primes, random primes, and non-singularity of a matrix   232
7.9   Inversion of structured integer matrices 233
7.10 Conclusion 236
7.11 Notes 236
7.12 Exercises 237

Conclusion   **241**

Bibliography   **243**

Index   **271**

Preface

Research in mathematics can be viewed as the search for some hidden keys that open numerous scientific locks. In this book, we seek key techniques and ideas to unify various computations with matrices and polynomials, which are the oldest subjects in mathematics and computational mathematics and the backbone of modern computations in the sciences, engineering, and communication.

Four millennia ago Sumerians wrote the solutions of polynomial equations on clay tablets. Later the ancient Egyptians did so on papyrus scrolls [B40], [Bo68]. Likewise, the solution of linear systems of equations, the most popular matrix operation, can also be traced back to ancient times.

How frequently do these subjects enter our life? Much more frequently than one commonly suspects. For instance, every time we turn on our radio, TV, or computer, convolution vectors (polynomial products) are computed. Indeed, all modern communication relies on the computation of convolution and the related operations of discrete Fourier, sine, and cosine transforms. Experts also know that most frequently the practical solution of a scientific or engineering computational problem is achieved by reduction to matrix computations.

Why are our two subjects put together? What do matrices and polynomials have in common?

Mathematicians know that matrices and polynomials can be added and multiplied together. Moreover, matrices endowed with certain structures (for instance, matrices whose entries are invariant in their shifts in the diagonal or antidiagonal directions) share many more common features with polynomials. Some basic operations with polynomials have an equivalent interpretation in terms of operations with structured matrices. A number of examples of this duality is covered in the book.

What is the impact of observing the correlation between matrices and polynomials? Besides the pleasure of having a unified view of two seemingly distinct subject areas, we obtain substantial improvement in modern computations. Structured matrices often appear in computational applications, in many cases along with polynomial computations. As a demonstration, in Sections 3.7–3.9 we cover a simple application of structured matrices to loss-resilient encoding/decoding and their well-known correlation to some celebrated problems of rational interpolation and approximation.

Applied linear algebra versus computer algebra

What is amazing — in view of the close ties between polynomial and structured matrix computations — is that applied mathematicians largely study the two subjects quite independently. The community of applied linear algebra, which studies computations with matrices and benefits from these studies, interacts very little with the computer algebra community which studies computations with polynomials and benefits from those studies. They constitute two distinct groups of people with distinct traditions, principles, and methods of working.

Matrix computation (applied linear algebra) people rely on numerical computation with rounding to a fixed precision. This enables faster computation using a small amount of computer memory. One must, however, take special care about rounding errors. The propagation of such errors in computational processes should be restricted to keep the output meaningful. This requirement has led to the advancement of error and perturbation analysis, approximation theory, and various techniques of algorithm design that stabilize computations numerically.

Computer algebra people are successfully exploiting and exploring an alternative path: error-free symbolic computations. This path requires more computer time and memory and thus is more expensive than numerical computation. The main advantage of symbolic computations is having completely reliable output. Various advanced mathematical tools have been developed to support this direction. Typical important examples are the transition to computations in finite fields, the Chinese remainder algorithm, and the p-adic Newton–Hensel lifting algorithms. Computations with polynomials make up much of computer algebra.

Polynomial and structured matrix computations combined

One of our goals is to reveal the correlation between computations with polynomials and structured matrices, continuing the line of the survey paper [P92a] and the book [BP94]. The expected impact includes better insight into both subjects and the unification of successful techniques and algorithms developed separately for each.

We study this correlation and its impact quite systematically in Chapters 2 and 3. This enables us to cover a substantial part of computer algebra using structured matrices. We observe close ties between structured matrices and the Nevanlinna–Pick and Nehari problems of rational interpolation and approximation. These celebrated algebraic problems allow numerical solution via reduction to matrix computations. Thus they may serve as a natural bridge between polynomial and matrix computations.

The displacement rank approach

Apart from unifying the study of matrices and polynomials, we focus on the design of effective algorithms unified over various classes of structured matrices. Our basic tool is the *displacement rank approach* to computations with structured matrices. The idea is to represent these matrices by their displacements, that is, the images of special displacement operators applied to the matrices. The displacements are defined by only small number of parameters, and the matrices can be recovered easily from their displacements.

The displacement rank approach consists of compression and decompression stages (the back and forth transition between matrices and their displacements) with dramatically simplified computations in-between (memory and time intensive computations with matrices are replaced with dramatically simplified operations with their displacements).

Some history and the displacement transformation approach to unification

The displacement rank approach to computations with structured matrices became celebrated after the publication of the seminal paper [KKM79], where some basic underlying results were presented for the class of Toeplitz-like matrices. Subsequent extensions to other classes of structured matrices followed [HR84], [GKK86], [GKKL87]. This was first done separately for each type of matrix structure, but in the paper [P90] we pointed out that the four most important classes of structured matrices, having structures of Toeplitz, Hankel, Vandermonde, and Cauchy types, are closely related to each other via their associated displacement operators.

For each of these four classes of matrices, we showed sample transformations into the three other classes by transforming the associated displacement operators and displacements. We proposed using such techniques systematically as a means of extending any efficient algorithm, developed for one of these four classes, to the other three classes. Later the approach proved to be effective for practical computations, such as the direct and iterative solution of Toeplitz/Hankel-like linear systems of equations (via the transformation of the displacement operators associated with their coefficients matrices to the ones of Cauchy type) [H95], [GKO95], [H96], [KO96], [HB97], [G97], [HB98], [Oa], numerical polynomial interpolation and multipoint evaluation [PLST93], [PZHY97], [PACLS98], [PACPS98], and algebraic decoding [OS99/OS00, Section 6]. In Chapters 4 and 5, we show applications of this approach for avoiding singularities in matrix computations and for accelerating numerical rational interpolation (Nevanlinna–Pick and Nehari problems). The list of applications of our approach is far from exhausted. The value of the method was widely recognized but its best known historical account in [G98] omits the source paper [P90]. Recent rediscovery of the method and techniques in [OS99/00], Section 6, caused even more confusion.

Symbolic operator approach to unification

An alternative and complementary method for unifying matrix structure is the *symbolic operator approach*. This approach also unifies numerical and algebraic algorithms. The matrix is associated with a largely unspecified displacement operator, and the high level description of the algorithms covers operations with displacements assuming their cited basic properties. This is the main framework for our presentation of some highly effective algorithms. The algorithms are completly specifiedwhen a structured input matrix, an associated displacement operator, and the rules of algebraic and numerical implementation are fixed. The specification is not our primary goal in this book, but we give some comments. We frequently combine the symbolic operator approach with displacement transformation techniques; we specify the results on the matrix structures. We also comment on the impact of our algorithms on the problems of computer algebra and numerical rational interpolation and approximation.

Unified superfast algorithms

How large is the gain of the displacement rank approach, based on shifting from matrices to their displacements?

An $n \times n$ structured matrix M has n^2 entries, all of which may be distinct, but its displacement may have small rank r, say equal to 1, 2 or 3, even where n is very large. Then the matrix can be represented with many fewer from rn to $2rn$ parameters.

Similarly, the solution of a structured linear system of n equations, $M\mathbf{x} = \mathbf{b}$, can also be accelerated dramatically. *Classical* Gaussian elimination uses the order of n^3 operations but has *fast* versions for structured matrices, running in time of the order of n^2, where numerical stabilization with pivoting is also incorporated [H95], [GKO95], [G97]. In Chapters 5–7, we cover *unified superfast algorithms*, which support the running time bound of the order of $r^2 n \log^2 n$ and which we present in a unified way for various matrix structures. The latter time bound is *optimal* up to the (small) factor of $r \log^2 n$.

Two technically distinct classes of superfast algorithms are shown to complement each other. Divide-and-conquer algorithms are covered in Chapter 5, and Newton's iteration in Chapters 6 and 7. In both cases the algorithms can be applied numerically, with rounding to a fixed finite precision, and symbolically (algebraically), with infinite precision. We do not develop these implementations but give some comments and refer the reader to http://comet.lehman.cuny.edu/vpan/newton for a numerical version of Newton's iteration for structured matrices.

The presented unified superfast algorithms are immediately extended to many important computational problems having ties to structured matrices. We cover some applications to computer algebra and the Nevanlinna–Pick and Nehari problems of numerical rational computations. The correlation to structured matrices enables better insight into both areas and an improvement of numerical implementation of known solution algorithms. For the Nevanlinna–Pick and Nehari problems, their reduction to computations with structured matrices is the only known way to yield a superfast and therefore nearly optimal (rather than just fast) solution, which is unified for several variations of these problems.

Our presentation

Our primary intended readership consists of researchers and algorithm designers in the fields of computations with structured matrices, computer algebra, and numerical rational interpolation, as well as advanced graduate students who study these fields. On the other hand, the presentation in the book is elementary (except for several results that we reproduced with pointers to source papers or books), and we only assume superficial knowledge of some fundamentals from linear algebra. This should make the book accessible to a wide readership, including graduate students and new researchers who wish to enter the main disciplines of our study: computations with structured matrices and polynomials. Examples, tables, figures, exercises of various levels of difficulty, and sample pseudocodes in Section 2.15 should facilitate their efforts.

Actually, the author was quite attracted by a chance to present advanced hot topics and even new results in a structured form. Most of the material is covered with proofs, derivations, and technical details, but we completely avoid long proofs. Wherever we omit them, we supply the relevant bibliography in the Notes at the end of each chapter. In particular, this applies to Newton's numerical iteration (Chapter 6) and the reduction of Nevanlinna–Pick and Nehari problems to matrix computations (Sections 3.8 and 3.9). Other topics are much more self-contained. We hope that the inclusion of details and some well-known definitions and basic results will not turn off more advanced readers who may just skip the elementary introductory parts of the book. Similarly, many readers may focus on the description of the main algorithms and skip some details of their analysis and the complexity estimates.

In the Notes, we cite the most relevant related works, reflecting also the earlier non-unified versions of the presented algorithms. The reader should be able to trace the most imporant related works from the cited bibliography. We apologize for any omission, which is inavoidable because of the huge and rapidly growing number of publications on computations with structured matrices.

To keep our presentation unified, we omit the eigenproblem for structured matrices and do not study some important matrix classes (such as multilevel matrices, banded matrices, infinite but finitely generated structured matrices, and block banded block Toeplitz matrices with Toeplitz blocks), whose treatment relies on distinct techniques. We give only brief comments on the important issues of numerical stability and parallel implementation of the presented algorithms, sending the reader to the relevant bibliography, and we leave the topic of data structures to more introductory texts.

Chapters 2 and 3 and the first sections of Chapter 7 overlap with some expository material on polynomial and Toeplitz matrix computations in the book [BP94], but we extend this material substantially, present it more clearly and more systematically, and show the correlation between computations with polynomials and structured matrices more extensively. Otherwise, most of our presentation is from journals and proceedings articles. Furthermore, several new unpublished results are included. Chapters 4–7 unify and extend scanty preceding works mostly devoted to some specified classes of structured matrices. Chapter 5 has a very preliminary exposition in the Proceedings paper [OP98] and a more developed Proceedings version in [P00]. The Proceedings papers [PR01], [PRW00], and [P01] are the predecessors of the first five sections of Chapter 6; they have no substantial overlap with the last six sections of the chapter. Chapter 7 overlaps with [P92] and [P00b]. In the last three chapters and also in Sections 3.6, 3.7, and 4.4, we present several novel techniques and algorithms and yield new record estimates for the arithmetic and Boolean complexity of some fundamental computations with structured matrices as well as for the cited problems of rational interpolation and approximation.

Selective reading

Despite our goal of a unified study, we structure the book to encourage selective reading of chapters and even sections. This makes the book accessible to graduate student, yet less boring for experts with specialized interests. Guiding graphs, titles, and figures help the selection. In particular, we give the following guiding graph for selective reading of chapters.

Guiding Graph for Selective Reading of Chapters

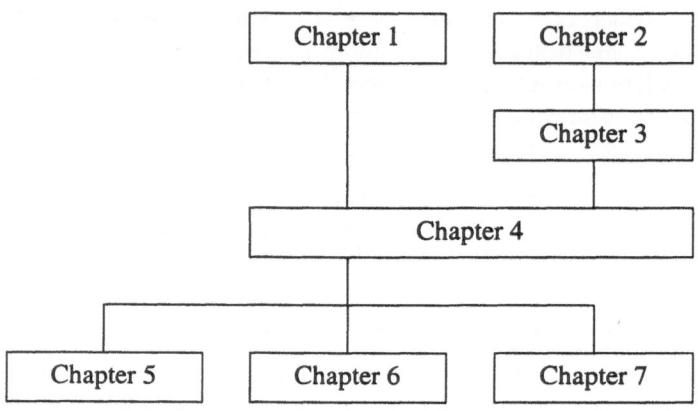

Let us also give some guiding comments (also see Section 1.10). In Chapters 2, 3, and 7, we assume an interest on the part of the readers in computer algebra problems. Passing through Chapters 2 and 3, these readers can be motivated to study the superfast algorithms of Chapter 5 but may decide to skip Chapter 6. On the contrary, readers who come from applied linear algebra and have no particular interest in computer algebra may focus on Chapters 4–6 and skip Chapters 2 and 3, except for some basic facts, the estimates collected in Table 1.2, and the definitions reproduced in Section 4.1. Furthermore, either Chapters 5 or 6 can be read independently of one another. In the beginning of Chapters 3–7 we display graphs for selective reading within each of these chapters. We propose that all sections of Chapters 1 and 2 be read in succession.

Chapter 1 outlines our main subjects and should serve as an introduction to the book. Technically, however, it can be skipped, except for some basic results, examples, and definitions in Sections 1.3–1.5, for which we give pointers whenever we use them.

Those readers whose interests are restricted to the tangential Nevanlinna–Pick and matrix Nehari problems of rational interpolation and approximation or to the loss-resilient encoding/decoding may start their reading with the respective sections of Chapter 3 and then if necessary may follow pointers and cross references to other parts of the book.

Summary of the book

Let us briefly summarize what we cover:

1. Unification of studies in the areas of:

    (a) computations with structured matrices and polynomials,

    (b) computer algebra and numerical linear algebra,

    (c) matrix structures of various classes,

    (d) structured matrices and Nevanlinna–Pick and Nehari problems.

2. Fundamental techniques:

    (a) displacement rank approach,

    (b) algorithmic transformation techniques,

    (c) divide-and-conquer method and recursive matrix factorization,

    (d) Newton's iteration in numerical and algebraic versions.

3. Superfast and memory efficient algorithms for several fundamental problems, including new algorithms and new derivations and analysis of some known algorithms for

    (a) structured matrix computations,

    (b) numerical rational interpolation and approximation,

    (c) loss-resilient encoding/decoding,

    (d) other areas of computer algebra.

Several unpublished results are included, in particular in Chapters 5–7 and in Sections 3.6, 4.4, and 4.6.2. (See more details in the Notes following each chapter).

The unification approach reflects the personal taste of the author whose primary interests include numerical linear algebra, polynomial computations, and algorithm analysis. He tried to focus on the key ideas and techniques of these areas. He believes, however, that the unification approach is beneficial for the subjects covered, since it provides for a deeper understanding of the power and the deficiencies of the solution algorithms.

This book should invite researchers (and perhaps some novices) to explore some subjects further. In particular, the following areas seem to be widely open to new research and/or implementation of recent algorithms:

- applications of the displacement transformation approach, including randomized transformations,

- analysis and implementation of the Newton-Structured Iteration and its generalizations and variations, including scaled versions, extensions to other residual correction methods, and particularly heuristic variations,

- complexity analysis and numerical analysis of the algorithms of Chapters 4 and 5 and their implementation and applications to specific classes of structured matrices,

- elaboration and implementation of the presented superfast algorithms for the solution of specific problems of encoding/decoding, computer algebra, and rational interpolation and approximation,

- parallel implementation of the presented algorithms (see [BP94, Chapter 4] and [P96], [P00b]).

Taking into account scientific and applied interest in topics covered, this book should fill a substantial void in the market. On the other hand, we hope that the elementary presentation will attract new people to the subjects and will help unify the efforts of researchers in several covered areas.

Acknowledgements

My involvement in the subjects of this book was motivated by my joint research with M. Abutabanjeh, A. Atinkpahoun, D. A. Bini, D. Bondyfalat, S. Branham, H. Brönnimann, H. Cebecioğlu, Z. Chen, J. W. Demmel, O. Dias (Tiga), I. Z. Emiris, Z. Galil, L. Gemignani, Y. Han, X. Huang, E. Kaltofen, M.-H. Kim, M. Kunin, J. Laderman, E. I. Landowne, K. Li, Y. Lin-Kriz, E. Linzer, W. L. Miranker, B. Mourrain, B. Murphy, V. Olshevsky, S. Pion, F. P. Preparata, S. Providence, Y. Rami, J. H. Reif, R. E. Rosholt, O. Ruatta, A. Sadikou, R. Schreiber, H. X. Sha, D. Shallcross, I. Sobze, C. Stewart, S. R. Tate, X. Wang, Y. Yu, and A. Zheng. I am grateful to all of them and to many other senior and junior colleagues, for their stimulating interest in my work, (p)reprints, support, discussions, and/or helpful comments on the subjects of this book, in particular to A. B. Borodin, B. Codenotti, S. A. Cook, R. M. Corless, G. Cybenko, D. Dobkin, W. Eberly, C. M. Fiduccia, P. Flajolet, A. Galligo, J. von zur Gathen, G. Heinig, T. Kailath, I. Kaporin, D. E. Knuth, Y. N. Lakshman, L. A. Levin, R. J. Lipton, L. Lovász, R. C. Melville, A. M. Odlyzko, J. Renegar, M.-F. Roy, B. Salvy, I. Shparlinski, M. Shub, S. Smale, H. J. Stetter, J. F. Traub, E. E. Tyrtyshnikov, M. Van Barel, P. Van Dooren, G. Villard, S. M. Watt, and S. Winograd. I specially thank Dario A. Bini for long time collaboration, Ioannis Z. Emiris and Bernard Mourrain for productive years of joint research on algebraic computation, Vadim Olshevsky for his enthusiasm about the papers [PZ00] and [PZHY97], suggestion of their speedy submission, his most valuable part in our joint effort on their extension, and advising me on the Nevanlinna–Pick and Nehari problems, both Georg Heinig and Vadim Olshevsky for their early appreciation of the displacement transformation method and successful demonstration of its power, and Erich Kaltofen and Joachim von zur Gathen for many informative discussions on algebraic complexity.

Very encouraging for me were the invitations to present the material of Chapters 4-6 at the Joint 1999 AMS/IMS/SIAM Summer Research Conference "Structured matrices in Operator Theory, Numerical Analysis, Control, Signal and Image Processing," organized by V. Olshevsky, in Boulder, Colorado, June 27–July 1, 1999; at the structured matrix session of G. Heinig and S. Serra Capizzano of the Second Conference on Numerical Analysis and Applications, organized by P. Yalamov in Rousse, Bulgaria, in June 2000; of Chapters 4, 5, and 7 at the Annual IMACS/ACA Conference, organized by N. Vassiliev in St. Petersburg, Russia, in June 2000; and of Chapters 4 and 6 at the P. Dewilde's session at the 14th International Symposium on Mathematical Theory of Networks and Systems in Perpignan, France, in June 2000; acceptance of my papers [P00] on Chapters 4 and 5 and [P00c] on Section 3.7 by the Program Committees of the ACM-SIAM SODA 2000 and the ACM-SIGSAM ISSAC 2000, respectively; the reception of my talks by the audiences at all these meetings; and the invitations to speak at the minisymposium on fast matrix algorithms (organized by S. Chandrasekaran and M. Gu) at the SIAM Annual Meeting in San Diego, California, in July 2001; at the Joint 2001 AMS/IMS/SIAM Summer Research Conference on Fast Algorithms, organized by V. Olshevsky in S. Hadley, Massachusetts, in August 2001; and at the special session on structured matrices organized by D. Bini and T. Kailath at the Joint AMS-UMI Conference in Pisa, Italy, in June 2002. T. Kailath's invitation to participate in the volume [KS99] revived my interest in Newton's iteration.

Besides this ample encouragement, writing of this book was supported by NSF Grant CCR 9732206, PSC CUNY Awards 61393-0030 and 62435-0031, and George N. Shuster Fellowship. Xinmao Wang and Hülya Cebecioğlu provided superb assistance with preparation of the book. I am grateful to Ioannis Z. Emiris and Robert C. Melville for prompt but careful review of the draft of the book and most valuable comments; to Anh Nguyen, Samer Salame, and particularly Elliott I. Landowne, Hülya Cebecioğlu, and Xinmao Wang for pointing out several typos in the manuscript; to Brian Murphy and Rhys Rosholt for many helpful comments on the style of the presentation in the original drafts of the Preface and Chapter 1; to Mikhail Kunin and Rhys Rosholt for carrying out the experiments on Newton's iteration, and to Brian Murphy, Rhys Rosholt, Samer Salame, Alexandre Ushakov, and Xinmao Wang for their assistance with writing pseudocodes. I greatly appreciate full cooperation of Birkhäuser/Springer, in particular most helpful collaboration with Ann Kostant, the timely liaison service of Tom Grasso and Nick Burns, the beautiful design of the book cover by Mary Burgess, and the excellent job of the copyeditor. It was both most vital and pleasant for me to have ample unconditional support of my research at Lehman College and the Graduate Center of CUNY as well as my home.

Glossary of mathematical notation

Notation	Explanation
$\mathbb{F}$	an arbitary field
$\mathbb{C}$	the field of complex numbers
$\mathbb{R}$	the field of real numbers
$M \in \mathbb{F}^{m \times n}$	an $m \times n$ matrix with entries from field $\mathbb{F}$
$M = (m_{i,j})_{i,j=0}^{m-1,n-1}$	$m \times n$ matrix with the (i, j)-th entry $m_{i,j}$
$\mathbf{v} \in \mathbb{F}^{k \times 1}$	a k-dimensional column vector with coordinates from filed $\mathbb{F}$
$\mathbf{v} = (v_i)_{i=0}^{k-1}$	k-dimensional vector with the i-th coordinate v_i, $i = 0, \ldots, k-1$
$S = \{s_1, \ldots, s_k\}$	the set of k elements $s_1, \ldots, s_k$ (not necessarily distinct)
$(W_1, \ldots, W_k)$	$1 \times k$ block matrix with blocks $W_1, \ldots, W_k$
$(\mathbf{w}_1, \ldots, \mathbf{w}_k)$	$n \times k$ matrix with the columns $\mathbf{w}_1, \ldots, \mathbf{w}_k$, where n is the dimension of the column vectors
$W^T, \mathbf{v}^T$	the transpose of a matrix W, the transpose of a vector $\mathbf{v}$
$W^*, \mathbf{v}^*$	the Hermitian (conjugate) transpose of a matrix, the Hermitian (conjugate) transpose of a vector
$\mathbf{e}_i$	the i-th coordinate vector
$I = I_n$	the $n \times n$ identity matrix
$0, O$	null matrices
$D(\mathbf{v}), \text{diag}(\mathbf{v})$	the diagonal matrix with a vector $\mathbf{v}$ defining its diagonal
$T = (t_{i-j})_{i,j=0}^{n-1}$	an $n \times n$ Toeplitz matrix

Z, Z_0	an $n \times n$ matrix with the (i, j)-th entry 1 for $i = j + 1$ and 0 for all other pairs i, j, the unit lower triangular Toeplitz matrix
Y_{00}	$Z + Z^T$
Y_{11}	$Z + Z^T + e_0 e_0^T + e_{n-1} e_{n-1}^T$
Z_f	$Z + f e_0 e_{n-1}^T$, the unit f-circulant matrix
$Z(\mathbf{v}), Z_0(\mathbf{v})$	$\sum_{i=0}^{n-1} v_i Z^i$, the lower triangular Toeplitz matrix with the first column $\mathbf{v}$
$Z_f(\mathbf{v})$	$\sum_{i=0}^{n-1} v_i Z_f^i$, the f-circulant matrix with the first column $\mathbf{v}$
$H = (h_{i+j})_{i,j=0}^{n-1}$	an $n \times n$ Hankel matrix
J	$(e_{n-1}, \ldots, e_0)$, the $n \times n$ unit Hankel matrix, the reflection matrix
$V(\mathbf{t}) = (t_i^j)_{i,j=0}^{n-1}$	the $n \times n$ Vandermonde matrix with the second column $\mathbf{t}$
$C(\mathbf{s}, \mathbf{t}) = (\frac{1}{s_i - t_j})_{i,j=0}^{n-1}$	the $n \times n$ Cauchy matrix defined by two vectors $\mathbf{s}$ and $\mathbf{t}$
$K(M, \mathbf{v}, k)$	$(M^i \mathbf{v})_{i=0}^{k-1}$, the $n \times k$ Krylov matrix with columns $M^i \mathbf{v}$, $i = 0, \ldots, k-1$
L	linear operator
A, B	operator matrices
G, H	generator matrices, $GH^T = L(M)$
$\nabla_{A,B}$	a Sylvester type displacement operator
$\nabla_{A,B}(M)$	$AM - MB$, Sylvester type displacement of M
$\Delta_{A,B}$	a Stein type displacement operator
$\Delta_{A,B}(M)$	$M - AMB$, Stein type displacement of M
ω, ω_n	a primitive n-th root of 1
$\mathbf{w}$	the vector $(\omega^i)_{i=0}^{n-1}$ of the n-th roots of 1
Ω	$(\omega^{ij})_{i,j=0}^{n-1}$, the scaled matrix of the DFT
$\text{DFT}(\mathbf{p})$	$(\sum_{i=0}^{n-1} p_i \omega_n^{ik})_{k=0}^{n-1}$, the discrete Fourier transform of a vector $\mathbf{p}$
DFT_n	computation of $\text{DFT}(\mathbf{p})$
$p(x) \bmod u(x)$	$p(x)$ modulo $u(x)$, the remainder of the division of univariate polynomials $p(x)$ by $u(x)$

F_u	the Frobenius (companion) matrix of a monic polynomial $u(x)$
A_u	the algebra of polynomials modulo a monic polynomial $u(x)$
trace(M)	$\sum_i m_{i,i}$, the trace of a matrix $M = (m_{i,j})$
det(M)	the determinant of a matrix M
rank(M)	the rank of a matrix M
log	logarithm
gcd(u, v)	greatest common divisor of polynomials $u(x)$ and $v(x)$
deg	degree
$u_t(x)$	$\prod_{i=0}^{n-1}(x - t_i)$, the monic polynomial with roots t_i
$\mathbf{u_t}$	the coefficient vector of the polynomial $u_t(x) - x^n$
$\lfloor x \rfloor, \lceil x \rceil$	two integers closest to a real x, $\lfloor x \rfloor \leq x \leq \lceil x \rceil$
n_+	$2^{\lceil \log_2 n \rceil}$
$M(n)$	arithmetic cost (in ops) of multiplication of two polynomials modulo x^{n+1}
$v_{r,n}(L)$ $m_{r,r_1,n}(L)$	arithmetic cost of multiplication of an $n \times n$ matrix by a vector and by a matrix (r and r_1 are the lengths of the L-generators for the matrices involved)
S and C	the matrices of the Discrete Sine and Cosine transforms (see Tables 3.1-3.5)
$\mathbf{Q}$	$(Q_i(x))_{i=0}^{n-1}$, a polynomial basis
$V_{\mathbf{Q}}(\mathbf{z})$	$(Q_i(x_j))_{i,j=0}^{n-1}$, a polynomial Vandermonde matrix
$M^{(k)}$	the $k \times k$ northwestern (leading principal) submatrix of a matrix M
$S^{(k)}(M) = S(M^{(k)}, M)$	the Schur complement of the block $M^{(k)}$ in a matrix M
$\sigma_j = \sigma_j(M)$	the j-th largest singular value of a matrix M
λ_j	an eigenvalue of a matrix M
$\|M\|_l$	the l-norm of a matrix M for $l = 1, 2, \infty$
$\|M\|_F$	the Frobenius norm of a matrix M
$v^- = v_{r,l}(L^{-1})$	the l-norm of a linear operator L^{-1} (l stands for $1, 2, \infty$ or F)
$\kappa(M) = \text{cond}_2(M)$	$\sigma_1(M)/\sigma_r(M)$, where $r = \text{rank}(M)$, the condition number of M
$e_{l,i}$	error norm $\|M^{-1} - X_i\|_l$

$\hat{e}_{2,i}$ $\|M^{-1} - Y_i\|_2$

$r_{l,i}$ residual norm $\|I - MX_i\|_l$

$\hat{r}_{2,i}$ $\|I - MY_i\|_2$

$\mu(q)$ $O((q \log q) \log \log q)$, the bit-operations cost of an
                               arithmetic operation modulo $2^q + 1$ over the integers

List of Tables

1.1 Four classes of structured matrices 3
1.2 Parameter and flop count for matrix representation and multiplication by a vector 4
1.3 Pairs of operators $\nabla_{A,B}$ and structured matrices 6
1.4 Operator matrices for the matrix product (Sylvester type operators). . 11

3.1 The matrices of eight forward and inverse discrete trigonometric transform ($\mu_0 = \mu_n = 1/\sqrt{2}$, $\mu_i = 1$ for $0 < i < n$.) 108
3.2 Polynomials $Q_k(y)$, $k = 0, \ldots, n - 1$ (T_i and U_i are the i-th degree Chebyshev polynomials of the first and second kinds.) 109
3.3 Polynomials $Q_n(x)$ and vectors $\mathbf{z}$ of their zeros 110
3.4 Factorization of the matrices T_Q of trigonometric transforms into the product of diagonal matrices W_Q and polynomial Vandermonde matrices $V_Q = V_Q(\mathbf{z})$ 111
3.5 Associated tridiagonal confederate matrices H_Q 112

4.1 Four classes of structured matrices 118
4.2 Parameter and flop count for representation of a structured matrix and for its multiplication by a vector 118
4.3 Some pairs of operators $\nabla_{A,B}$ and associated structured matrices . . . 120
4.4 Upper bounds on the growth of the L-rank in the transformations with Vandermonde multipliers. 146

6.1 Upper bounds on the 2-norms of the inverse displacement operators. . 188
6.2 Upper bounds on the Frobenius norms of the inverse displacement operators 189
6.3 Lower and upper bounds for the Frobenius norms of the inverse displacement operators. 189
6.4 Results for the tridiagonal Toeplitz matrices M of class 1 with the entries on the main diagonal equal to 4 and on the first super and subdiagonals equal to 1. 205
6.5 Results for the tridiagonal Toeplitz matrices M of class 1 with the entries on the main diagonal equal to 2 and on the first super and subdiagonals equal to -1. 205

6.6   The matrices $M = \left(\frac{1}{|i-j|}\right)_{i,j}$ of class 2 206

6.7   Random Toeplitz matrices M of class 3, $n = 100$ (non-homotopic iteration) 206

6.8   Symmetric indefinite Toeplitz matrices $\left(\begin{smallmatrix} 0 & M \\ M^T & 0 \end{smallmatrix}\right)$ for a random Toeplitz matrix M of class 3, $n = 100$ 206

6.9   Symmetric positive definite Toeplitz matrices $M^T M$ for a random Toeplitz matrix M of class 3, $n = 100$ 207

6.10 Results for random symmetric positive definite Toeplitz matrices M of class 4, $n=50$ 207

6.11 Results for random symmetric positive definite Toeplitz matrices M of class 4, $n=100$ 208

6.12 Results for random symmetric positive definite Toeplitz matrices M of class 4, $n=150$ 208

6.13 Results for random symmetric positive definite Toeplitz matrices M of class 4, $n=200$ 209

6.14 Results for random symmetric indefinite Toeplitz matrices M of class 5, $n=50$ 209

6.15 Results for random symmetric indefinite Toeplitz matrices M of class 5, $n=100$ 209

6.16 Results for random symmetric indefinite Toeplitz matrices M of class 5, $n=150$ 210

6.17 Results for random symmetric indefinite Toeplitz matrices M of class 5, $n=200$ 210

6.18 Results for random symmetric indefinite Toeplitz matrices M of class 5, $n=250$ 210

6.19 Results for random symmetric indefinite Toeplitz matrices M of class 5, $n=300$ 211

6.20 Results for homotopic and nonhomotopic processes applied to symmetric positive definite matrix obtained from the original by premultiplying with its transpose. 211

6.21 Maximal values of the residual norms for non-homotopic processes with the original, symmetrized double-sized, and symmetrized positive definite matrices, respectively. 212

List of Figures

2.1 Correlation among polynomial and matrix computations I.   34

2.2 Correlation among polynomial and matrix computations II.   50

3.1 Fan-in process, $m_j^{(h+1)} = m_{2j}^{(h)} m_{2j+1}^{(h)}$ 75

3.2 Fan-out process, $r_j^{(h)} = r_{\lfloor j/2 \rfloor}^{(h+1)} \bmod m_j^{(h)} = v(x) \bmod m_j^{(h)}$ 75

3.3 Correlation among polynomial and matrix computations III.   83

3.4 Correlation among polynomial and matrix computations IV.   89

3.5 Correlation among polynomial and matrix computations V.   93

3.6 Correlation of loss-resilient encoding decoding to structured matrix computations.   95

Chapter 1

Computations with Structured Matrices. Introduction

Summary   In this chapter we partly repeat but substantially extend the introductory description of our two main subjects sketched in the preface, that is, the subjects of unified superfast algebraic and numerical computations with structured matrices and the displacement rank approach. The latter subject is covered in some technical detail in Sections 1.2–1.5. The rest of the chapter is a non-technical but extensive introduction to unified superfast algorithms for computation with structured matrices. Further elaboration and technical details are given in Chapters 4–7.

1.1   Application areas, our subjects and objectives

Structured matrices (such as circulant, Toeplitz, Hankel, Frobenius, Sylvester, subresultant, Bezout, Vandermonde, Cauchy, Loewner, and Pick matrices) are omnipresent in algebraic and numerical computations in the various sciences, engineering, communication, and statistics.

Among the areas of application, we single out computations with polynomials and rational functions. In the survey [P92a] and the book [BP94], the correlation between fundamental computations with polynomials and structured matrices was examined systematically and demonstrated by many examples. In the next two chapters, we recall and extend these studies. Polynomial computation is a substantial part of computer algebra. Our exposition should provide a better insight into this domain and motivate the interest of computer algebraists in structured matrices. As a by-product, we show a simple improvement over the known methods for deterministic loss-resilient encoding/decoding based on the application of structured matrices.

On the other hand, structured matrices (particularly the ones with structure of Cauchy type) are closely related to the celebrated problems of numerical rational interpolation and approximation such as Nevanlinna–Pick and Nehari problems. Various effective approaches involving deep mathematical tools were proposed for the solution of these problems. Computationally, these approaches are similar to each other and to the MBA approach of recursive factorization of structured matrices, due to Morf [M74], [M80], and Bitmead and Anderson [BA80]. The papers [BGR90], [BGR90a], [GO93], and [GO94b] studied the reduction of the State-Space solution of the tangential and tangential boundary Nevanlinna–Pick problems and the matrix Nehari problem to the inversion or recursive factorization of the associated Cauchy-like matrices. In [OP98] a dramatic acceleration of the known solution algorithms for these problems was proposed based on this association. We outline and extend this work in Sections 3.8, 3.9, and Chapter 5.

Studying the close correlation between computations with polynomials and rational functions and computations with structured matrices should benefit both areas. For multiplication, division, multipoint evaluation, interpolation, and other basic computational problems with polynomials, we give the solution in terms of both polynomials and structured matrices. In some cases, the two resulting algorithms are computationally identical, that is, they compute the same output and intermediate values but have two different derivations and representations — one in terms of polynomials and the other in terms of structured matrices. Besides providing two distinct insights into the subject, having two approaches helps in finding various amendments, extensions and generalizations. In many cases, appealing to polynomial algorithms enables us to compute the solution faster, whereas the structured matrix approach leads to improving the numerical stability of the algorithms. For the Nevanlinna–Pick and Nehari problems, the structured matrix approach enables both unification of the solution algorithms and their dramatic acceleration.

Thus, we have two related goals of algorithm design:

1. To devise effective algorithms for polynomial and rational computations

and

2. To devise effective algorithms for computations with structured matrices.

The latter goal is achieved in Chapters 4–7, based on the reduction of computations with structured matrices essentially to multiplication of such matrices by vectors and their recursive compression. Nearly linear time algorithms for multiplication by vectors of Toeplitz, Hankel, Vandermonde, and Cauchy matrices ultimately reduce the task to polynomial multiplication, and further to discrete Fourier, sine and/or cosine transforms. These algorithms can be found in Chapters 2 and 3, which cover polynomial and rational computations. On the other hand, one of the results of our study in Chapters 2 and 3 is the reduction of the most fundamental polynomial and rational computations to computations with structured matrices, that is, the reduction of goal 1 to goal 2.

The compression algorithms (covered in Sections 4.6 and 4.7) ultimately reduce the compression problem again to multiplication of structured matrices by vectors and therefore again to discrete Fourier, sine and/or cosine transforms. These transforms are very fast and allow an effective parallel implementation; therefore, so do most computations with structured matrices. We briefly comment on the parallel complexity of the resulting algorithms in Sections 4.6 and 4.7 and in Chapters 6 and 7 but do not delve into the detailed study of the large subject of parallel computing, to which we refer the reader to [L92], [BP94, Chapter 4], [Q94], [P96], [P00b], [LP01], and references therein.

1.2   Four features of structured matrices

Table 1.1 represents the four most popular classes of structured matrices, and many other classes and subclasses are well studied, such as rationally generated Toeplitz and Hankel matrices, f-circulant Toeplitz matrices, banded Toeplitz matrices, block Toeplitz matrices, block matrices with Toeplitz, Hankel, or Vandermonde blocks; Sylvester, subresultant, Frobenius (companion), Bezout, Pick, Loewner, and polynomial Vandermonde matrices.

Structured matrices M

a) can be represented with only a small number of parameters,

b) can be multiplied by vectors fast, in nearly linear time,

c) have close algorithmic correlation to computations with polynomials and rational functions, in particular to their multiplication, division, interpolation, and multipoint evaluation, and

d) are naturally associated with linear *displacement operators* L (typically, incorporating the operators of shift and scaling) and can be recovered easily from the operators L and image matrices $L(M)$, called the *displacements* of M, which have smaller rank r.

Table 1.1: Four classes of structured matrices

Toeplitz matrices $(t_{i-j})_{i,j=0}^{n-1}$

$$\begin{pmatrix} t_0 & t_{-1} & \cdots & t_{1-n} \\ t_1 & t_0 & \ddots & \vdots \\ \vdots & \ddots & \ddots & t_{-1} \\ t_{n-1} & \cdots & t_1 & t_0 \end{pmatrix}$$

Hankel matrices $(h_{i+j})_{i,j=0}^{n-1}$

$$\begin{pmatrix} h_0 & h_1 & \cdots & h_{n-1} \\ h_1 & h_2 & \iddots & h_n \\ \vdots & \iddots & \iddots & \vdots \\ h_{n-1} & h_n & \cdots & h_{2n-2} \end{pmatrix}$$

Vandermonde matrices $(t_i^j)_{i,j=0}^{n-1}$

$$\begin{pmatrix} 1 & t_0 & \cdots & t_0^{n-1} \\ 1 & t_1 & \cdots & t_1^{n-1} \\ \vdots & \vdots & & \vdots \\ 1 & t_{n-1} & \cdots & t_{n-1}^{n-1} \end{pmatrix}$$

Cauchy matrices $\left(\frac{1}{s_i-t_j}\right)_{i,j=0}^{n-1}$

$$\begin{pmatrix} \frac{1}{s_0-t_0} & \cdots & \frac{1}{s_0-t_{n-1}} \\ \frac{1}{s_1-t_0} & \cdots & \frac{1}{s_1-t_{n-1}} \\ \vdots & & \vdots \\ \frac{1}{s_{n-1}-t_0} & \cdots & \frac{1}{s_{n-1}-t_{n-1}} \end{pmatrix}$$

Feature a) can be immediately seen for the matrices of Table 1.1. Indeed, Toeplitz, Hankel, Vandermonde, and Cauchy matrices are completely defined by the vectors $(t_i)_{i=1-n}^{n-1}$, $(h_i)_{i=0}^{2n-2}$, $\mathbf{t} = (t_i)_{i=0}^{n-1}$, and $\mathbf{s} = (s_i)_{i=0}^{n-1}$ and $\mathbf{t}$, respectively. In Table 1.2 and Chapters 2 and 3, we show features b) and c) for the same matrices. In the next sections and in Chapters 4–7, we use feature d) for specified linear operators L to extend the cited four basic classes having features a) and b) and manipulate efficiently the resulting structured matrices.

Table 1.2: Parameter and flop count for matrix representation and multiplication by a vector

Matrices M	Number of parameters per an $m \times n$ matrix M	Number of flops for computing $M\mathbf{v}$
general	mn	$2mn - n$
Toeplitz	$m + n - 1$	$O((m + n)\log(m + n))$
Hankel	$m + n - 1$	$O((m + n)\log(m + n))$
Vandermonde	m	$O((m + n)\log^2(m + n))$
Cauchy	$m + n$	$O((m + n)\log^2(m + n))$

1.3 Displacements (some history, definitions and examples)

Compared to general matrices, computations with structured matrices, if performed properly, require much less computer time and memory space. This is an important reason for the popularity of structured matrices. The key words here are "if performed properly," and this is a major focus of the book.

To properly perform computations with structured matrices, we explore and exploit the **displacement rank approach**. In presenting this, we follow and extend the seminal paper [KKM79], which defined matrices with the structures of Toeplitz type as the ones having small displacement rank.

Initially, the approach was intended for more restricted use. The paper [KKM79] "introduced the concept of displacement ranks, which serve as a measure of how 'close' to Toeplitz a given matrix is". According to the related paper [FMKL79, page 33], the idea was to underline the "shift invariance" (or "stationarity" or "homogeneity") feature of the matrices of "Toeplitz or displacement type, that is, of the form

$$R = [r_{i-j}], \quad 0 \le i, j \le N."$$

The idea, however turned out to be even deeper, more powerful, and more general than it may have seemed initially, and so we assume a more general *modern interpretation of the displacement of a matrix M as the image L(M) of an appropriate linear displacement operator L* applied to the matrix M and revealing its structure. Formally, we associate $m \times n$ structured matrices (having entries in a fixed field $\mathbb{F}$ or, say in the complex number field $\mathbb{C}$) with linear operators $L : \mathbb{F}^{m \times n} \to \mathbb{F}^{m \times n}$ of *Sylvester type*, $L = \nabla_{A,B}$:

$$L(M) = \nabla_{A,B}(M) = AM - MB,$$

and *Stein type*, $L = \Delta_{A,B}$:

$$L(M) = \Delta_{A,B}(M) = M - AMB,$$

for fixed pairs $\{A, B\}$ of *operator matrices*.

Theorem 1.3.1. $\nabla_{A,B} = A\Delta_{A^{-1},B}$ *if the operator matrix* A *is non-singular, and* $\nabla_{A,B} = -\Delta_{A,B^{-1}}B$ *if the operator matrix* B *is non-singular.*

For operator matrices we most frequently use the matrices Z_f, Z_f^T, and $D(\mathbf{v})$, where

Z_f is the unit f-circulant matrix,

$$
Z_f = \begin{pmatrix} 0 & & & f \\ 1 & \ddots & & \\ & \ddots & \ddots & \\ & & 1 & 0 \end{pmatrix}, \tag{1.3.1}
$$

f is any scalar,

Z_f^T is the transpose of Z_f, and

$D(\mathbf{v})$ is a diagonal matrix with diagonal entries $v_0, \ldots, v_{n-1}$,

$$
D(\mathbf{v}) = \operatorname{diag}(v_i)_{i=0}^{n-1} = \begin{pmatrix} v_0 & & & \\ & v_1 & & \\ & & \ddots & \\ & & & v_{n-1} \end{pmatrix}. \tag{1.3.2}
$$

Here and hereafter, the blank areas in matrix representation are assumed to be filled with zero entries.

The operators of Sylvester and Stein types can be transformed easily into one another if at least one of the two associated operator matrices is non-singular.

With each class of $m \times n$ structured matrices M, we associate linear operators $L = \nabla_{A,B}$ and $L = \Delta_{A,B}$ for which the rank r of the displacement $L(M)$ (called the *L-rank* or *displacement rank* of the matrix M) remains relatively small as the size of M grows large, say $r = O(1)$ or $r = o(\min\{m, n\})$. Examples 1.3.2–1.3.5 and Table 1.3 show some specific choices of operators for the four basic classes of structured matrices and for some related matrices.

Example 1.3.2. Toeplitz matrices. Let $T = (t_{i-j})_{i,j=0}^{n-1}$. Then $\nabla_{Z_1,Z_0}(T) =$

$$
Z_1 T - T Z_0 = \begin{pmatrix} t_{n-1} & t_{n-2} & \cdots & t_0 \\ t_0 & t_{-1} & \cdots & t_{1-n} \\ \vdots & \vdots & \ddots & \vdots \\ t_{n-2} & t_{n-1} & \cdots & t_{-1} \end{pmatrix} - \begin{pmatrix} t_{-1} & \cdots & t_{1-n} & 0 \\ \vdots & \ddots & \vdots & \vdots \\ t_{n-3} & \cdots & t_{-1} & 0 \\ t_{n-2} & \cdots & t_0 & 0 \end{pmatrix}
$$

$$
= \begin{pmatrix} t_{n-1}-t_{-1} & \cdots & t_1-t_{1-n} & t_0 \\ 0 & \cdots & 0 & t_{1-n} \\ \vdots & & \vdots & \vdots \\ 0 & \cdots & 0 & t_{-1} \end{pmatrix} \text{ is a rank 2 matrix.}
$$

Table 1.3: Pairs of operators $\nabla_{A,B}$ and structured matrices

operator matrices		class of structured	rank of
A	B	matrices M	$\nabla_{A,B}(M)$
Z_1	Z_0	Toeplitz and its inverse	≤ 2
Z_1	Z_0^T	Hankel and its inverse	≤ 2
$Z_0 + Z_0^T$	$Z_0 + Z_0^T$	Toeplitz+Hankel	≤ 4
$D(t)$	Z_0	Vandermonde	≤ 1
Z_0	$D(t)$	inverse of Vandermonde	≤ 1
Z_0^T	$D(t)$	transposed Vandermonde	≤ 1
$D(s)$	$D(t)$	Cauchy	≤ 1
$D(t)$	$D(s)$	inverse of Cauchy	≤ 1

Example 1.3.3. Hankel matrices. Let $H = (h_{i+j})_{i,j=0}^{n-1}$. Then $\nabla_{Z_1,Z_0^T}(H) =$

$$Z_1 H - H Z_0^T = \begin{pmatrix} h_{n-1} & h_n & \cdots & h_{2n-2} \\ h_0 & h_1 & \cdots & h_{n-1} \\ \vdots & \vdots & \ddots & \vdots \\ h_{n-2} & h_{n-1} & \cdots & h_{2n-3} \end{pmatrix} - \begin{pmatrix} 0 & h_0 & \cdots & h_{n-2} \\ 0 & h_1 & \cdots & h_{n-1} \\ \vdots & \vdots & \ddots & \vdots \\ 0 & h_{n-1} & \cdots & h_{2n-3} \end{pmatrix}$$

$$= \begin{pmatrix} h_{n-1} & h_n - h_0 & \cdots & h_{2n-2} - h_{n-2} \\ h_0 & 0 & \cdots & 0 \\ \vdots & \vdots & & \vdots \\ h_{n-2} & 0 & \cdots & 0 \end{pmatrix} \text{ is a rank 2 matrix.}$$

Example 1.3.4. Vandermonde matrices. Let $V = (t_i^j)_{i,j=0}^{n-1}$. Then we have

$$\nabla_{D(t),Z_0}(V) = (t_i \ \cdots \ t_i^{n-1} \ t_i^n)_{i=0}^{n-1} - (t_i \ \cdots \ t_i^{n-1} \ 0)_{i=0}^{n-1} = (0 \ \cdots \ 0 \ t_i^n)_{i=1}^{n-1};$$

this is a rank 1 matrix.

Example 1.3.5. Cauchy matrices. Let $C = \left(\frac{1}{s_i - t_j}\right)_{i,j=0}^{n-1}$. Then we have

$$\nabla_{D(s),D(t)}(C) = \left(\frac{s_i}{s_i - t_j}\right)_{i,j=0}^{n-1} - \left(\frac{t_j}{s_i - t_j}\right)_{i,j=0}^{n-1} = \begin{pmatrix} 1 & \cdots & 1 \\ \vdots & & \vdots \\ 1 & \cdots & 1 \end{pmatrix};$$

this is a rank 1 matrix.

    Operator matrices vary depending on the structure of the matrix M: shift operator matrices for the structures of Toeplitz and Hankel type change into scaling operator matrices for the Cauchy structure and into pairs of shift and scaling operator matrices for the structures of Vandermonde type.

    Besides the matrices Z_f in (1.3.1) and $D(v)$ in (1.3.2), Table 1.3 shows the operator matrix $Z_0 + Z_0^T$. Other known operator matrices are naturally associated with various other matrix structures. The next example is a more advanced extension of Toeplitz structure. (The example is given only for demonstration. We do not use it in this book.)

Example 1.3.6. Block Toeplitz matrices M.

$$M = \begin{pmatrix} M_0 & M_{-1} & \cdots & M_{1-q} \\ M_1 & M_0 & \ddots & \vdots \\ \vdots & \ddots & \ddots & M_{-1} \\ M_{p-1} & \cdots & M_1 & M_0 \end{pmatrix}, \text{ each block } M_i \text{ is an } s \times t \text{ matrix.}$$

Let $A = \begin{pmatrix} & 0_s \\ I_{(p-1)s} & \end{pmatrix}$, $B = \begin{pmatrix} & I_{(q-1)t} \\ 0_t & \end{pmatrix}$, that is, let A and B be 2×2 block antidiagonal matrices (of sizes $ps \times ps$ and $qt \times qt$, respectively). Here, the antidiagonal blocks $I_{(p-1)s}$, $I_{(q-1)t}$, 0_s, and 0_t are identity blocks of sizes $(p-1)s \times (p-1)s$ and $(q-1)t \times (q-1)t$ and null blocks of sizes $s \times s$ and $t \times t$, respectively. Then $\text{rank}(\Delta_{A,B}(M)) \le s + t$. This bound is easily verified based on the observation that $I_{(p-1)s} = I_{p-1} \otimes I_s$, $I_{(q-1)t} = I_{q-1} \otimes I_t$, where $\otimes$ denotes the Kronecker product. $\square$

The reader is referred to Table 1.3, Example 1.4.1 in Section 1.4, Section 1.5, and the first five sections of Chapter 4 for further specific examples and basic properties of operators associated with structured matrices.

We conclude this section with the next definition, which summarizes several concepts most relevant to our study of structured matrices.

Definition 1.3.7. For an $m \times n$ matrix M and an associated operator L, the value $r = \text{rank}(L(M))$ is said to be its L-rank r, also called its *displacement rank*. If this value r is small relative to $\min(m,n)$, say, if $r = O(1)$ or $r = o(\min\{m,n\})$ as m and n grow large (see Remark 1.3.8 below), then we call the matrix L-*like* and *having structure of type L*. In the case where the operator L is associated with Toeplitz, Hankel, Vandermonde, or Cauchy matrices M, we call an L-like matrix *Toeplitz-like, Vandermonde-like*, or *Cauchy-like matrix*, respectively, and we say that the matrix has the structure of *Toeplitz, Hankel, Vandermonde*, or *Cauchy type*, respectively.

Remark 1.3.8. To formalize the basic concepts of L-like matrices and the related concepts of Definitions 1.3.7 and 1.5.7 in Section 1.5, we may define "small" values of the rank in Definition 1.3.7 as the values that remain bounded by a constant independent of m and n as both dimensions m and n grow large. Alternative formalizations are also possible, for instance, where "small" means $O(\log^d(m+n))$ for a fixed positive d or for $d = O(1)$ or where "small" means $o(m+n)$.

1.4 COMPRESS, OPERATE, DECOMPRESS

The displacement rank approach requires using the operators L with two *key features*:

1) The displacement rank of a structured matrix M, that is, $\text{rank}(L(M))$, must be *small* relative to the size of M (so the displacement can be *compressed*).

2) Operations with structured matrices can be performed *much faster* via operations with their displacements, with the application of appropriate techniques.

3) Matrices can be *recovered (decompressed) easily* from their displacements.

Feature 1) is reflected in Table 1.3 for some sample classes of structured matrices. Due to feature 1), the displacement $L(M)$ (as any matrix of small rank r) can be represented with a short *generator* defined by only a small number of parameters. In particular, the singular value decomposition (SVD) of the $m \times n$ matrix $L(M)$, having mn entries (as well as the eigendecomposition if the matrix is Hermitian or real symmetric) gives us its *orthogonal generator* that consists of two *generator matrices* G and H having orthogonal columns and sizes $m \times r$ and $n \times r$, respectively, that is, a total of $(m + n)r$ entries (see Section 4.6).

Features 1) and 3) combined enable compact representation of a structured matrix. The following important example demonstrates the key feature 3).

Example 1.4.1. Cauchy-like matrices. Consider the operator $L = \nabla_{D(s), D(t)}$,

$$L(M) = D(s)M - M D(t), \tag{1.4.1}$$

where all components of the vectors **s** and **t** are distinct. For a fixed matrix M, let its displacement $L(M)$ be represented by a pair of $n \times l$ matrices $\{G, H\}$ such that

$$L(M) = \sum_{i=1}^{l} \mathbf{g}_i \mathbf{h}_i^T = G H^T, \tag{1.4.2}$$

$$G = \begin{pmatrix} g_{11} & \cdots & g_{1l} \\ g_{21} & \cdots & g_{2l} \\ \vdots & & \vdots \\ g_{n1} & \cdots & g_{nl} \end{pmatrix} = (\mathbf{g}_1 \ \cdots \ \mathbf{g}_l), \quad H = \begin{pmatrix} h_{11} & \cdots & h_{1l} \\ h_{21} & \cdots & h_{2l} \\ \vdots & & \vdots \\ h_{n1} & \cdots & h_{nl} \end{pmatrix} = (\mathbf{h}_1 \ \cdots \ \mathbf{h}_l),$$

$$\tag{1.4.3}$$

$l \geq r = \text{rank}\, L(M)$. It is immediately verified that (1.4.1)–(1.4.3) are satisfied by the matrix

$$M = \sum_{k=1}^{l} D(\mathbf{g}_k) C(\mathbf{s}, \mathbf{t}) D(\mathbf{h}_k), \tag{1.4.4}$$

where $C = C(\mathbf{s}, \mathbf{t})$ is the Cauchy matrix of Example 1.3.5.

Conversely, (1.4.1)–(1.4.3) imply (1.4.4) because the operator $L(M)$ in (1.4.1) is surely non-singular, that is, the matrix equation $L(M) = 0$ implies immediately that $M = 0$. Indeed, $(L(M))_{i,j} = (s_i - t_j)(M)_{i,j}$, and, by assumption, $s_i \neq t_j$ for all pairs (i, j).

(1.4.4) *bilinearly expresses* a matrix M via the columns of its *L-generator (or displacement generator)* $\{G, H\}$ *of length* l. (Note that the L-generator has a total of $2ln$ entries — much less than n^2 if $l \ll n$.) We also have an alternative bilinear expression for a matrix M in (1.4.4) via the rows $\mathbf{u}_i^T$ of G and the rows $\mathbf{v}_j^T$ of H:

$$M = \left(\frac{\mathbf{u}_i^T \mathbf{v}_j}{s_i - t_j} \right)_{i,j=0}^{n-1}.$$

The latter expression is dual to (1.4.4). For $\mathbf{u}_i^T = (x_i, -1)$, $\mathbf{v}_j^T = (1, y_j)$, this expression covers the class of *Loewner matrices* $\left(\frac{x_i - y_j}{s_i - t_j} \right)_{i,j}$ that are encountered in the study of rational interpolation. If the length l of the L-generator $\{G, H\}$ for the matrix M in (1.4.1)–(1.4.3) is small relative to n, then, according to Definition 1.3.7, we say that M is a Cauchy-like matrix.   □

The reader is referred to Sections 4.3–4.5 for further demonstration of the key feature 3). This feature enables simple extension of a compressed (that is, memory efficient) representation of the displacement $L(M)$ to a compressed representation of the matrix M, that is, it enables simple inversion of the operator L. (In Section 4.5, the approach works even where the operator L^{-1} is not invertible and the matrix $L^{-1}(L(M))$ is undefined on some of its entries.) Typically for the customary displacement operators L, the inverse operator L^{-1} is linear where it is defined, and M's entries are expressed bilinearly via the columns of the generator matrices G and H in (1.4.2). Furthermore, based on these expressions, multiplication of the matrix M by a vector is easily reduced to multiplication by vectors of r or $2r$ rudimentary structured matrices such as f-circulant ones (defined in Section 2.6), Vandermonde, or (as in the above example) Cauchy matrices. This computation is simple because it amounts essentially to only a constant or logarithmic number of discrete Fourier or sine and/or cosine transforms (see Sections 2.4, 2.6, 3.1, and 3.6).

The following flowchart represents the displacement rank approach to computations with structured matrices:

COMPRESS, OPERATE, DECOMPRESS.

Our study elaborates this flowchart for some of the most fundamental computations *unified over various classes of structured matrices*. That is, the algorithms are described for any input matrix represented by its displacement of smaller rank, where the displacement operator is treated as *symbolic*, with unspecified operator matrices. Nevertheless the description is sufficiently complete to allow immediate application of the algorithms to any specified pair of a structured matrix and an associated displacement operator.

The COMPRESS stage exploits the key features 1), as explained in the previous section. The key feature 3) of the displacement rank approach is exploited at the DECOMPRESS stage (e.g., one may rapidly recover the solution to a structured linear system of equations from a short displacement generator of the inverse of the structured input matrix or of its largest non-singular submatrix). We revisit these two stages and study them systematically in Chapter 4. Our next section supports the OPERATE stage.

1.5   Basic operations with compressed structured matrices

At the OPERATE stage, computations with structured matrices are reduced to operations with their short generators. We have the following simple results.

Theorem 1.5.1. *For two pairs of scalars a and b and matrices M and N of the same size and any linear operator L (in particular, for $L = \nabla_{A,B}$ and $L = \Delta_{A,B}$ for any pair of matrices A and B), we have $L(aM + bN) = aL(M) + bL(N)$.*

Theorem 1.5.2. *For any triple of matrices A, B, and M of compatible sizes, we have*

$$(\nabla_{A,B}(M^T))^T = -\nabla_{B^T,A^T}(M),$$
$$(\Delta_{A,B}(M^T))^T = \Delta_{B^T,A^T}(M).$$

Theorem 1.5.3. *Let M be a non-singular matrix. Then*

$$\nabla_{B,A}(M^{-1}) = -M^{-1}\nabla_{A,B}(M)M^{-1}.$$

Furthermore,

$$\Delta_{B,A}(M^{-1}) = BM^{-1}\Delta_{A,B}(M)B^{-1}M^{-1},$$

if B is a non-singular matrix, whereas

$$\Delta_{B,A}(M^{-1}) = M^{-1}A^{-1}\Delta_{A,B}(M)M^{-1}A,$$

if A is a non-singular matrix.

Proof. The claimed equations can be immediately verified by inspection. Let us also show a proof with derivation. Pre- and post-multiply the equation $\nabla_{A,B}(M) = AM - MB$ by M^{-1} and obtain the first equation of Theorem 1.5.3. Pre-multiply it by B^{-1} and then interchange B and B^{-1} throughout to obtain the second equation. Post-multiply it by A^{-1} and then interchange A and A^{-1} throughout to obtain the last equation of Theorem 1.5.3.   □

Theorem 1.5.3 expresses the displacement of the inverse of a structured matrix M via the displacement of M, the operator matrices A and B, and the products of the inverse with $2r$ specified vectors, where $r = \text{rank}(L(M))$ for $L = \nabla_{A,B}$ or $L = \Delta_{A,B}$. More specialized and more explicit expressions are known in the important special case where M is a Toeplitz or Hankel matrix (see (2.11.4), (4.2.10), and (4.2.11), Remark 4.4.11, Theorem 2.1.5, and Exercise 2.24, as well as the Notes to this section).

Theorem 1.5.4. (See Table 1.4.) *For any 5-tuple $\{A, B, C, M, N\}$ of compatible matrices, we have*

$$\nabla_{A,C}(MN) = \nabla_{A,B}(M)N + M\nabla_{B,C}(N),$$
$$\Delta_{A,C}(MN) = \Delta_{A,B}(M)N + AM\nabla_{B,C}(N).$$

Furthermore,

$$\Delta_{A,C}(MN) = \Delta_{A,B}(M)N + AMB\Delta_{B^{-1},C}(N),$$

if B is a non-singular matrix, whereas

$$\Delta_{A,C}(MN) = \Delta_{A,B}(M)N - AM\Delta_{B,C^{-1}}(N)C,$$

if C is a non-singular matrix.

Table 1.4: Operator matrices for the matrix product (Sylvester type operators).
The matrix in the first line of each column represents either one of two factors or
their product; the two other lines show a pair of the associated operator matrices.

M	N	MN
A	B	A
B	C	C

Proof. $\nabla_{A,C}(MN) = AMN - MNC = (AM - MB)N + M(BN - NC)$. This proves
the first equation of Theorem 1.5.4. Similarly, we verify the second equation. Now,
Theorem 1.3.1 leads us to the next two equations.   □

Definition 1.5.5. Operator pairs $(\nabla_{A,B}, \nabla_{B,C})$, $(\Delta_{A,B}, \nabla_{B,C})$, $(\Delta_{A,B}, \Delta_{B^{-1},C})$, and
$(\Delta_{A,B}, \Delta_{B,C^{-1}})$, allowing application of Theorem 1.5.4, are called *compatible*.

Now represent the matrices A, B, M, $\nabla_{A,B}(M)$, and $\Delta_{A,B}(M)$ as 2×2 block matrices with blocks $A_{i,j}$, $B_{i,j}$, $M_{i,j}$, $\nabla_{i,j}$, and $\Delta_{i,j}$, respectively, having compatible sizes
(for i, $j = 0, 1$). We can easily verify the next result.

Theorem 1.5.6.

$$\nabla_{A_{ii},B_{jj}}(M_{ij}) = \nabla_{ij} - R_{i,j},$$

$$\Delta_{A_{ii},B_{jj}}(M_{ij}) = \Delta_{ij} + S_{i,j},$$

where

$$R_{i,j} = M_{i,1-j}B_{1-j,j} - A_{i,1-i}M_{1-i,j},$$
$$S_{i,j} = A_{i,i}M_{i,1-j}B_{1-j,j} + A_{i,1-i}M_{1-i,j}B_{j,j} + A_{i,1-i}M_{1-i,1-j}B_{1-j,j},$$

for i, $j = 0, 1$.

Remark 1.5.7. The expressions of Theorem 1.5.6 are effective where the matrices
$A_{i,1-i}$, $B_{j,1-j}$ and, consequently, $R_{\nabla,i,j}$ and $R_{\Delta,i,j}$ have small rank (relative to the
matrix size) for all pairs (i, j), i, $j \in \{0, 1\}$. (For formal meanings of "small ranks,"
see Remark 1.3.8.) If this property holds for a pair of operator matrices A and B, we
call such matrices and the associated operators $L = \nabla_{A,B}$ and $L = \Delta_{A,B}$ *nearly block
diagonal*. All operator matrices of Table 1.3 are nearly block diagonal.

Remark 1.5.8. Computations with compressed structured matrices remain efficient as
long as they are done properly to keep the length of the generators small. Generally,
this requires special techniques because the length grows when Theorems 1.5.1, 1.5.4,
and 1.5.6 are applied.

1.6 Two classes of superfast algorithms. The case of Toeplitz-like matrices

A key word for the algorithms in this book is *"superfast"*. Some of the most fundamental computations with structured matrices (such as the solution of a non-singular
structured linear system of n equations) can be accelerated to achieve solution in linear
arithmetic time, that is, $O(n)$ for $n \times n$ matrices (up to logarithmic or polylogarithmic
factors). For comparison, the *classical* algorithms running in arithmetic time of the
order of n^3 and *fast* algorithms using time of the order of n^2 either ignore the structure
or proceed with its limited use, respectively. All superfast algorithms that we cover
require linear memory space (that is, $O(n)$ versus at least n^2 space required in the case
where the algorithms deal with unstructured dense matrices). *In our unifying presentation, we apply the nomenclature "superfast" to all algorithms running in linear time
(up to polylogarithmic factors), including algorithms for polynomial computations.*

The presented superfast algorithms are *nearly optimal* in using computer memory
and time. Indeed, the representation of structured input matrices and polynomials requires the order of n memory space where n parameters are used. Furthermore, each
arithmetic operation has two operands and thus involves at most two input parameters.
This simple argument implies the information lower bounds of the order of n on the
arithmetic time cost.

In Chapters 5–7, we cover two classes of superfast algorithms: *the divide-and-
conquer direct algorithms* and *Newton's iterative algorithms*. Both are well known
for computations with general matrices, but the displacement rank approach enables
dramatic acceleration (from cubic to nearly linear arithmetic running time) in the case
where the input matrices are structured.

The algorithms were first developed for Toeplitz-like matrices. Let us now briefly
recall this development. In subsequent sections, we comment on numerical and algebraic versions of the resulting algorithms and on the general techniques for the extension of the same superfast algorithms to various classes of structured matrices.

In [M74], [M80], and [BA80] the displacement rank approach is combined with
the simple but powerful construction in the paper [S69] for recursive triangular factorization of a matrix. The resulting divide-and-conquer algorithm solves a non-singular
Toeplitz or Toeplitz-like linear system of n equations by using $O(n \log^2 n)$ flops (arithmetic operations). That is, the MBA algorithm is superfast versus the classical solution
of a linear system of n equations by Gaussian elimination, which uses the order of n^3
flops, and the fast algorithms that solve a Toeplitz linear system in $O(n^2)$ flops.

Less widely known (particularly until publication of the Chapter [PBRZ99] in
[KS99]) were the papers [P92], [P92b], [P93], and [P93a], where superfast Newton's
iteration for numerical inversion of a well conditioned Toeplitz-like matrix was developed based on the displacement rank approach. Newton's iteration is strongly stable
numerically and allows much more effective parallel implementations (essentially reduced to FFT's) than does any alternative superfast algorithm.

Each step of the iteration is reduced to performing matrix multiplication twice,
which use $O(n \log n)$ flops if the original Toeplitz-like matrix structure is maintained
throughout. To maintain the structure, it was proposed in the papers [P92b], [P93], and

[P93a] to represent Toeplitz-like matrices via Singular Value Decompositions (SVD's) of their displacements (which defines *orthogonal displacement generators*) and to truncate the smallest singular values to compress these generators. The construction has resulted in superfast algorithms for well conditioned input matrices, for which polylogarithmic number of Newton's steps (each using nearly linear number of flops) ensure convergence. An alternative technique for compression of the generators of structured matrices in the process of Newton's iteration has been elaborated later on.

In Chapters 5–7, we cover all these constructions not only for Toeplitz-like matrices but in a *unified way* for various other classes of structured matrices. The unified algorithms remain superfast, and other features are extended from the Toeplitz-like case. Let us briefly compare the two groups of algorithms.

The divide-and-conquer algorithms are direct and easier to analyze. They are convenient for error-free computations with singular matrices, say for the computation of the rank and null space of a matrix, but they also have numerically stable versions.

Newton's iteration has several other attractive features. Its implementation is simpler than the implementation of the divide-and-conquer algorithms. The iteration is strongly stable and enables computation of the numerical rank and numerical generalized inverse of a matrix, which is important in applications to signal processing. Newton's iteration is also a fundamental tool of computer algebra, which supports effective variation of computer precision in the error-free computation of the output. Both numerical and algebraic versions allow most effective parallel implementations, which is superior to the divide-and-conquer approach. Newton's iteration is a highly developed subject but still intensively researched with promise of further substantial improvements.

In spite of their cited attractive features, superfast algorithms are not always the methods of choice in practice. Other criteria such as the overhead constants hidden in the "O" notation, simplicity of implementing the algorithm, efficiency of its parallelization, and numerical stability may give an edge to fast algorithms.

Leaving this question open in every special case, we find it at least conceptually useful to focus our current study on unified superfast methods based on the displacement rank approach. In the most important case of Toeplitz/Hankel-like input, however, the superfast algorithms in this book comply quite well with the listed criteria. Preliminary testing of modified structured versions of Newton's iteration has confirmed its efficiency (see Section 6.11), whereas practical implementations have already been reported for superfast divide-and-conquer algorithms [VK98], [VHKa].

1.7 Unified algorithms and algorithmic transformations

As we mentioned, the displacement rank approach was first proposed for matrix structures of Toeplitz type with the initial goal of obtaining fast (running in arithmetic time $O(n^2)$) algorithms for the computation of the Gohberg–Semencul type expressions for the inverse R^{-1} of an $n \times n$ non-singular Toeplitz-like matrix R (see Theorems 1 and 2 of [KKM79] and a proof of the latter theorem in [FMKL79]). Gradually, the approach was extended to the study of matrix structures of other types. The cited superfast algorithms (that is, the MBA algorithm and Newton's iteration) have been extended to these matrix classes as well.

The extensions relied on two approaches. First, some simple techniques for superfast transformation of the four classes of structured matrices of Toeplitz, Hankel, Cauchy, and Vandermonde types into each other were shown in the paper [P90] based on the transformation of the associated displacement operators and displacements. (Toeplitz-to-Hankel and Hankel-to-Toeplitz transformations are actually trivial, but we cite them for the sake of completeness.) The paper proposed using such superfast transformations to extend successful algorithms developed for one class of structured matrices to other classes. The MBA algorithm and Newton's iteration can be immediately extended from input matrices with Toeplitz type structure to various other classes of structured matrices in this way. This *displacement transformation approach* (or shortly *transformation approach*) turned out to be effective in several areas, in particular for direct and iterative solution of Toeplitz/Hankel-like linear systems of equations (via the transformation of the displacement operators associated with their coefficients matrices to the operators of Cauchy type). In Chapters 4 and 5, we show applications of this approach for avoiding singularities in matrix computations and accelerating numerical rational interpolation (Nevanlinna–Pick and Nehari problems). (See Sections 3.6, 4.8, and 4.9, Remark 5.1.1, and Example 5.5.1.)

Later on, an alternative way to unifying structured matrix algorithms was developed, based on replacing all or some of the shift operators by scaling operators in MBA and Newton's constructions. This immediately transformed the original constructions devised for Toeplitz-like input matrices to cover the Cauchy-like and Vandermonde-like inputs. In the next important step, displacement operators with specific (shift and scaling) operator matrices were replaced by symbolic displacement operators. The generalization enabled the design of the unified superfast MBA and Newton algorithms for all classes of structured matrices whose associated operators had the key features 1)–3) listed in Section 1.4. This covered all popular classes of structured matrices. Furthermore, the latter approach (which we call *the symbolic operator approach*) enabled superfast computations even with some useful but so far unused matrices having rather weird structures, such as the skew-Hankel-like matrices in Example 5.5.1 in Chapter 5. Under this approach, the algorithms reduce computations with structured matrices essentially to a sufficiently short sequence of operations of multiplication of such matrices by vectors. The algorithms and their analysis apply to any matrix associated with an unspecified readily invertible displacement operator (viewed as a symbolic operator). Such algorithms remain effective as long as the matrix can be multiplied by a vector fast and has a displacement of a smaller rank. This is the main framework for our high level description of algorithms in the book. The description unifies several classes of structured input matrices as well as algebraic and numerical implementation methods. The choices of the input matrix, displacement operator, and algebraic or numerical techniques are specified in the subsequent implementation stage. In spite of our focus on using the symbolic operator approach, we wish to stress two points.

1) The unification is not universal in our study. We either do not or very briefly cover some important and intensively studied classes of structured matrices (see Section 1.9).

2) The displacement transformation and the symbolic operator approaches comple-

ment each other as allies. In some cases, transformations are simple, cause no numerical stability problems (e.g., from Toeplitz-like to Cauchy-like matrices or in symbolic computations), and enable us to exploit various algorithmic advantages of specific matrix structures. For example, Toeplitz/Hankel-like matrices are generally multiplied by a vector faster than Vandermonde-like and Cauchy-like matrices. Therefore we may accelerate some Vandermonde-like and Cauchy-like computations by transforming them into Toeplitz/Hankel-like computations. Likewise, Cauchy and Cauchy-like matrices have a similar advantage over Vandermonde-like matrices due to the Fast Multipole Algorithm of [GR87], which effectively multiplies a Cauchy matrix by a vector, so the transformations from Vandermonde to Cauchy-like matrices improve the known methods for polynomial interpolation and multipoint evaluation (see Notes). On the other hand, Cauchy-like matrices (unlike Toeplitz/Hankel-like matrices) preserve their structure in row and column interchange; this enables structure preserving pivoting and therefore fast and numerically stable Gaussian elimination. Displacement transformation enables extension of these nice properties to the Toeplitz/Hankel-like input (see Section 4.9).

1.8   Numerical and algebraic versions of structured matrix algorithms. Flops and ops

The symbolic operator approach allows both numerical and algebraic (symbolic) implementations of structured matrix algorithms. Let us next briefly recall general implementation rules and then comment on the structured matrix case.

1.8.1   Numerical implementation

In numerical implementation, with rounding to a fixed finite precision, one generally asks how much the output can be corrupted by rounding errors, in other words, how meaningful is the output. This leads us to the well-studied subjects of numerical conditioning and stability. We only briefly touch upon them and only in conjunction with our topics. For more detailed coverage, see [W63], [W65], [S74], [A78], [CdB80], [BP94, Section 3.3 and Appendix 3A], [GL96], [Hi96], and [BPa]. On numerical stability of computations with Toeplitz and Hankel matrices, see [C80], [B85], [FZ93], [F94]. On more recent studies exploiting and extending the transformation approach of the paper [P90], see [GKO95], [G98], [Oa].

Let us give some brief introductory comments. Partition the subject into two parts:

a) *conditioning of the computational problem*, measured by its *condition number κ*, that is, by the ratio of two bounds — on the input perturbation and the resulting output error

and

b) *numerical stability of the solution algorithm*, measured by the ratio e_I/e_C of two output error bounds caused by rounding to a fixed finite precision — one, e_I

caused by rounding the input values and another, e_C caused by rounding also all other values computed by the algorithm.

The study of conditioning and stability requires specification of the input. Therefore, in this study we should first descend down to earth from our high level symbolic description of unified algorithms and specify the input matrix and the associated displacement operator. Let us next comment on issues a) and b).

a) A computational problem and the matrix defining it are called *ill conditioned* if the condition number κ is large and is called *well conditioned* otherwise. The concepts of "large," "ill" and "well" are not well defined but usually still help in the discussion about numerical behavior. In some cases, we have a clear answer: the number κ is large where it is infinite or has the order of 2^N where N is the input size of the problem. Computation of the greatest common divisor of two polynomials and polynomial interpolation are examples of ill conditioned problems.

Among matrix computations, we are most interested in the computation of the inverse or generalized inverse of a matrix and the solution of a linear system of equations. For these problems, the condition numbers can be calculated as the products of the norms of the input matrix and its generalized inverse or as the maximal ratio of two singular values of the input matrix (see Definition 6.1.1 in Chapter 6).

How large are the condition numbers of structured matrices? There are many well conditioned structured matrices, such as the identity or discrete Fourier transform matrices. A random Toeplitz/Hankel, Vandermonde, or Cauchy matrix, however, is likely to be badly ill conditioned, and this holds also where the matrix is positive definite [G62], [GI88], [Gau90], [T94], [TZa], [FOa].

b) Like conditioning, numerical stability is also defined based on the informal concept of "large." That is, the algorithm is numerically stable if the ratio e_I/e_C is not large and numerically unstable otherwise. Note that the value $\log_2(e_I/e_C)$ shows the number of bits of the output lost in the result of rounding to the fixed finite precision throughout the entire computation.

Estimating the ratio e_I/e_C and $\log_2(e_I/e_C)$ may seem to be a formidable task. In many cases, however, the task is dramatically simplified based on the advanced techniques of numerical analysis (see the bibliography cited in the beginning of this section). These techniques, including Wilkinson's backward error analysis, have been particularly well developed for matrix computations.

Numerical stability is known to be strong for Newton's iteration for general matrices [PS91]. Based on the techniques of Section 6.6 of Chapter 6 and the papers [P92b], [P93], [PR01], [PRW00], and [PRWa], one should be able to estimate perturbations of the values computed numerically at the stages particular to the displacement rank approach, although no such analysis has been published so far. On the other hand, the MBA algorithm is prone to numerical stability problems for the general input of Toeplitz/Hankel type but becomes (weakly) stable if the input matrix is positive definite [B85].

1.8.2 Algebraic implementation

To avoid problems of numerical conditioning and numerical stability, one may shift to error-free symbolic computations. This is a realistic method supported by several packages and libraries of subroutines and by various advanced algebraic techniques. In Sections 3.5 (see Remark 3.5.12) and 7.7, we encounter the sample techniques of the *Chinese Remainder Algorithm* for integers and the *Newton–Hensel lifting*. Presently, symbolic implementations of matrix computations are typically outperformed (in terms of the computer time and computer memory involved) by numerical implementations with rounding to the single or double IEEE precision (provided, of course, that numerical implementation produces a desired uncorrupted output). This situation should motivate further efforts for improving the symbolic approach [Ba], perhaps by taking the path of creating more specialized packages of subroutines. There is also some hope for synergy of the combination of techniques of both numerical and symbolic computations where they are applied towards a common goal. During the last decade this idea was quite popular among the designers of algebraic algorithms [P98a]. In several cases the resulting algorithms successfully exploit structured matrix computations [CGTW95], [P95a], [P96a], [P97], [P98], [MP98], [MP00], [Pa], [EPa].

1.8.3 Unification and separation of numerical and algebraic versions of algorithms

Our symbolic operator presentation of the MBA algorithm enables both numerical and algebraic implementations. That is, we define the algorithm over an arbitrary field $\mathbb{F}$. In the case of algebraic implementation, we work either in an abstract field or in a specific finite field such as $\mathrm{GF}(p)$, the field of integers modulo a prime p. In the numerical case, we formally work in the fields of rational, real, or complex numbers and truncate each computed value to a fixed finite precision or to the two-level (single-double) precision.

For operations in the chosen field, we use the abbreviation *ops*. (Ops are *flops* in the floating point numerical computations with the complex, real, or rational numbers.)

The two implementations may require distinct specification even at a high level of their description which we adopt. In such cases we supply some further details. For numerical and symbolic versions of the MBA algorithm, we show two distinct special techniques for countering possible singularities (see Sections 5.5–5.7 of Chapter 5). On the other hand, numerical and algebraic versions of Newton's iteration are presented separately, in Chapters 6 and 7, respectively.

1.8.4 Computational complexity estimates

The bound of $O(n \log^2 n)$ or $O(nr^2 \log^2 n)$ flops involved in the MBA numerical algorithm for an $n \times n$ Toeplitz-like input matrix is extended to the bound of $O(r^2 v(n) \log n)$ ops in the symbolic MBA algorithm, provided that r is the displacement rank of an $n \times n$ input matrix and $v(n)$ ops are sufficient to multiply a matrix of this class by a vector (for $r = 1$). In our study, $v(n)$ varies from $O(n \log n)$ to $O((n \log^2 n) \log \log n)$ depending on the class of matrices and field of constants allowed. We may estimate the *bit-operation complexity* by summing the bit-operations over all ops involved, provided we are given the precision of performing each op (see Exercise 2.9 in Chapter 2).

Similar comments apply to the other algorithms of the book. The flop estimate of $O(n \log^d n)$ per step of the *Newton–Toeplitz Iteration* can be extended to the bound of $O((rv(n) \log^{d-1} n) \log \log n)$ ops per step for several other classes of structured input matrices over any field. In the numerical version of this iteration, a polylogarithmic number of steps is still sufficient provided that the structured input matrix is well conditioned. In the algebraic version, the number of steps may grow to n but the bit-operation cost remains nearly optimal provided that the input matrix is filled with integers and the exact output values are required. That is, the algorithm remains superfast at the bit-operation level.

For *parallel implementation* of the algorithms, we only supply some simple estimates that immediately follow from the known results on polynomial multiplication and FFT. For Newton's iteration on n arithmetic processors, parallel arithmetic time decreases by the factor of n versus the sequential time. This is because the computation is ultimately reduced to polynomial multiplication modulo $x^{O(n)}$ and/or FFT, which allow parallel acceleration by the factor of n (see Remarks 2.4.5 and 2.4.6 in Section 2.4). The product of the parallel arithmetic time and processor bounds remains at the level of sequential arithmetic time bound, which characterizes nearly optimal parallelization. The MBA algorithm and its extensions allow parallel acceleration only by a polylogarithmic factor.

The same asymptotic estimates for the computational cost hold where MBA and Newton's algorithms are applied to a singular matrix M. In this case the output sets include the rank of M and a short generator for its null space basis and if the matrix is non-singular, then also for its inverse. Newton's numerical iteration also outputs the Moore–Penrose generalized inverse M^+. The sequential time bound grows by roughly the factor of n where Newton's algebraic iteration outputs the characteristic polynomial of M.

1.9   Limitations of the unification

Our unified study still omits several highly important classes of structured matrices, in particular, banded and, more generally, diagonal+semi-separable matrices [EG97], [EG97a], [EG98], [EG99], [EG99a]; multivariate Bezoutians and multilevel matrices [H70a], [GL83], [PT84], [HR84], [LT85], [FP88], [FD89], [HF89], [W90], [GKZ94], [T96], [MP98], [MP00]; infinite but finitely generated matrices [DvdV98], [BM99]; polynomial Vandermonde matrices [H90], [KO96], [KO97]; confluent (Cauchy-like) matrices [BGR90]; and some highly important subclasses in the class of Toeplitz/Hankel-like matrices with additional structure such as (block) banded (block) Toeplitz matrices [BP88], [BP91a], [BM99], [BM01], [BPa], rationally generated Toeplitz matrices [GS84], [J89], [CN96], [CN99], and circulant matrices [D79].

Furthermore, for these classes of matrices, the techniques we present require extension, complementation, or replacement by other methods. In particular, effective computations with diagonal+semi-separable matrices require their association with non-linear operators [EG97], [EG97a], [EG98], [EG99], [EG99a]. On the other hand, only a part of our study can be extended to multivariate Bezoutians and multilevel matrices M. They possess properties a)–c) of structured matrices listed in Section 1.2 but not property d) — their recovery from the displacements $L(M)$ is not simple, and super-fast inversion of these matrices is a hard open problem [MP00]. On the contrary, in the case of matrices with additional structure, other techniques may supersede ours. For instance, the preconditioned conjugate gradient method is superior for some important classes of rationally generated Toeplitz matrices [CN96], [CN99], [KO96], whereas for (block) banded matrices with Toeplitz blocks, the matrix structure is used most effectively in combination with the powerful techniques of the block cyclic reduction [BM99], [BPa]. Furthermore, the power of our unified algorithms may be substantially enhanced when they are applied to a specific class of structured matrices. In particular, see [VK98], [VHKa] on practical benefits of modification of the divide-and-conquer algorithms for Toeplitz linear systems where the correlation to rational interpolation on the complex unit circle is exploited.

1.10   Organization of the book

We cover the cited unified algorithms in some detail, particularly, the divide-and-conquer algorithms. In the next two chapters, we cover the correlation of computations with structured matrices and some fundamental polynomial and rational computations. Apart from limited application in Chapter 7, *the material of Chapters 2 and 3 is not used in subsequent chapters* except for some basic facts and definitions (mostly from Section 2.1) and the results on multiplication of structured matrices by vectors displayed already in Table 1.2. Graphs in the beginning of Chapters 3–7 provide a guide to optional selective reading. In Chapter 4, we recall and extend various fundamental techniques for computations with structured matrices partly outlined in the present chapter. In Chapter 5, we elaborate on the unified superfast extension of the MBA algorithm. This includes specifications for a singular input and computations in finite fields, which are required in some important applications, for instance, to computer algebra and algebraic codes. In Chapters 6 and 7, which can be read independently of Chapter 5, we describe unified numerical and algebraic versions of Newton iteration for the inversion of structured matrices. In Section 3.7, we show application of structured matrices to loss-resilient encoding/decoding. In Sections 3.8 and 3.9, we cover applications to the tangential Nevanlinna–Pick and matrix Nehari problems.

     Every chapter is followed by several exercises of various levels of difficulty.

1.11   Notes

Section 1.1   The most popular are structured matrices of Toeplitz/Hankel type. They are encountered, for example, in the shift register and linear recurrence computation [Be68], [M75], inverse scattering [BK87], adaptive filtering [K74], [Ha96], the modeling of stationary and non-stationary processes [KVM78], [K87], [L-AK84], [L-AKC84], [KAGKA89], numerical study of Markov chains [S94], [BM99], [BPa], the solution of partial differential and integral equations [B85], [C47/48], [KLM78], [KVM78], and algebraic computations [H70], [BT71], [GY79], [BGY80], [P91], [P92], [BP94], [P97], [EP97], [MP98], [MP00], [BPa]. There is also substantial interest in the matrix structures of Vandermonde and Cauchy types because of their correlation to polynomial and rational interpolation and multipoint evaluation [GY79], [BGY80], [F85], [D89], [D89a], [BGR90], [FF90], [PLST93], [FHR93], [H95], [PZHY97], [S97], [PACLS98], [PACPS98], [OP98], [FFGK98], [S98], [OP99], [OS00b] and their applications to algebraic codes [MS77], [WB94], [P00c], [OS00], [Pb], particle and many-body simulation [A85], [A86], [C73], [D83], [HE81], integral equations [R85], [JO89], the study of galaxies [A63], [B88], conformal maps [T86], and the Riemann zeta function [OS88].

The advanced techniques for the Nevanlinna–Pick and Nehari problems of rational computations include the band extension method of H. Dym and I. Gohberg, the Buerling–Lax theorem approach of K. Glover and of J. Ball, I. Gohberg, and L. Rodman, the de Branges reproducing kernel method of H. Dym, and the lifting-of-commutant method of C. Fojas (see [D89], [D89a], [BGR90], [FF90], [S97], [FFGK98], and [S98], and the bibliography therein).

Section 1.2 There is huge literature on various specific classes of structured matrices; see, e.g., [D79], [BGR90], [BGR90a], [H90], [BP94], [CN96], [KO97], [OP98], [KS99], [MP00], and the bibliography therein.

Section 1.3 [KKM79] was the most influential paper for the development of the displacement rank approach. The proof of one of its two theorems was given in [FMKL79]. Among the most important works preceding [KKM79] and [FMKL79], we single out [GS72] and [M74], and we cite [HR84], [K91], [KS95], and [KS99] for historical accounts and bibliography of the early period of the study of this subject. The nomenclature "Vandermonde-like" is sometimes used for a distinct class of matrices [H88], [H90], [L96], which is close to what we define by (3.2.7) and call polynomial Vandermonde matrices, using the nomenclature of [KO96], [KO97]. For simplicity we mostly restrict our study to square structured matrices. The extension to the rectangular case is usually straightforward, many of its details are covered in [PWa]. Example 1.3.6 is from [PWb].

Section 1.4 Example 1.4.1 is due to [HR84], [GO94], [H95]. On application of Loewner matrices to rational interpolation, see [F85].

Section 1.5 Theorems 1.5.1–1.5.4 and 1.5.6 are taken from [P99], [P00]. Theorem 1.5.3 was used in [BP93] in the Toeplitz-like case; it was probably used earlier. Theorem 1.5.4 was used in [CKL-A87] in the Toeplitz-like case and in [P90] in the general case of Stein type operators. Numerous inversion formulae have been published for Toeplitz/Hankel matrices [GS72], [GF74], [H70a], [H79], [FMKL79], [BGY80], [T81], [HR84], [LT85], [AG89], [T90], [Gad90], [LDC90], [AG91], [HR98], and [BMa].

Section 1.6 Fast Toeplitz solvers using $O(n^2)$ flops were proposed in [L47], [D59], and [T64]. The first superfast solution of Toeplitz systems in $O(n \log^2 n)$ flops, due to [BGY80], does not apply to the Toeplitz-like case and is prone to numerical stability problems [B85]. Also well known are the superfast Toeplitz-like solvers proposed in the 1980's and 1990's, but technically they develop the ideas of I. Schur in the early 1900's (see [M81], [dH87], [AGr88], [KS95], [KS99], and the bibliography therein).

Section 1.7   The displacement rank was defined for the matrices of Hankel type already in [KKM79], for the matrices of Vandermonde and Cauchy types in [HR84], [GKK86], [GKKL87], [P90], [FHR93], [BP94], [GO94], [H95], [GKO95], [KO96], [PZ00]. The displacement transformation approach (due to [P90]) was applied to the solution of Toeplitz/Hankel-like linear systems of equations (via the reduction of the associated displacement operators to the operators of Cauchy type) in [H95], [GKO95], [H96], [KO96], [HB97], [G97], [HB98], [Oa]; numerical polynomial interpolation and multipoint evaluation in [PLST93], [PZHY97], [PACLS98], [PACPS98]; and algebraic decoding in [OS99]/[OS00], Section 6 (also see the Notes to Sections 4.8 and 4.9). The extension of the MBA and Newton algorithms to Cauchy-like and Vandermonde-like matrices is due to [PZHD97], [PACPS98], [PZACP99], and [PZ00] (also see the Notes to Section 5.1). The symbolic operator approach for the MBA and Newton constructions were developed in [OP98], [P99], [P00], [PR01], [PRW00], and [PRWa] (see more in the Notes to Sections 5.1 and 6.3-6.9).

1.12   Exercises

1.1   (a) Compute the displacements $\nabla_{Z_e, Z_f}(T)$, $\nabla_{Z_e^T, Z_f^T}(T)$, $\Delta_{Z_e^T, Z_f}(T)$, and $\Delta_{Z_e, Z_f^T}(T)$ for a Toeplitz matrix $T = (t_{i-j})_{i,j=0}^{n-1}$. Represent each as the sum of r outer products of pairs of vectors, for $r = 2$.

  Do the same for the following displacements:

    (b) $\nabla_{Z_e, Z_f^T}(H)$, $\nabla_{Z_e^T, Z_f}(H)$, $\Delta_{Z_e, Z_f}(H)$, and $\Delta_{Z_e^T, Z_f^T}(H)$ for a Hankel matrix $H = (h_{i+j})_{i,j=0}^{n}$,

    (c) $\nabla_{Z_0 + Z_0^T, Z_0 + Z_0^T}(H+T)$ for a Hankel matrix $H = (h_{i+j})_{i,j=0}^{n-1}$ and a Toeplitz matrix $T = (t_{i-j})_{i,j=0}^{n-1}$ (in which case $r = 4$),

    (d) $\nabla_{D(s), Z_f}(V(s))$, $\nabla_{D^{-1}(t), Z_f^T}(V(t))$, $\Delta_{D(s), Z_f^T}(V(s))$, $\Delta_{D^{-1}(t), Z_f}(V(t))$ for Vandermonde matrices $V(s) = (s_i^j)_{i,j=0}^{n-1}$ and $V(t) = (t_i^j)_{i,j=0}^{n-1}$, $t_i \neq 0$ for all i (in which case $r = 1$),

    (e) $\Delta_{D^{-1}(s), D(t)}(C)$, where $s_i \neq 0$ for all i, and $\Delta_{D(s), D^{-1}(t)}(C)$, where $t_j \neq 0$ for all j for a Cauchy matrix $C = \left(\frac{1}{s_i - t_j}\right)_{i,j=0}^{n-1}$ (in which case $r = 1$). Note that $C = \left(\frac{1}{(s_i + a) - (t_j + a)}\right)_{i,j}$ for any scalar a, so the pair of vectors s and t can be replaced by the pair $s + a\mathbf{l}$ and $t + a\mathbf{l}$ for $\mathbf{l} = (1)_{i=0}^{n-1} = (1, 1, \ldots, 1)^T$ and for any scalar a.

1.2   (a) Give upper bound on $\text{rank}(\nabla_{Z_e, Z_f}(B))$ for a $k \times l$ block matrix B with Toeplitz blocks.

(b) Let $k = l = 2$, and specify the expressions for the matrices $R_{i,j}$ and $S_{i,j}$ in Theorem 1.5.6.

(c), (d), and (e) Do parts (a) and (b) where the blocks of the block matrix B are Hankel, Vandermonde, and Cauchy matrices, respectively, associated with the operators in Examples 1.3.3–1.3.5, respectively.

1.3 Prove that $\text{rank}(\nabla_{Z_0, Z_0^T}(T_0 T)) \leq 4$ for two Toeplitz matrices T and T_0.

1.4 Give further examples of displacement operators $L = \nabla_{A,B}$ and $L = \Delta_{A,B}$ such that $\text{rank}(L(M))$ is small for

(a) Toeplitz matrices,

(b) Hankel matrices,

(c) Toeplitz+Hankel matrices,

(d) Vandermonde matrices.

1.5 (a) Show a row and column permutation that transforms the block Toeplitz matrix M in Example 1.3.6 into an $s \times t$ block matrix $\widehat{M}$ with $p \times q$ Toeplitz blocks.

(b) Estimate $\text{rank}(\nabla_{Z,Z}(\widehat{M}))$.

1.6 Verify Theorems 1.5.1, 1.5.2, and 1.5.6.

1.7 Verify that all operators of Table 1.3 are nearly block diagonal (see Remark 1.5.7).

1.8 Give upper bounds on

(a) $\text{rank}(L(aM + bN))$ in terms of $\text{rank}(L(M))$ and $\text{rank}(L(N))$,

(b) $\text{rank}(L(MN))$ in terms of $\text{rank}(L_1(M))$ and $\text{rank}(L_2(N))$ for $L = \nabla_{A,C}$, $L_1 = \nabla_{A,B}$, and $L_2 = \nabla_{B,C}$.

(c) The same as in part (b) but for the operators $L = \Delta_{A,C}$, $L_1 = \Delta_{A,B}$, and $L_2 = \Delta_{B,C^{-1}}$.

1.9 Compare the ranks in each of the following pairs of displacements:

(a) $\Delta_{A,B}(M)$ and $\Delta_{B,A}(M^{-1})$ where the matrix M and at least one operator matrix A or B are non-singular,

(b) $\nabla_{A,B}(M)$ and $\nabla_{B,A}(M^{-1})$ where the matrix M is non-singular,

(c) $\nabla_{A,B}(M)$ and $\nabla_{B^T,A^T}(M^T)$,

(d) $\Delta_{A,B}(M)$ and $\Delta_{B^T,A^T}(M^T)$.

1.10 Extend Tables 1.3 and 1.4 to the case of Stein type operators.

Chapter 2

Toeplitz/Hankel Matrix Structure and Polynomial Computations

Summary Toeplitz matrices, Hankel matrices, and matrices with similar structures (such as Frobenius, Sylvester, and subresultant matrices) are probably the most studied structured matrices. Among their numerous important areas of application, we select fundamental polynomial computations, including multiplication, and the Extended Euclidean Algorithm, together with their several extensions and applications. In this chapter we reveal the correlation among computations with polynomials and structured matrices of Toeplitz and Hankel types (see Figures 2.1 and 2.2) and show superfast algorithms for these computations. In the next chapter we similarly study matrices of Vandermonde and Cauchy types. Apart from the introductory material of the next section and the estimates for the arithmetic cost of multiplying Toeplitz, Hankel, Vandermonde, and Cauchy matrices by vectors, the results of these two chapters are little used in Chapters 4–7. Some sample pseudocodes for the algorithms of Sections 2.4 and 2.5 are collected in Section 2.15.

2.1 Definitions and preliminaries

Let us recall and extend the relevant definitions and basic results of Chapter 1. We assume computations in a fixed field $\mathbb{F}$ that contains all constants involved into our computations. Depending on the applications, this can be the field $\mathbb{C}$ of complex numbers or the Galois field $GF(p)$ of integers modulo a fixed prime.

Definition 2.1.1. W^T, W^*, $\mathbf{v}^T$, and $\mathbf{v}^*$ are the *transposes* and the *Hermitian (conjugate) transposes* of a matrix W and a vector $\mathbf{v}$, respectively. $(W_1, \ldots, W_n)$ is the $1 \times n$ block matrix with the blocks $W_1, \ldots, W_n$. $D(\mathbf{v}) = \mathrm{diag}(\mathbf{v}) = \mathrm{diag}(v_0, \ldots, v_{n-1})$ being an $n \times n$ *diagonal* matrix in (1.3.2). $\mathbf{e}_i$ is the i-th coordinate vector, having its i-th coordinate 1 and all other coordinates 0, so $\mathbf{e}_0 = (1, 0, \ldots, 0)^T$. $\mathbf{1} =$

$(1, 1, \ldots, 1)^T = \sum_i \mathbf{e}_i$. The null matrices are denoted by 0 and O, and the $n \times n$ null matrix is also denoted by 0_n and O_n. I and I_n denote the $n \times n$ *identity* matrix, $I = \begin{pmatrix} 1 & & \\ & \ddots & \\ & & 1 \end{pmatrix} = (\mathbf{e}_0, \ldots, \mathbf{e}_{n-1})$. $J = \begin{pmatrix} & & 1 \\ & \ddots & \\ 1 & & \end{pmatrix} = (\mathbf{e}_{n-1}, \ldots, \mathbf{e}_0)$ is the $n \times n$ *reflection* matrix. (Here and hereafter, a blank space in matrix representation is assumed to be filled with zeros.) $Z_f = (\mathbf{e}_1, \ldots, \mathbf{e}_{n-1}, f\mathbf{e}_0)$ is the *unit f-circulant* matrix of equation (1.3.1), $Z_0 = Z = (\mathbf{e}_1, \ldots, \mathbf{e}_{n-1}, \mathbf{0})$ is the *unit lower triangular Toeplitz* matrix. $Z_f(\mathbf{u}) = \sum_i u_i Z_f^i$ is an *f-circulant* matrix for a vector $\mathbf{u} = (u_i)_{i=0}^{n-1}$ (see (2.6.1) in Section 2.6). $Z_0(\mathbf{u}) = Z(\mathbf{u}) = \sum_i u_i Z^i$ is a *lower triangular Toeplitz matrix*. $K(M, \mathbf{v}, k) = (M^i \mathbf{v})_{i=0}^{k-1}$ is the $n \times k$ *Krylov matrix* defined by the triple of a natural k, an $n \times n$ matrix M, and a vector $\mathbf{v}$ of dimension n. The same triple $(M, \mathbf{v}, k)$ defines the *Krylov space* of vectors with the basis formed by the columns of the matrix $K(M, \mathbf{v}, k)$. $K(M, \mathbf{v})$ denotes the matrix $K(M, \mathbf{v}, n)$.

Definition 2.1.2. ω_n is a primitive n-th root of 1, i.e., $\omega_n^n = 1$, $\omega_n^s \neq 1$, $s = 1, 2, \ldots, n - 1$; e.g., $\omega_n = \exp(2\pi \sqrt{-1}/n)$ in the complex number field $\mathbb{C}$. $\mathbf{w}_n = (\omega_n^i)_{i=0}^{n-1}$ is the vector of all the n-th roots of 1. $\Omega_n = (\omega_n^{ij})_{i,j=0}^{n-1}$ is the $n \times n$ matrix of the **discrete Fourier transform (DFT)**, which is a special case of Vandermonde matrices (see Table 1.1). *The discrete Fourier transform of a vector* $\mathbf{p}$ of dimension n is the vector $DFT(\mathbf{p}) = \Omega_n \mathbf{p}$. DFT_n denotes the computation of this vector provided that the vectors $\mathbf{p}$ and $\mathbf{w}_n$ are given (the vector $\mathbf{w}_n$ can be immediately obtained in $n - 2$ ops if just ω_n, a primitive n-th root of 1, is given). We write $DFT = DFT_n$, $\omega = \omega_n$, $\mathbf{w} = \mathbf{w}_n$, $\Omega = \Omega_n$ where n is known from the context. The **inverse DFT problem** is the problem of computing the vector $\mathbf{p} = \Omega^{-1} \mathbf{v}$ satisfying the equation $\Omega \mathbf{p} = \mathbf{v}$ for two given vectors $\mathbf{v} = (v_i)_{i=0}^{n-1}$ and $\mathbf{w}_n$.

Hereafter, $\lceil x \rceil$ and $\lfloor x \rfloor$ denote two integers closest to a real x such that $0 \leq \lceil x \rceil - x < 1$, $0 \leq x - \lfloor x \rfloor < 1$. For an integer k, we write

$$k_+ = 2^{\lceil \log_2 k \rceil}, \tag{2.1.1}$$

that is, k_+ is the smallest integral power of 2 exceeding $k - 1$.

The following simple results can be easily verified (see Exercise 1.10).

Theorem 2.1.3. *For the $n \times n$ matrix Z_e and any scalar e, we have $Z_e^n = eI$, $JZ_eJ = Z_e^T$. For $f \neq 0$, we have $Z_{1/f}^T = Z_f^{-1}$.*

Theorem 2.1.4. $J^2 = I$, $J\mathbf{v} = (v_{n-1-i})_{i=0}^{n-1}$, *and* $JD(\mathbf{v})J = D(J\mathbf{v})$, *for any vector* $\mathbf{v} = (v_i)_{i=0}^{n-1}$.

Theorem 2.1.5. *Pre-multiplication by the matrix J (as well as post-multiplication) transforms any Hankel matrix into a Toeplitz matrix and vice-versa.*

Because of the latter theorem, Hankel matrices differ from Toeplitz matrices only by the reflection permutation of their rows (or columns). So we explicitly specify only Toeplitz computations in this chapter.

Theorem 2.1.6.

$$\sum_{j=0}^{n-1} \omega_n^{ij} = \begin{cases} n & \text{if } n \text{ divides } i \\ 0 & \text{otherwise} \end{cases}$$

or, equivalently,

$$n\Omega^{-1} = \Omega^* = (\omega^{-ij})_{i,j=0}^{n-1}.$$

Theorem 2.1.7. ω^{-1} *is a primitive n-th root of* 1*, that is,* Ω^* *is also a matrix of DFT.*

Theorem 2.1.8. $K(Z_f, \mathbf{v}) = Z_f(\mathbf{v})$ *is an f-circulant matrix,* $K(Z_f^T, \mathbf{v}) = J Z_f(J\mathbf{v})$*,* $K(D(\mathbf{v}), 1) = (v_i^j)_{i,j=0}^{n-1}$ *is a Vandermonde matrix, for the* $n \times n$ *matrices* Z_f*,* $Z_f(\mathbf{v})$*,* J*,* $D(\mathbf{v})$*, and* $K(M, \mathbf{v})$ *of Definition 2.1.1, any vector* $\mathbf{v} = (v_i)_{i=0}^{n-1}$*, and any scalar* f*.*

As arranged in Section 1.8, our abbreviation "op" stands for a "field operation" (or a "flop" in numerical computations).

2.2 Fast Fourier transform (polynomial version)

A straightforward computation of the vector DFT($\mathbf{p}$) requires $2n^2 - n$ ops, provided that the $n \times n$ matrix Ω and a vector $\mathbf{p}$ of dimension n are given. Let us show a *superfast solution*, using $O(n \log n)$ ops, provided that n is a power of 2, that is, $n = 2^h$, h is a positive integer.

The DFT($\mathbf{p}$) is the vector of the values $p(\omega^j) = \sum_{i=0}^{n-1} p_i \omega^{ij}$ of a polynomial $p(x) = \sum_{i=0}^{n-1} p_i x^i$, so that the DFT is equivalent to the evaluation of a polynomial $p(x)$ on the node set $\{1, \omega, \dots, \omega^{n-1}\}$. Here is the **polynomial version of the FFT algorithm**, where $n = 2^h$.

Split $p(x)$ as follows:

$$p(x) = p_0(y) + x p_1(y), \quad \text{where } y = x^2,$$
$$p_0(y) = p_0 + p_2 x^2 + \dots + p_{n-2} x^{n-2},$$
$$p_1(y) = x(p_1 + p_3 x^2 + \dots + p_{n-1} x^{n-2}).$$

This reduces the DFT$_n$ for $p(x)$ to two problems of DFT$_{n/2}$ (for $p_0(y)$ and $p_1(y)$) at a cost of multiplying $p_1(y)$ by x n times for $x = \omega^i$, $i = 0, 1, \dots, n-1$, and of the pairwise addition of the n output values to $p_0(\omega^{2i})$. Since $\omega^{i+n/2} = -\omega^i$ for even n, we perform multiplication only $n/2$ times, that is, $f(n) \le 2f(n/2) + 1.5n$ provided that $f(k)$ ops are sufficient for DFT$_k$. By splitting the polynomials $p_0(y)$ and $p_1(y)$ recursively, we obtain the following estimate.

Theorem 2.2.1. *For* $n = n_+ = 2^h$ *and for a positive integer* h *(see (2.1.1)), the DFT$_n$ requires* $f(n) \le 1.5nh = 1.5n \log_2 n$ *ops.*

The next result can be obtained based on a converse divide-and-conquer process [BP94, p.12] or as a simple corollary of Theorems 2.1.6, 2.1.7, and 2.2.1.

Theorem 2.2.2. *For* $n = n_+$ *(see (2.1.1)), the inverse DFT$_n$ problem can be solved by using* $n + 1.5n \log_2 n$ *ops.*

2.3   Fast Fourier transform (matrix version)

A matrix version of FFT leads essentially to the same algorithm. Based on this version, the algorithm has been proved to be stable numerically.

In practice, the numerical implementation of DFT_n (for $n = n_+$ being a power of 2) relies on the following recursive factorization of the matrix Ω_n, called the **binary split fast Fourier transform (FFT) algorithm**:

$$\Omega_{2k} = P_{2k} \begin{pmatrix} \Omega_k & \\ & \Omega_k \end{pmatrix} \begin{pmatrix} I_k & \\ & D_{2k} \end{pmatrix} \begin{pmatrix} I_k & I_k \\ I_k & -I_k \end{pmatrix}. \qquad (2.3.1)$$

Here $D_{2k} = \operatorname{diag}(\omega_{2k}^i)_{i=0}^{k-1}$, $\Omega_2 = \left(\begin{smallmatrix} 1 & 1 \\ 1 & -1 \end{smallmatrix} \right)$, and P_{2k} is the $2k \times 2k$ odd/even permutation matrix, such that $P_{2k}(\mathbf{u}) = \mathbf{v}$ where $\mathbf{u} = (u_i)_{i=0}^{2k-1}$, $\mathbf{v} = (v_i)_{i=0}^{2k-1}$, $v_i = u_{2i}$, $v_{i+k} = u_{2i+1}$, $i = 0, 1, \ldots, k-1$. Computationally, this algorithm is identical to the FFT algorithm of the previous section, that is, both algorithms compute exactly the same output and intermediate values (and therefore both use $1.5n \log_2 n$ ops).

The matrix representation provides a customary framework for the backward error analysis, which shows numerical stability of this algorithm.

Theorem 2.3.1. *Let d and k be positive integers. Let $n = n_+ = 2^k$ (see (2.1.1)). Let $\mathbf{p}$ be an n-dimensional complex vector. Let $DFT(\mathbf{v})$ denote the forward DFT of a vector $\mathbf{v}$. Finally, let $FFT_d(\mathbf{p})$ denote the vector computed by the binary split FFT algorithm applied to a vector $\mathbf{p}$ and implemented in the floating point arithmetic with d bits, $d \geq 10$. Then we have*

$$FFT_d(\mathbf{p}) = DFT(\mathbf{p} + \mathbf{e}(\mathbf{p}, n))$$

where $\mathbf{e}(\mathbf{p}, n))$ is the error vector,

$$\|\mathbf{e}(\mathbf{p}, n)\| \leq ((1 + \rho 2^{-d})^k - 1)\|\mathbf{p}\|, \quad 0 < \rho < 4.83,$$

and $\|\mathbf{p}\| = \|\mathbf{p}\|_2 = (\sum_i |p_i|^2)^{1/2}$ denotes the Euclidean norm of a vector $\mathbf{p} = (p_i)_i$. Moreover,

$$\|\mathbf{e}(\mathbf{p}, n)\| \leq 5k2^{-d}\|\mathbf{p}\| \text{ for any } n \leq 2^{2^{d-6}}.$$

By observing that $\|\Omega_n \mathbf{p}\| = \sqrt{n}\|\mathbf{p}\|$, we deduce the next result.

Corollary 2.3.2. *Under the notation of Theorem 2.3.1, we have*

$$\|FFT_d(\mathbf{p}) - DFT(\mathbf{p})\| \leq 5\sqrt{n}(\log n)2^{-d}\|\mathbf{p}\|$$

provided that $n = n_+ \leq 2^{2^{d-6}}$.

The condition $n \leq 2^{2^{d-6}}$ actually allows much larger values for n than the values customarily used in all practical computations.

2.4 Polynomial multiplication (with extensions)

Problem 2.4.1. Polynomial multiplication. Given the coefficients of two polynomials, $u(x) = \sum_{i=0}^{m} u_i x^i$ and $v(x) = \sum_{i=0}^{n} v_i x^i$, compute the coefficients of the polynomial $p(x) = \sum_{i=0}^{m+n} p_i x^i = u(x)v(x)$. (The coefficient vector $(p_i)_{i=0}^{m+n}$ of the polynomial $p(x)$ is also customarily called by *the convolution* of the coefficient vectors $(u_i)_{i=0}^{m}$ and $(v_i)_{i=0}^{n}$ of the two polynomials $u(x)$ and $v(x)$, and the problem is also called the **computation of the convolution**.)

The straightforward solution algorithm uses $9(m+1)(n+1) - m - n - 1$ ops. For $m = n$, this is $9(n+1)^2 - 2n - 1$. A non-trivial algorithm multiplies a pair of polynomials of degrees less than n by using $M(n)$ ops where

$$2n \leq M(n) = O((n \log n) \log \log n) \qquad (2.4.1)$$

and can be applied over any field or even over any ring of constants.

Let us show a superfast algorithm that relies on the *evaluation/interpolation techniques* of A. L. Toom, 1963. The solution assumes computations in the field containing a primitive 2^h root of 1 for $h = \lceil \log_2(m+n+1) \rceil$, that is, $h = \lceil \log_2(2n+1) \rceil$ for $m = n$, and then supports the complexity bound of $M(n)$ ops for

$$M(n) = O(n \log n). \qquad (2.4.2)$$

Solution: Write $h = \lceil \log_2(m+n+1) \rceil$, $N = (m+n+1)_+ = 2^h$.

1. Compute the values $U_j = \sum_i u_i \omega_N^{ij}$, $V_j = \sum_i v_i \omega_N^{ij}$, $j = 0, 1, \ldots, N-1$, of the polynomials $u(x)$ and $v(x)$ at the N-th roots of 1 (that is, twice perform DFT$_N$).

2. Compute the values $P_j = U_j V_j$, $j = 0, 1, \ldots, N-1$ (by using N ops).

3. Compute and output the values $(p_j)_{j=0}^{N-1} = \frac{1}{N} \Omega_N^*(P_j)_{j=0}^{N-1}$ (that is, perform the inverse DFT$_N$). □

Simple counting shows that this evaluation/interpolation technique uses $O((m + n+1) \log(m+n+1))$ ops provided that a primitive N-th root of 1 or all the N-th roots of 1 are available.

Let us next relate the resulting algorithm and cost bounds (2.4.1), (2.4.2) to **multiplication of an $(n+1) \times (n+1)$ Toeplitz matrix by a vector.**

Indeed, the vector equation

$$
\begin{pmatrix}
u_0 & & & O \\
\vdots & \ddots & & \\
\vdots & & \ddots & u_0 \\
u_m & & \ddots & \vdots \\
& \ddots & & \vdots \\
O & & & u_m
\end{pmatrix}
\begin{pmatrix}
v_0 \\
\vdots \\
v_n
\end{pmatrix}
=
\begin{pmatrix}
p_0 \\
\vdots \\
\vdots \\
p_m \\
\vdots \\
p_{m+n}
\end{pmatrix}
\qquad (2.4.3)
$$

is equivalent to the polynomial equation

$$\left(\sum_{i=0}^{m} u_i x^i\right)\left(\sum_{i=0}^{n} v_i x^i\right) = \sum_{i=0}^{m+n} p_i x^i. \qquad (2.4.4)$$

Consequently, the coefficients $p_0, \ldots, p_{m+n}$ of the latter polynomial product form the vector on the right-hand side of (2.4.3), which is a product of an $(m + n + 1) \times (n + 1)$ Toeplitz matrix T by a vector. (2.4.3) reduces Problem 2.4.1 of polynomial multiplication (2.4.4) (computation of the convolution) to Toeplitz-by-vector multiplication. Let us also show the converse reduction. Consider (2.4.3) for $m = 2n$ where the Toeplitz matrix T has size $(3n + 1) \times (n + 1)$ but is not general (it has zero entries at and around two corners). The $(n + 1) \times (n + 1)$ submatrix formed by the rows in the middle — from the $(n + 1)$-th to the $(2n + 1)$-th — is a general Toeplitz matrix, however. Its multiplication by a vector is a subproblem of the convolution problem and can be done in $O(M(n))$ ops. Hereafter, various polynomial computations are reduced to polynomial multiplication, and their complexity is usually expressed in terms of $M(n)$.

The correlation between Toeplitz-by-vector multiplication and convolution is very close because it comes merely from equating the coefficients of all powers x^i on both sides of (2.4.4). Nevertheless, some differences become clear immediately in various extensions and generalizations.

1) The evaluation/interpolation scheme can be immediately extended to *multiplication of several polynomials*, with no simple counterpart in the matrix version of the convolution algorithm.

2) *Some computations with integers are related closely to polynomial computations.* Indeed, any non-negative integer can be written in binary form as the value of a polynomial $\sum_i b_i x^i$ at $x = 2$ where the b_i's are 0's or 1's or, more generally, as the value of this polynomial at $x = 2^h$ where the b_i's are integers in the range from 0 to $2^h - 1$. The latter observation, due to [FP74], *enables immediate interpolation to a polynomial $p(x) = \sum_{i=0}^{n} p_i x^i$ with bounded non-negative integer coefficients p_i, $0 \le p_i < 2^h$, $i = 0, \ldots, n$, from its single value v at $x = 2^h$ for a sufficiently large h.* In an extension allowing also negative integer coefficients p_i, $-2^{h-1} \le p_i < 2^{h-1}$, first recover the polynomial $p(x) + 2^{h-1} \sum_{i=0}^{n-1} x^i$ from its value $v + 2^{h-1} n$, and then obtain the polynomial $p(x)$. These techniques (called **binary segmentation**) enable a natural extension of some operations with polynomials having integer coefficients to the same operations with integers and vice-versa; the techniques apply to multiplication, division, the computation of the greatest common divisors, and the Chinese Remainder Algorithm. Similar correlation between computations with integers and matrices was also observed, but this is a more subtle approach. Its exploitation requires some extra care and so far has not utilized Toeplitz type structures.

3) The superfast algorithms of this section have been extended to multivariate polynomial multiplication. The extensions do not rely on correlation to Toeplitz matrices.

Let us conclude this section with two important extensions of superfast convolution algorithms.

Problem 2.4.2. Generalized DFT. Given the values $q_0, q_1, \ldots, q_{k-1}$, a, a^{-1}, and $a^{-1/2}$, compute the values $r_h = \sum_{j=0}^{k-1} q_j a^{hj}$ for $h = 0, 1, \ldots, k-1$. (For $a = \omega_k$, this is DFT$(\mathbf{q})$, $\mathbf{q} = (q_i)_{i=0}^{k-1}$.) As in the FFT algorithm, we may rewrite $r_h = r_{h,0} + a r_{h,1}$, $r_{h,0} = \sum_i q_{2i}(a^2)^{hi}$, $r_{h,1} = \sum_i q_{2i+1}(a^2)^{hi}$, and this would enable us to relax the assumption that the input includes the value $a^{-1/2}$.

Solution: Write

$$r_h = a^{-h^2/2} \sum_{j=0}^{k-1} a^{(j+h)^2/2}(a^{-j^2/2}q_j), \quad h = 0, 1, \ldots, k-1.$$

Substitute $m = k-1$, $n = 2k-2$, $p_{n-h} = r_h a^{h^2/2}$, $u_j = a^{-j^2/2}q_j$ for $j \le m$, $u_j = 0$ for $j > m$, $v_s = a^{(n-s)^2/2}$ and rewrite the expressions for r_h as follows:

$$p_i = \sum_{j=0}^{i} u_j v_{i-j}, \quad i = m, m+1, \ldots, n.$$

Now, the generalized DFT problem has been reduced to a subtask of the convolution task, that is, of Problem 2.4.1, which can be solved in $O(M(k))$ ops. □

Problem 2.4.3. Taylor's shift of the variable. Given a scalar s and the coefficients $r_0, r_1, \ldots, r_n$ of a polynomial $r(x) = \sum_{i=0}^{n} r_i x^i$, compute the coefficients of the polynomial

$$q(y) = r(y+s) = \sum_{h=0}^{n} r_h (y+s)^h = \sum_{g=0}^{n} q_g(s) y^g,$$

or, equivalently, compute at $x = s$ the values of the polynomial $r(x)$ and all scaled derivatives $r^{(g)}(s)/g! = q_g(s)$ for $g = 1, \ldots, n-1$.

Solution: Expand the powers $(y+s)^h$ and obtain that

$$q_g(s) = \sum_{h=g}^{n} r_h s^{h-g} \frac{h!}{g!(h-g)!}, \quad g = 0, 1, \ldots, n.$$

Substitute $p_{n-g} = g! q_g(s)$, $u_{n-h} = h! r_h$, $v_k = s^k/k!$, $j = n-h$ and $i = n-g$ and arrive at the following expressions (which are a part of the convolution Problem 2.4.1):

$$p_i = \sum_{j=0}^{i} u_j v_{i-j}, \quad i = 0, 1, \ldots, n.$$

Superfast solution (using $O(M(n))$ ops) follows. □

By combining the solutions of Problems 2.4.1 and 2.4.2, we obtain superfast algorithms that use $O(n \log n)$ ops to compute the values $p(r^i + t) = \sum_{j=0}^{n} p_j (r^i + t)^j$ of a polynomial $p(x)$ at the points $x = r^i + t$, $i = 0, 1, \ldots, n$, for any set $\{p_0, \ldots, p_n, t, r^{-1/2}\}$. Furthermore, for any a, b, and r, we may fix s and t such that $(r^i + s)^2 + t = r^{2i} + 2ar^i + b$, $i = 0, 1, \ldots, n$. Therefore, we obtain the following result.

Theorem 2.4.4. *For any set of scalars $c \neq 0$, $a, b, p_0, \ldots, p_n, t$, the values of the polynomial $p(x) = \sum_{i=0}^{n} p_i x^i$ at the points $t_i = ct^{2i} + 2at^i + b$, $i = 0, 1, \ldots, n$, can be computed in $O(M(n))$ ops.*

Remark 2.4.5. The FFT, convolution, and consequently Toeplitz-by-vector multiplication algorithms allow effective *parallelization* using arithmetic time $O(\log n)$ on $O(M(n)/\log n)$ processors. For computing in the complex field, we have the bound of arithmetic time $O(\log n)$ on n processors. These bounds are basic for estimating the parallel arithmetic complexity of the algorithms of this book.

Remark 2.4.6. The stated estimates for parallel complexity of FFT are not complete. They do not cover the time-complexity of *processor communication* and *synchronization*. The same applies to parallel complexity estimates that we state for all other algorithms in this book because these algorithms are ultimately reduced to FFT's (or FDST/FDCT's, that is, fast discrete sine/cosine transforms) either directly or via polynomial multiplication (convolution), equivalent to multiplication of a Toeplitz matrix by a vector. The time-complexity of FFT (as well as FDST/FDCT) covering processor communication and synchronization has been thoroughly estimated under most realistic models of parallel computing. The extension of the estimates to the algorithms in this book, however, would have moved the reader into a huge domain, which is completely different from our present subjects, both in spirit and techniques. As opposed to the book, the study of processor communication and synchronization depends heavily on the specific computers on which the algorithms are implemented and therefore on technological progress in the computer industry. On the other hand, the comparative efficiency of parallel versions of the algorithms in this book is already sufficiently well represented by arithmetic complexity estimates. The latter estimates definitely favor Newton's iteration over the divide-and-conquer approach, and this conclusion does not change when the communication/synchronization cost is also taken into account.

2.5   Polynomial division and triangular Toeplitz matrix inversion

Problem 2.5.1. Inversion of a $K \times K$ non-singular triangular Toeplitz matrix.
Let $v_0 \neq 0$. Compute the vector $Z^{-1}(\mathbf{v})\mathbf{e}_0$, where $Z(\mathbf{v}) = \sum_{i=0}^{K-1} v_i Z^i$ for a vector $\mathbf{v} = (v_i)_{i=0}^{K-1}$ is a triangular Toeplitz matrix in Definition 2.1.1.

Problem 2.5.2. Computing the reciprocal of a polynomial modulo a power of x.
Given a positive integer K and a polynomial $v(x) = \sum_{i=0}^{K-1} v_i x^i$, $v_0 \neq 0$, compute a polynomial $t(x) = \sum_{i=0}^{K-1} t_i x^i$ such that

$$t(x)v(x) = 1 \bmod x^K. \tag{2.5.1}$$

Solution of Problems 2.5.1 and 2.5.2: First observe the *equivalence of Problems 2.5.1 and 2.5.2.* Equate the coefficients of x^i on both sides of the latter equation for all i, that $Z^K = 0$, and deduce that $Z(\mathbf{t})Z(\mathbf{v}) = I$ where $\mathbf{t} = (t_i)_{i=0}^{K-1}$, $\mathbf{v} = (v_i)_{i=0}^{K-1}$, and $Z(\mathbf{t}) = \sum_{i=0}^{K-1} t_i Z^i$. Then reverse the argument. This proves the equivalence and also shows that $Y = Z^{-1}(\mathbf{v}) = Z(\mathbf{t})$ is a lower triangular Toeplitz matrix completely defined by its first column vector $Y\mathbf{e}_0$.

Here is one of the known algorithms for a superfast computation of the vector $Y\mathbf{e}_0$ in $O(M(K))$ ops for $M(n)$ in (2.4.1), (2.4.2). Observe the following block representation:

$$ Y = \begin{pmatrix} T & 0 \\ T_0 & T_1 \end{pmatrix}^{-1} = \begin{pmatrix} T^{-1} & 0 \\ -T_1^{-1}T_0T^{-1} & T_1^{-1} \end{pmatrix}. $$

Here, T_0 is a Toeplitz matrix, T is a $\lceil K/2 \rceil \times \lceil K/2 \rceil$ triangular Toeplitz matrix, and T_1 is its $\lfloor K/2 \rfloor \times \lfloor K/2 \rfloor$ submatrix, so T_1^{-1} is a submatrix of the matrix T^{-1}. Let us write $h = \lceil \log_2 K \rceil$. ($T_1 = T$, $T_1^{-1} = T^{-1}$ for even K.) To compute the vector $Y\mathbf{e}_0$, it remains to compute the vectors $\mathbf{x} = T^{-1}\mathbf{e}_0$ (which defines the matrix T^{-1} and its submatrix T_1^{-1}), $\mathbf{y} = T_0\mathbf{x}$, and $-T_1^{-1}\mathbf{y}$. By results of the previous section, the two latter steps require at most $cM(2^{h-1})$ ops for some constant c. Let $I(s)$ ops be sufficient for the inversion of an $s \times s$ triangular Toeplitz matrix. Then we obtain that

$$ I(K) \leq I(2^h) \leq 2cM(2^{h-1}) + 2^{h-1} + I(2^{h-1}). $$

Continue recursively and obtain that

$$ \begin{aligned} I(K) \leq I(2^h) &\leq 2cM(2^{h-1}) + 2^{h-1} + I(2^{h-1}) \\ &\leq 2c(M(2^{h-1}) + M(2^{h-2})) + 2^{h-1} + 2^{h-2} + I(2^{h-2}) \\ &\leq \sum_{i=0}^{h-1}(2cM(2^i) + 2^i) + 1 \\ &< 2^h + 2cM(2^h). \end{aligned} $$

This completes the solution of Problems 2.5.1 and 2.5.2 at the arithmetic cost of $M(n)$. $\qquad\square$

Problems 2.5.1 and 2.5.2 are closely related to the following problem.

Problem 2.5.3. Polynomial division. For two polynomials $u(x) = \sum_{i=0}^{m} u_i x^i$ and $v(x) = \sum_{i=0}^{n} v_i x^i$, $v_n = 1$, $m \geq n$, compute *the quotient* $q(x) = \sum_{i=0}^{m-n} q_i x^i$ and *the remainder $r(x)$* of their division defined uniquely by the relations

$$ u(x) = q(x)v(x) + r(x), \quad \deg r(x) < n. \tag{2.5.2} $$

If $r(x) = 0$, we say that the polynomial $v(x)$ *divides* $u(x)$ or that $v(x)$ is a *divisor* of $u(x)$. We call two polynomials *co-prime* if they have only a constant common divisor.

The straightforward solution of the polynomial division problem requires $2Kn$ ops, for $K = m - n + 1$. The cost bound turns into $2(n+1)^2$ for $m = 2n + 1$. Let us show superfast solution using $O((m/K)M(K))$ ops.

Solution: First equate the coefficients of the leading powers x^i, $i = m, m-1, \ldots, n$, on both sides of the equation in (2.5.2) and obtain a non-singular upper triangular Toeplitz linear system of K equations,

$$Z^T(\mathbf{p})\mathbf{q} = (u_i)_{i=n}^m.$$

Here

$$Z(\mathbf{p}) = \sum_{j=0}^{K-1} p_j Z^j$$

is the matrix of Definition 2.1.1, $\mathbf{q} = (q_i)_{i=0}^{m-n} = (Z^T(\mathbf{p}))^{-1}(u_i)_{i=n}^m$, $\mathbf{p} = (p_j)_{j=0}^{m-n}$, $p_j = (v_{n-j})$ for $j = 0, \ldots, m-n$; $p_{n-i} = v_i = 0$ for $i < 0$, $v_n = 1$. Now, we solve Problem 2.5.1 to obtain the matrix $(Z^T(\mathbf{p}))^{-1} = (Z^{-1}(\mathbf{p}))^T$, then compute the quotient vector $\mathbf{q} = (Z^T(\mathbf{p}))^{-1}(u_i)_{i=n}^m$ in $O(M(K))$ ops, and finally compute the remainder $r(x)$ in $O((m/K)M(K))$ ops, by using the equation in (2.5.2) reduced modulo x^n or by using its matrix version obtained by equating the coefficients of x^i, $i = 0, 1, \ldots, n-1$, on both sides of (2.5.2).   □

The latter algorithms for polynomial division and reciprocals were first discovered in their seemingly different but equivalent polynomial versions. Also several other dual pairs of equivalent polynomial and matrix algorithms for polynomial division are derived differently, but compute the same intermediate and output values.

We conclude this section with two sample applications of the solution of a triangular Toeplitz linear system and polynomial division.

Problem 2.5.4. Computing the power sums of polynomial roots. Unlike the roots r_j of a polynomial

$$x^n + \sum_{i=0}^{n-1} p_i x^i = \prod_{j=1}^n (x - r_j) \qquad (2.5.3)$$

their power sums $s_k = \sum_{j=1}^n r_j^k$ can be rationally expressed via the coefficients. Namely, we have the following system of Newton's identites,

$$\sum_{i=0}^{k-1} p_{n-i} s_{k-i} + k p_{n-k} = 0, \quad k = 1, 2, \ldots, n, \qquad (2.5.4)$$

$$\sum_{i=0}^n p_{n-i} s_{n+k-i} = 0, \quad k = 1, 2, \ldots, m-n, \qquad (2.5.5)$$

for any $m > n$. A simple derivation of (2.5.5) is shown at the beginning of Section 2.7.)

Solution: The n equations (2.5.4) form a unit triangular Toeplitz linear system, from which we may compute the power sums $s_1, \ldots, s_n$ by using $O(M(n))$ ops. Substitute these values s_i into the first n equations of system (2.5.5) to obtain a unit triangular Toeplitz linear system for the next portion of the power sums $s_{n+1}, \ldots, s_{2n}$ and then recursively continue this computation to obtain the values $s_1, \ldots, s_K$ by using a total of $O((K/n)M(n))$ ops for any K. $\qquad\square$

The converse problem of computing the coefficients $\{p_i\}_{i=0}^{n-1}$ where the power sums are given is a little more involved but has a superfast solution, whose complexity ranges from $O(M(n))$ to $O(M(n)\log n)$ ops depending on the field $\mathbb{F}$ of constants.

Problem 2.5.5. Generalized Taylor expansion. Given the coefficients of two polynomials $u(x)$ of degree N and $t(x)$ of degree n, let m denote $\lceil (N+1)/n \rceil - 1$ and compute the coefficients of the polynomials $s_i(x)$, $i = 0, 1, \ldots, m$, of degrees less than n, defining the Taylor expansion of $u(x)$ as a polynomial in $t(x)$,

$$u(x) = \sum_{i=0}^{m} s_i(x)(t(x))^i.$$

Solution: Again apply the divide-and-conquer method assuming without loss of generality that $m + 1 = (N+1)/n = 2^h$ is an integral power of 2. Compute the integer $k = \lceil m/2 \rceil$, divide the polynomials $u(x)$ by $(t(x))^k$, and obtain that $u(x) = q(x)(t(x))^k + r(x)$. This reduces the original problem to the computation of the generalized Taylor expansions of two polynomials $q(x)$ and $r(x)$ of degrees less than $nk = (N+1)/2$. Recursively, apply this algorithm and note that only the powers $(t(x))^J$ for $J = 2^j$, $j = 1, 2, \ldots, h - 1$, have to be computed (we may compute them by means of *repeated squaring* or by means of the evaluation and interpolation on the sets of roots of 1). Then the desired generalized Taylor expansion of the polynomial $u(x)$ is computed in $m + 1$ operations of polynomial division. All these operations together require at most $\sum_{j=1}^{h-1} 2^j D((N+1)/2^j) = O(M(N)\log m)$ ops with a small overhead constant hidden in the "O" notation. The latter bound dominates the arithmetic cost $O(M(N))$ of repeated squaring. Here, $D(s) = O(M(s))$ denotes the cost of the division with a remainder of two polynomials of degrees of at most s. $\qquad\square$

If $t(x) = x - a$ we arrive at Problem 2.4.3 of Taylor's shift of the variable.

2.6 The algebra of f-circulant matrices. Diagonalizations by DFT's

In this section, we extend Problems 2.5.2 and 2.5.3 but start with definitions of independent interest. For any fixed f, the $n \times n$ matrix Z_f in (1.3.1) and Definition 2.1.1 generates the *algebra of f-circulant matrices*,

$$Z_f(\mathbf{v}) = K(Z_f, \mathbf{v}) = \sum_{i=0}^{n-1} v_i Z_f^i = \begin{pmatrix} v_0 & fv_{n-1} & \cdots & fv_1 \\ v_1 & v_0 & & \vdots \\ \vdots & & \ddots & fv_{n-1} \\ v_{n-1} & \cdots & v_1 & v_0 \end{pmatrix}; \qquad (2.6.1)$$

each matrix is defined by its first column $\mathbf{v} = (v_i)_{i=0}^{n-1}$ and by a scalar f (see Definition 2.1.1 and Theorem 2.1.8). *Circulant, skew-circulant,* and *lower triangular Toeplitz*

Figure 2.1: Correlation among polynomial and matrix computations I. ("P" is an abbreviation for "Problem," the arrows indicate reductions of problems to each other.)

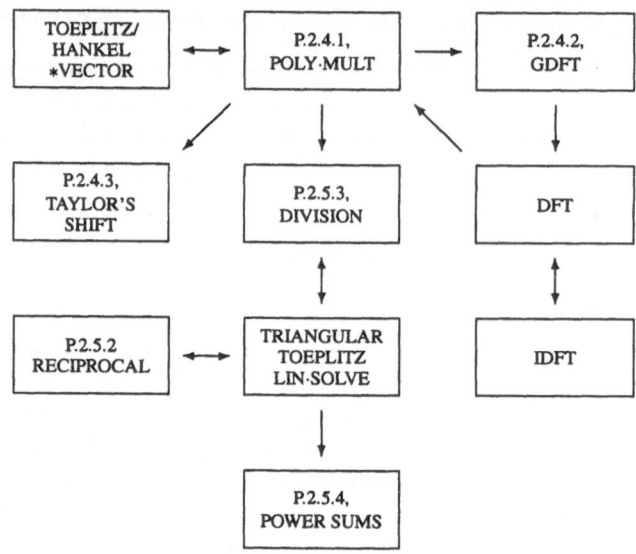

matrices are three subclasses of f-circulant matrices where $f = 1$, $f = -1$, and $f = 0$, respectively. f-circulant matrices form a subclass of Toeplitz matrices. On the other hand, every $n \times n$ Toeplitz matrix $T = (t_{i-j})_{i,j=0}^{n-1}$ can be uniquely represented as the sum

$$T = Z_e(\mathbf{u}) + Z_f(\mathbf{v}) \qquad (2.6.2)$$

for any fixed triple (e, f, t_{-n}), $e \neq f$, where $u_i + v_i = t_i$, $eu_i + fv_i = t_{i-n}$, $i = 0, 1, \ldots, n - 1$.

The matrix Z_f is f-*potent of order* n, that is,

$$Z_f^n = fI, \qquad (2.6.3)$$

and this immediately implies the next results, which we used for $f = 0$ in the previous section.

Theorem 2.6.1. *The algebra* $\{Z_f(\mathbf{v})\}$ *of* f-*circulant matrices having size* $n \times n$ *is isomorphic to the algebra* $\{p(x) \bmod (x^n - f)\}$ *of polynomials modulo* $x^n - f$ *where the first column* $\mathbf{v} = (v_i)$ *of a matrix* $Z_f(\mathbf{v})$ *is mapped into a polynomial* $\sum_i v_i x^i$ *and vice-versa.*

Corollary 2.6.2. *Let* $u(x) = \sum_i u_i x^i$, $v(x) = \sum_i v_i x^i$, $u(x)v(x) \bmod (x^n - f) = p(x) = \sum_i p_i x^i$, $q(x)t(x) \bmod (x^n - f) = 1$, $q(x) = \sum_i q_i x^i$, $t(x) = \sum_i t_i x^i$. *Then* $Z_f(\mathbf{u})Z_f(\mathbf{v}) = Z_f(\mathbf{p})$, $Z_f(\mathbf{q}) = Z_f^{-1}(\mathbf{t})$ *where* $\mathbf{u} = (u_i)$, $\mathbf{v} = (v_i)$, $\mathbf{p} = (p_i)$, $\mathbf{q} = (q_i)$, $\mathbf{t} = (t_i)$.

Corollary 2.6.3. *The products and inverses of f-circulant matrices are f-circulant matrices.*

The next superfast algorithms for multiplication and inversion of f-circulant matrices of size $n \times n$ use $O(M(n))$ ops for $M(n)$ in (2.4.1), (2.4.2). By the above results, these operations are equivalent to the problems of polynomial multiplication and division modulo $x^n - f$, which we already solved for $f = 0$. Since a square f-circulant matrix is fully defined by its first column, it is sufficient for us to compute this column of the f-circulant product or inverse.

To perform f-circulant multiplication and division efficiently for $f \neq 0$, we rely on the following diagonalization of the matrices $Z_f(\mathbf{v})$ in (2.6.1) by a *similarity transformation*, that is, by a transformation $W^{-1}Z_f(\mathbf{v})W$ for a non-singular matrix W. (Note that it is sufficient to prove this diagonalization for the generator Z_f of the algebra and see extensions in Section 3.11.)

Theorem 2.6.4. *We have*

$$Z_1(\mathbf{v}) = \Omega^{-1} D(\Omega \mathbf{v}) \Omega. \tag{2.6.4}$$

More generally, for any $f \neq 0$, we have

$$Z_{f^n}(\mathbf{v}) = U_f^{-1} D(U_f \mathbf{v}) U_f \tag{2.6.5}$$

where

$$U_f = \Omega D(\mathbf{f}), \quad \mathbf{f} = (f^i)_{i=0}^{n-1}, \tag{2.6.6}$$

and the matrices $D(\mathbf{u})$ and $\Omega = \Omega_n = (\omega_n^{ij})_{i,j=0}^{n-1}$ are defined in Definitions 2.1.1 and 2.1.2, so $D(\mathbf{f}) = I$, $U_f = \Omega$ where $f^n = 1$.

Proof. Pre-multiply both sides of (2.6.4) by Ω and (2.6.5) by U_f and verify the resulting equations by inspection. □

(2.6.4)–(2.6.6) reduce multiplication of the matrix $Z_f(\mathbf{v})$ by a vector essentially to three DFT_n, and similarly for the inverse matrix $Z_f^{-1}(\mathbf{v})$ where $f \neq 0$. If our computations are over a field with the n_+-th roots of 1 (see (2.1.1)), we may apply the FFT algorithm and achieve the overall cost bound of $O(n \log n)$ ops for these operations. Due to (2.6.2), *the bound $O(n \log n)$ is extended to multiplication of $n \times n$ Toeplitz matrices and, consequently, by Theorem 2.1.5, to the same operation for Hankel matrices.*

An inspection shows that the Toeplitz/Hankel-by-vector multiplication based on Theorems 2.6.4 and 2.1.5 and (2.6.2) uses FFT's of smaller sizes and, consequently, uses fewer ops than the evaluation-interpolation solution in Section 2.4. Practical numerical computations routinely rely on the former approach, and a satisfactory analysis of numerical stability has been developed for it. Algebraically, the approach involves roots of 1, which is a restriction on the field of constants, versus the evaluation-interpolation approach of Toom. On the other hand, Toom's approach can be adjusted to yield the equivalent dual polynomial versions of the matrix algorithms for multiplication of two polynomials modulo $x^n - 1$ and modulo $x^n + 1$ [BP94, pages 13–14] and, therefore, for multiplication of a Toeplitz matrix by a vector. The solution of Problems 2.5.2 and 2.5.3 for computing polynomial reciprocal and the inverse of a triangular

Toeplitz matrix are covered by neither Theorem 2.6.4 (note its assumption $f \neq 0$) nor the evaluation-interpolation method, but the situation is changed when we seek (arbitrary precision) approximate solutions. Consider the following pair of dual algorithms for *any precision approximate polynomial division.*

a) [B84]. For a small positive ϵ, invert the matrix $Z_\epsilon(\mathbf{v})$ by applying Theorem 2.6.4 to approximate first the matrix $(Z(\mathbf{v}))^{-1}$ and then the quotient polynomial $q(x)$.

b) [PLS92]. Extend the evaluation/interpolation techniques to polynomial division with vanishing remainder. Then extend these techniques to any pair of input polynomials $u(x)$ and $v(x)$ in (2.5.2) by choosing very large values as the nodes of evaluation. More precisely, choose these values to be the scaled K-th roots of 1, $N\omega_K^i$, $i = 0, \ldots, K-1$ for $K = (m - n + 1)_+$ (see (2.1.1)) and for a large value N. This makes the remainder nearly vanish, so

$$(q_i)_{i=0}^{K-1} \approx \text{diag}((N^{-i})_{i=0}^{K-1})\Omega_K^{-1}(t_j)_{j=0}^{K-1} \text{ where } t_j = \frac{u(N\omega_K^j)}{v(N\omega_K^j)}, \ j = 0, \ldots, K-1.$$

The two latter dual algorithms do not give us an exact solution but save some ops versus the exact solution algorithms and have a simpler structure allowing for more effective parallel implementation.

2.7   Algebra of polynomials modulo a polynomial and the Frobenius matrix algebra

In the two previous sections, we computed a polynomial reciprocal modulo a polynomial $u(x)$ in the cases of a monomial $u(x) = x^K$ or a binomial $u(x) = x^n - f$. Let us extend our study to the case of any fixed polynomial $u(x)$ by using matrix representation and more advanced techniques.

With a monic polynomial $u(x) = x^m + \sum_{i=0}^{m-1} u_i x^i = \prod_{j=1}^n (x - r_j)$, associate the $m \times m$ *Frobenius (companion) matrix,*

$$F_u = \begin{pmatrix} & & & -u_0 \\ 1 & & & -u_1 \\ & \ddots & & \vdots \\ & & 1 & -u_{m-1} \end{pmatrix} = Z - \mathbf{u}e_{m-1}^T, \quad \mathbf{u} = (u_i)_{i=0}^{m-1},$$

whose characteristic polynomial, $\det(xI - C)$, equals $u(x)$. By virtue of the Cayley-Hamilton theorem, we have $u(F_u) = 0$. Consequently,

$$F_u^i u(F_u) = 0, \quad \text{trace}(F_u^i u(F_u)) = 0, \quad i = 0, 1, \ldots$$

where $\text{trace}(F_u) = \sum_{j=1}^n r_j$, and therefore, $s_k = \sum_{j=1}^n r_j^k = \text{trace}(F_u^k)$, $k = 0, 1, \ldots$. The latter equations imply (2.5.5) for $p(x) = u(x)$.

In the previous section, we covered the case where $F_u = Z_f$, that is, $u_0 = -f$, $u_1 = \ldots = u_{m-1} = 0$. By extending Theorem 2.6.1, we obtain

Theorem 2.7.1. *The algebra generated by a matrix F_u for any fixed vector $\mathbf{u} = (u_i)_{i=0}^{m-1}$ is isomorphic to the algebra A_u of polynomials reduced modulo the polynomial $u(x) = x^m + \sum_{i=0}^{m-1} u_i x^i$ where a polynomial $f(x) = \sum_{i=0}^{m-1} f_i x^i \in A_u$ is mapped into the matrix $F_u(\mathbf{f}) = \sum_{i=0}^{n-1} f_i F_u^i$, and vice-versa.*

Likewise, Theorem 3.11.1 of Section 3.11 extends Theorem 2.6.4.

The polynomials in the algebra A_u have degree of at most $m-1$ and can be expressed in any basis formed by any m linearly independent polynomials of degree less than m. The straightforward choice is the monomial basis $\{1, x, \ldots, x^{m-1}\}$, but better numerical stability of computations in the algebra A_u is achieved by using *Horner's basis*

$$\{h_{m-i}(x) = x^{m-i} + u_{m-1}x^{m-i-1} + \ldots + u_i, \quad i = 1, \ldots, m\}.$$

Theorem 2.7.2. *Let $H\mathbf{y} = \mathbf{f}$, $\mathbf{f} = (f_i)_{i=0}^{m-1}$, $\mathbf{y} = (y_i)_{i=0}^{m-1}$, $H = (h_{i,j})_{i,j=0}^{m-1}$, $h_{i,j} = u_{i+j+1}$, for $i + j < m$, $h_{i,j} = 0$ for $i + j \geq m$, $i, j = 0, 1, \ldots, m-1$. Then we have $\sum_{i=0}^{m-1} f_i x^i = \sum_{i=0}^{m-1} y_i h_i(x)$ for the polynomials $h_i(x)$ of Horner's basis (defined above), that is, the transition between Horner's and monomial bases is defined by the triangular Hankel matrix H.*

The following result specifies a simple structure of matrices generated by the matrix F_u.

Theorem 2.7.3. *We have the equation*

$$F_u(\mathbf{f}) = Z(\mathbf{f}) - Z(\mathbf{u})Z^T(Z\mathbf{y})$$

where the vector $\mathbf{y} = H^{-1}\mathbf{f}$ is defined in Theorem 2.7.2.

The computations in the quotient algebra A_u (and, equivalently, in the *Frobenius algebra $A(F_u)$* generated by the matrix F_u) have applications to univariate polynomial root-finding and have further non-trival extensions to root-finding for a system of mutivariate polynomials.

A fundamental problem of computer algebra is the *computation of the reciprocal in the algebra A_u.* A special case where $u(x) = x^n - f$ was covered in the previous section.

Problem 2.7.4. Computing the reciprocal of a polynomial modulo a polynomial.
Compute a polynomial $t(x) = \sum_{i=0}^{m-1} t_i x^i$ that satisfies the following equations:

$$s(x)u(x) + t(x)v(x) = t(x)v(x) \bmod u(x) = 1 \qquad (2.7.1)$$

for two positive integers m and n, $m \geq n$, two fixed co-prime monic polynomials $u(x) = \sum_{i=0}^{m} u_i x^i$ and $v(x) = \sum_{i=0}^{n} v_i x^i$, $u_m = v_n = 1$, and an unknown polynomial $s(x) = \sum_{i=0}^{n-1} s_i x^i$.

For $u(x) = x^K$, (2.7.1) turns into (2.5.1) and Problem 2.5.4 turns into Problem 2.5.2.

Theorem 2.7.5. *$O(M(m) \log m)$ ops for $M(n)$ in (2.4.1), (2.4.2) are sufficient to solve Problem 2.7.4.*

Proof. Let us first reduce Problem 2.7.4 to structured matrix computations (in two ways).

Problem 2.7.4 is equivalent to the computation of the vector $\mathbf{t} = F_u^{-1}(\mathbf{v})\mathbf{e}_0$, $\mathbf{t} = (t_i)_{i=0}^{m-1}$, where $\mathbf{v} = (v_i)_{i=0}^{m-1}$, $v_i = 0$ for $i > n$, and one may exploit the structure of the matrix $F_u(\mathbf{v})$ defined by Theorem 2.7.3.

In a more customary reduction, we equate the coefficients on both sides of (2.7.1) to obtain the *Sylvester linear system* of $m + n$ equations,

$$S\begin{pmatrix} \mathbf{s} \\ \mathbf{t} \end{pmatrix} = \mathbf{e}_0, \qquad (2.7.2)$$

where $S = (U, V)$, $U = (u_{i-j})_{i=0,j=0}^{m+n-1,n-1}$, $V = (v_{i-j})_{i=0,j=0}^{m+n-1,m-1}$, $u_h = v_h = 0$ for $h < 0$, $u_h = 0$ for $h > m$, $v_h = 0$ for $h > n$, $\mathbf{s} = (s_i)_{i=0}^{n-1}$. The $(m+n) \times (m+n)$ matrix $S = S(u, v)$ is called the *Sylvester* or *resultant matrix* associated with the two polynomials $u(x)$ and $v(x)$. Its determinant, $\det S$ is called the *resultant* of the two polynomials $u(x)$ and $v(x)$. S is a 1×2 block matrix with Toeplitz blocks U and V, that is, it has a structure of Toeplitz type. We have $\det S = u_m^n v_n^m \prod_{i=1}^m \prod_{j=1}^n (x_i - y_j)$ provided that $u(x) = u_m \prod_{i=1}^m (x - x_i)$, $v(x) = v_n \prod_{j=1}^n (x - y_j)$. Therefore, the matrix S is non-singular (that is, $\det S \neq 0$) if and only if the polynomials $u(x)$ and $v(x)$ are co-prime. It follows that testing non-singularity of the matrix $S = S(u, v)$ as well as the matrix $F_u(\mathbf{v})$ in Theorem 2.7.3 is equivalent to testing whether the two polynomials $u(x)$ and $v(x)$ are co-prime. If so, the computation of the polynomial $(1/v(x)) \bmod u(x)$ is reduced to the computation of the vectors $S^{-1}\mathbf{e}_0$ or $F_u^{-1}(\mathbf{v})\mathbf{e}_0$.

In Chapter 5, we present superfast algorithms for testing non-singularity of structured matrices and for the inversion of non-singular structured matrices. The algorithms apply to the input matrices $F_u(\mathbf{f})$ in Theorem 2.7.3 and S in (2.7.2). This implies *superfast computation of reciprocals in the algebra* A_u, that is, a superfast solution of Problem 2.7.4. In Sections 2.10 and 2.11, we show other superfast algorithms for this problem. Both groups of algorithms support Theorem 2.7.5. □

Let us consider two simplifications of Problem 2.7.4.

Problem 2.7.6. Computing the reciprocal of a polynomial modulo a power of a polynomial. Solve Problem 2.7.4 in the case where the polynomial $u(x)$ has the form $(g(x))^b$, for a given polynomial $g(x)$ and an integer $b > 1$.

Solution: We reduce the problem to evaluating the reciprocal of the input polynomial $v(x)$ modulo $g(x)$ (rather than modulo the higher degree polynomial $u(x) = (g(x))^b$).

Algorithm 2.7.7. Computing the reciprocal of a polynomial modulo a power of a polynomial.

COMPUTATION:

1. Compute the coefficients of the polynomial $v_0(x) = v(x) \bmod g(x)$, where $\deg v_0(x) < \deg g(x) = m/b$.

2. Compute the coefficients of the polynomial $t(x) = t_0(x) = (1/v_0(x))$ mod $g(x)$, which is the solution to Problem 2.7.4 for $v(x) = v_0(x)$ and $u(x) = g(x)$ (recall that the polynomials $v(x)$ and $u(x)$ are co-prime and, therefore, the polynomials $v_0(x)$ and $g(x)$ are co-prime).

3. Compute the coefficients of two polynomials, $w(x) = 1 - v(x)t_0(x)$ mod $u(x)$ and $z(x) = 1/(v(x)t_0(x))$ mod $u(x) = 1/(1 - w(x))$ mod $u(x)$, by means of the following formula:

$$z(x) = \sum_{j=0}^{b-1} (w(x))^j \bmod u(x) = \prod_{h=1}^{\lceil \log_2 b \rceil} (1 + (w(x))^{2^h}) \bmod u(x).$$

(Here, we use the equations $1/(1-w) = 1+w+w^2+\cdots = (1+w)(1+w^2)(1+w^4)\cdots$, $w(x) = 1 - v(x)t_0(x) = 0$ mod $g(x)$, and consequently $(w(x))^b = 0$ mod $u(x)$ because $u(x) = (g(x)^b)$.)

4. Output the coefficients of the polynomials $t(x) = t_0(x)z(x)$ mod $u(x)$.

We perform $O(bM(m/b))$ ops at stage 1, $O(M(m/b)\log(1 + m/b))$ ops at stage 2, and $O(M(m))$ ops at stages 3 and 4. At all stages combined, we use $O(M(m/b)\log(1 + m/b) + M(m))$ ops versus $O(M(m)\log m)$ ops involved in the solution of Problem 2.7.4. This saves roughly the factor of min$\{b, \log m\}$ ops.

In the next important special case, the solution of Problems 2.7.4 and 2.7.6 is simplified further.

Problem 2.7.8. Computing the reciprocal of a polynomial modulo a shifted power. Solve Problems 2.7.4 and 2.7.6 in the special case where $g(x) = x - z$, $u(x) = (x-z)^m$, that is, compute the first m Taylor coefficients of the reciprocal polynomial $v(x) = (1/u(x))$ at $x = z$.

Solution: Apply the following algorithm.

Algorithm 2.7.9. Computing the reciprocal of a polynomial modulo a shifted power.

COMPUTATION: Successively compute the following polynomials:

1. $w(y) = v(x)$ for $y = x - z$,

2. $q(y) = (1/w(y))$ mod y^m,

3. $t(x) = q(x - z)$.

The algorithm uses $O(M(m))$ ops to compute two Taylor shifts of the variable and a reciprocal modulo a power y^m (see Problems 2.4.3 and 2.5.2).

Finally, here is a simple generalization of Problem 2.7.4.

Problem 2.7.10. Division modulo a polynomial. Given two co-prime monic poly-
nomials, $u(x)$ of degree m and $v(x)$ of degree n, and a monic polynomial $r(x) = x^k + \sum_{i=0}^{k-1} r_i x^i$, $n \leq m$, $k < m$, compute a unique polynomial $t(x) = \sum_{i=0}^{m-1} t_i x^i$ that
satisfies the equations

$$r(x) = s(x)u(x) + t(x)v(x) = t(x)v(x) \bmod u(x), \tag{2.7.3}$$

for an unknown polynomial $s(x) = \sum_{i=0}^{k+m-1} s_i x^i$.

Solution: Problem 2.7.4 is the special case where $r(x) = 1$. On the other hand,
Problem 2.7.10 is the problem of division in the algebra A_u and is immediately reduced
to Problem 2.7.4 when equation (2.7.1) is multiplied by $r(x)$ and the output polynomial
$t(x)$ is reduced modulo $u(x)$. ☐

2.8 Polynomial gcd and the (Extended) Euclidean Algorithm

The problems and algorithms in this section are related to Problems 2.7.4 and 2.7.10 as
well as to Toeplitz/Hankel matrix computations and are among the most fundamental
in computer algebra.

Problem 2.8.1. Computing a polynomial gcd. Given two polynomials,

$$u(x) = \sum_{i=0}^{m} u_i x^i \text{ and } v(x) = \sum_{i=0}^{n} v_i x^i, \quad m \geq n \geq 0, \ u_m v_n \neq 0, \tag{2.8.1}$$

compute their *greatest common divisor* (the *gcd*) $\gcd(u, v)$, that is, a common divisor
of $u(x)$ and $v(x)$ having the highest degree in x. (The gcd of two non-zero polynomials
$u(x)$ and $v(x)$ is unique up to within constant factors, or it can be assumed to be monic
and unique.)

Remark 2.8.2. $\mathrm{lcm}(u, v)$, the *least common multiple* (the *lcm*) of two polynomials
$u(x)$ and $v(x)$, that is, their common multiple having the smallest degree, equals
$u(x)v(x)/\gcd(u, v)$. Finally, observe that *any solution of the gcd Problem 2.8.1 can be
immediately extended to computing the gcd of several polynomials, $u_1(x), \ldots, u_k(x)$,*
because $\gcd(u_1, \ldots, u_{i+1}) = \gcd(\gcd(u_1, \ldots, u_i), u_{i+1})$ for $i = 2, 3, \ldots, k - 1$, and
similarly for the computation of $\mathrm{lcm}(u_1, \ldots, u_{i+1})$.

In this section we cover the classical solution algorithm, its superfast version, and
its important extension. In the next section, we show various applications, in particular
to rational function reconstruction, with some feedback to the computation of the gcd
(see Algorithm 2.9.6).

Solution I (matrix version): The gcd problem can be reduced to computations with
structured matrices (of Toeplitz/Hankel type) based on the following results (see more
in Sections 2.10 and 2.11).

Theorem 2.8.3. *Let $r(x) = \gcd(u, v)$. Then*

$$2k = 2 \deg r(x) = m + n - \text{rank}(S) \qquad (2.8.2)$$

for the matrix S in (2.7.2), and furthermore (2.7.3) hold for some co-factor polynomials $s(x)$ and $t(x)$ of degrees of at most $n - k - 1$ and $m - k - 1$, respectively.

Based on this theorem, one may immediately reduce Problem 2.8.1 to the solution of a structured linear system of equations whose coefficient matrix is an $(m + n - 2k) \times (m + n - 2k)$ submatrix of the matrix S. We cover and extend this approach in Section 2.10. Together with the algorithms in Chapter 5, this yields superfast solution algorithms.   □

Solution II (EA):   An alternative solution is via the classical algorithm.

Algorithm 2.8.4. Euclidean Algorithm (EA).

INPUT: Natural numbers m and n and the coefficients $u_0, \ldots, u_m, v_0, \ldots, v_n$ of two polynomials, $u(x) = \sum_{i=0}^{m} u_i x^i$ and $v(x) = \sum_{j=0}^{n} v_j x^j$, of degrees m and n, respectively.

OUTPUT: The coefficients of the polynomial $\gcd(u, v)$.

INITIALIZATION: Write $u_0(x) = u(x)$, $v_0(x) = v(x)$.

COMPUTATION:
    STAGE $i, i = 0, 1, \ldots, l - 1$. Compute the polynomial

$$v_{i+1}(x) = u_i(x) \bmod v_i(x) = u_i(x) - q_{i+1}(x)v_i(x), \qquad (2.8.3)$$

where $q_{i+1}(x)$ is the quotient polynomial and l is the smallest integer such that $v_l(x) \equiv 0$. Write $u_{i+1}(x) = v_i(x)$.
    Output $u_l(x) = \gcd(u, v)$.

The following equation,

$$\gcd(u_i(x), v_i(x)) = \gcd(u_{i+1}(x), v_{i+1}(x))$$

holds for all i and implies correctness of the algorithm. Observe that for each i the i-th stage amounts to computing the remainder of the division of polynomials $u_i(x)$ by $v_i(x)$ and $l \leq n + 1$, that is, there can be at most $n + 1$ stages of this algorithm since $\deg v_{i+1}(x) < \deg v_i(x)$.
    Hereafter, assume that $m = n$ (otherwise replace $u(x)$ or $v(x)$ by $u(x) + v(x)$). Then the algorithm involves $O(n^2)$ ops.
    For completeness, here is a superfast version of the EA using $O(M(n) \log n)$ ops for $M(n)$ in (2.4.1), (2.4.2), even though Solution I has a simpler derivation and better numerical stability (see Remark 2.8.11 at the end of this section).

Solution III (superfast EA): We start with preliminaries. The algorithm relies on the following matrix representation of (2.8.3):

$$\begin{pmatrix} u_i(x) \\ v_i(x) \end{pmatrix} = \tilde{Q}_i(x) \begin{pmatrix} u_{i-1}(x) \\ v_{i-1}(x) \end{pmatrix} = \hat{Q}_i(x) \begin{pmatrix} u(x) \\ v(x) \end{pmatrix},$$

$$\tilde{Q}_i(x) = \begin{pmatrix} 0 & 1 \\ 1 & -q_i(x) \end{pmatrix}, \quad \hat{Q}_0(x) = 1, \tag{2.8.4}$$

$$\hat{Q}_i(x) = \hat{Q}_{i-1}(x)\tilde{Q}_i(x) = \tilde{Q}_i(x)\tilde{Q}_{i-1}(x)\cdots\tilde{Q}_1(x),$$

$$i = 1,\ldots,l; \quad l \le n+1.$$

We also note the following helpful matrix equation proposed by W. Eberly,

$$\tilde{Q}_i(x) = \begin{pmatrix} 0 & 1 \\ 1 & -q_i(x) \end{pmatrix} = \begin{pmatrix} 0 & 1 \\ 1 & -q_{h_i,i} \end{pmatrix} \begin{pmatrix} 1 & -q_{h_i,i} \\ 0 & 1 \end{pmatrix} \cdots \begin{pmatrix} 1 & -q_{0,i} \\ 0 & 1 \end{pmatrix},$$

which holds provided that $q_i(x) = \sum_{k=0}^{h_i} q_{k,i} x^k$. Due to this matrix representation of Euclid's algorithm (EA), the computation of $\gcd(u, v)$ has been reduced to the computation of the first row of the matrix $\hat{Q}_l(x)$ and to multiplication of this row by the vector $(u(x), v(x))^T$, that is, to performing polynomial multiplication twice. Given the matrices $\tilde{Q}_i(x)$, $i = 1,\ldots,l$, the matrix $\hat{Q}_l(x)$ can be computed in $l - 1$ recursive steps (2.8.4) of 2×2 matrix multiplication.

Now write $d_i = \deg v_{i-1}(x) - \deg v_i(x)$, for $i = 1,\ldots,l-1$, and first assume that $d_i = 1$ for all i. Then, by inductive application of (2.8.3), verify the following *key observation: the matrix $\tilde{Q}_k(x)$ only depends on x and the k leading coefficients of each of the polynomials u(x) and v(x).* To relax the assumption that $d_i = 1$ for all i, we shift from $\tilde{Q}_k(x)$ to the matrices $Q_k(x)$, which we define by extending the sequence of triples $(u_i(x), v_i(x), \hat{Q}_i(x))$, $i = 1,\ldots,l$, so that the i-th triple is replaced by its d_i copies. Let $\{w_k(x), z_k(x), Q_k(x)\}$, $k = 1,\ldots,n$, denote the resulting sequence of triples.

Example 2.8.5. Consider a specific instance:

$$\begin{array}{lll} u(x) = u_0(x) = 8x^5 + 4x + 4, & v(x) = v_0(x) = 4x^5 + 2, & \\ u_1(x) = v_0(x) = 4x^5 + 2, & v_1(x) = 4x, & q_1(x) = 2, \\ u_2(x) = v_1(x) = 4x, & v_2(x) = 2, & q_2(x) = x^4, \\ u_3(x) = v_2(x) = 2, & v_3(x) = 0, & q_3(x) = 2x, \end{array}$$

$$\hat{Q}_1(x) = \tilde{Q}_1(x) = \begin{pmatrix} 0 & 1 \\ 1 & -2 \end{pmatrix}, \quad \tilde{Q}_2(x) = \begin{pmatrix} 0 & 1 \\ 1 & -x^4 \end{pmatrix}, \quad \tilde{Q}_3(x) = \begin{pmatrix} 0 & 1 \\ 1 & -2x \end{pmatrix},$$

$$\hat{Q}_2(x) = \tilde{Q}_1(x)\tilde{Q}_2(x) = \hat{Q}_1(x)\tilde{Q}_2(x),$$

$$\hat{Q}_3(x) = \tilde{Q}_1(x)\tilde{Q}_2(x)\tilde{Q}_3(x) = \hat{Q}_2(x)\tilde{Q}_3(x),$$

Under these assumptions, we have $d_1 = 4$, $d_2 = 1$,

$$\begin{array}{llll} Q_i(x) = \hat{Q}_i(x), & w_i(x) = u_1(x), & z_i(x) = v_1(x), & i = 1, 2, 3, 4, \\ Q_5(x) = \hat{Q}_2(x), & w_5(x) = w_2(x), & z_5(x) = v_2(x), & \\ Q_6(x) = \hat{Q}_2(x), & w_6(x) = w_3(x), & z_6(x) = v_3(x). & \end{array}$$

$\square$

Now, we have

$$(w_k(x), z_k(x))^T = Q_k(x)(u(x), v(x))^T,$$
$$\deg Q_k(x) \le n - \deg w_k(x) \le k, \quad k = 1, \ldots, n. \tag{2.8.5}$$

Due to re-enumeration and extension of the sequence of triples, the *key-observation* is extended to the case of any sequence d_i but for the matrices $Q_k(x)$ rather than $\tilde{Q}_k(x)$.

At this point, we assume that $n + 1$ is a power of 2 (otherwise, replace $u(x)$ by $x^d u(x)$ and $v(x)$ by $x^d v(x)$ where $d + n + 1$ is a power of 2 and $d \le n$) and specify a superfast version of Euclid's algorithm as follows:

Algorithm 2.8.6. Superfast EA.

INPUT: A natural number k and the coefficients of two polynomials $u(x)$ and $v(x)$ of degree $n = 2^k$.

OUTPUT: The coefficients of $\gcd(u, v)$ defined within a scalar factor.

INITIALIZATION: Evaluate the matrix $Q_1(x) = \tilde{Q}_1(x)$. (This involves just two leading coefficients u_m and v_n.)

COMPUTATION: Proceed recursively for $h = 0, 1, \ldots, \log_2(n + 1)$.

1. Assume that a procedure for the evaluation of the matrix $Q_H(x)$ is available for $H = 2^h$ (for $h = 0$ such a procedure amounts to the division of $u(x)$ by $v(x)$), and recursively define a procedure for the evaluation of the matrix $Q_{2H}(x)$ as follows:

   (a) Note that the entries of $Q_H(x)$ only depend on the H leading terms of each of the two polynomials $u(x)$ and $v(x)$. Keep these terms and also the next H terms of each of the polynomials and drop all other terms when applying the matrix equation (2.8.5) for $k = H$. Denote the resulting polynomials by $u^*(x)$ and $v^*(x)$. Compute the matrix $Q_H(x)$ and then the pair of polynomials $(u_H^*(x), v_H^*(x))^T = Q_H(x)(u^*(x), v^*(x))^T$.

   (b) Apply step (a) starting with the pair $(u_H^*(x), v_H^*(x))$, rather than with the pair $(u(x), v(x))$ in order to compute the matrix $Q_H^*(x)$. Note that the resulting matrix is the same as in the case where step (a) is applied starting with the pair $(w_H(x), z_H(x))$.

   (c) Compute the matrix $Q_{2H}(x) = Q_H^*(x)Q_H(x)$.

2. Compute the vector $(w_n(x), z_n(x))^T = Q_n(x)(u(x), v(x))^T$ and output the polynomial $w_n(x)$, which is a scalar multiple of $\gcd(u, v)$.

The h-th recursive step of Stage 1 amounts to a small number of operations of multiplication and division with polynomials having at most $3 * 2^h$ non-zero terms (which are the leading terms, since we drop the other terms) and thus only requires $O(M(2^h))$ ops, so the overall number of ops involved in the computations is $O(M(n) \log n)$.

Next, we recall a celebrated extension of the EA which is closely related to the rational function reconstruction problem.

Algorithm 2.8.7. Extended Euclidean Algorithm (EEA).

INPUT: Natural numbers $m, n, m \geq n$, and the coefficients of the polynomials $u(x)$ and $v(x)$ in (2.8.1).

OUTPUT: The coefficients of the polynomials $q_{i-1}(x)$, $r_i(x)$, $s_i(x)$, and $t_i(x)$, $i = 2, 3, \ldots, l$, where the integer l and the polynomials are defined below.

INITIALIZATION: Write $s_0(x) = t_1(x) = 1$, $s_1(x) = t_0(x) = 0$, $r_0(x) = u(x)$, $r_1(x) = v(x)$.

COMPUTATION: For $i = 2, 3, \ldots, l+1$, compute the coefficients of the polynomials $q_{i-1}(x), r_i(x), s_i(x), t_i(x)$ by using the following relations:

$$r_i(x) = r_{i-2}(x) - q_{i-1}(x)r_{i-1}(x), \tag{2.8.6}$$
$$s_i(x) = s_{i-2}(x) - q_{i-1}(x)s_{i-1}(x), \tag{2.8.7}$$
$$t_i(x) = t_{i-2}(x) - q_{i-1}(x)t_{i-1}(x). \tag{2.8.8}$$

Here $\deg r_i(x) < \deg r_{i-1}(x)$, $i = 2, 3, \ldots, l$, that is, we have $r_i(x) = r_{i-2}(x) \bmod r_{i-1}(x)$, and l is such that $r_{l+1}(x) \equiv 0$, that is, $r_{l-1}(x) = q_l(x)r_l(x)$, and therefore, $r_l(x) = \gcd(u(x), v(x))$.

Note that the sequence $\{r_0(x), r_1(x), r_2(x), \ldots, r_l(x)\}$ consists of the polynomials $u_i(x), v_i(x), i = 1, 2, \ldots, l$, computed by Algorithm 2.8.4.

Theorem 2.8.8. *The polynomials $r_i(x)$, $s_i(x)$, and $t_i(x)$ in (2.8.6)–(2.8.8) satisfy the following relations:*

$$s_i(x)r_0(x) + t_i(x)r_1(x) = r_i(x), \quad i = 1, \ldots, l, \tag{2.8.9}$$

$$\deg r_{i+1}(x) < \deg r_i(x), \qquad\qquad i = 1, \ldots, l-1,$$
$$\deg s_{i+1}(x) > \deg s_i(x) = \deg(q_{i-1}(x)s_{i-1}(x)) = \deg v(x) - \deg r_{i-1}(x),$$
$$i = 2, \ldots, l,$$
$$\deg t_{i+1}(x) > \deg t_i(x) = \deg(q_{i-1}(x)t_{i-1}(x)) = \deg u(x) - \deg r_{i-1}(x),$$
$$i = 2, \ldots, l,$$

where we write $\deg s_{l+1}(x) = \deg t_{l+1}(x) = +\infty$, $\deg r_{l+1} = -\infty$.

Proof. The theorem is deduced easily from (2.8.6)–(2.8.8) by induction on i.   □

Theorem 2.8.9. *For $i = 1, \ldots, l$, the polynomials $s(x) = s_i(x)$ and $t(x) = t_i(x)$ represent a unique solution to the relations $\deg(s(x)u(x)+t(x)v(x)-r_i(x)) < \deg r_i(x)$, $\deg s(x) < \deg v(x) - \deg r_i(x)$, $\deg t(x) < \deg u(x) - \deg r_i(x)$.*

Equations (2.8.6) imply that the $q_i(x)$ are the quotients and the $r_i(x)$ are the remainders of the polynomial division performed in the EA, so $\gcd(u(x), v(x))$ is a scalar multiple of $r_l(x)$. In some applications, equations (2.8.6) are replaced by their scaled versions, where the scaled sequence $r_0(x), \ldots, r_i(x)$ is called *the pseudo remainder sequence* of polynomials. All these polynomials have integer coefficients if $u(x)$ and $v(x)$ have integer coefficients.

Theorem 2.8.10. *Algorithm 2.8.7 (the EEA) involves $O(n^2)$ ops for $m = O(n)$, whereas the four polynomials $q_i(x)$, $r_i(x)$, $s_i(x)$, and $t_i(x)$ in (2.8.6)–(2.8.8) for any fixed $i \leq l$ can be computed by using $O(M(n) \log n)$ ops, for $M(n)$ in (2.4.1), (2.4.2).*

Proof. The EEA uses $O(m^2)$ ops to compute the coefficients of the polynomials $q_i(x)$, $r_i(x)$, $s_i(x)$, and $t_i(x)$ for all i. This estimate is obvious if $\deg q_i(x) = 1$ for all i, but it is not hard to obtain it in the general case as well since $\sum_i \deg q_i(x) \leq n$. For $m = O(n)$, the latter cost bound is optimal up to a constant factor, because $n(n + 1)/2$ ops is a lower bound on the worst case cost of computing the coefficients of the polynomials $r_2(x), \ldots, r_l(x)$, which form the *polynomial remainder sequence* for the input polynomials $u(x)$ and $v(x)$. Indeed, together these polynomials may have up to $n(n + 1)/2$ distinct coefficients.

On the other hand, the cited superfast version of the EA (as well as the superfast matrix algorithms in Sections 2.10 and 2.11) can be immediately extended to compute the four polynomials $q_i(x)$, $r_i(x)$, $s_i(x)$, and $t_i(x)$ in (2.8.6)–(2.8.8) for any fixed i by using $O(M(m) \log m)$ ops. $\square$

Remark 2.8.11. Algorithms 2.8.4, 2.8.6, and 2.8.7 are unstable numerically. These algorithms provide no reliable means for computing an *approximate gcd*. (For a pair of polynomials $u(x)$ and $v(x)$ and a positive ϵ, an approximate gcd (or an ϵ-gcd) is a common divisor of the largest degree for two polynomials, $u^*(x)$ and $v^*(x)$ that are allowed to deviate from the polynomials $u(x)$ and $v(x)$ within a fixed relative norm bound ϵ.) The computation of an approximate gcd has applications to network theory, linear systems, control theory, and computer aided design. Some algorithms for computing approximate gcds utilize the combination of the EEA with two matrix methods for the gcd that we describe in Sections 2.10 and 2.11.

The EEA is closely related to various subjects of computational mathematics such as (block) tridiagonalization of a matrix by the Lanczos algorithm, orthogonal polynomials, and the minimal realization problem [G92], [Par92]; the solution of linear Diophantine equations [GG99], and finite resolution computational geometry [GY86]. In the next sections, we study applications to rational function computations, reconstruction of a rational number from its residue, and Toeplitz/Hankel matrix computations. On the other hand, the development of the EA and the EEA presented above for polynomials can be extended to integers; some practical heuristic and randomized algorithms use binary segmentation (see Section 2.4) to reduce the computation of the gcd of two polynomials with integer coefficients to the gcd of integers [CGG84], [S85], [S88]. Applied to a pair of positive integers u and v, the EEA produces the **continued fraction expansion** of the rational number u/v, defined by the quotients $q_1, q_2, \ldots, q_l$ of the EEA, where $r_{l+1} = 0$. Furthermore, for any $h < l$, the continued fraction defined by $q_1, q_2, \ldots, q_{h+\delta}$ for $-1 \leq \delta \leq 1$ produces approximation p/q such that $|(u/v) - (p/q)| \leq 1/(\sqrt{5}q^2)$, p and q are integers, which is substantially better than the bound $0.5/q$ produced by the decimal or binary expansions of u/v.

2.9 Applications and extensions of the EEA: modular division, rational function reconstruction, Padé approximation, and linear recurrence span

Solution of Problems 2.7.4 and 2.7.10:   Clearly, the EEA solves Problem 2.7.4 on computing a reciprocal in the algebra A_u, that is, the reciprocal of a polynomial modulo a fixed polynomial $u(x)$. This is an immediate consequence of Theorems 2.8.8 and 2.8.9. Indeed, by virtue of these theorems, for three given polynomials $r_0(x)$, $r_1(x)$, and $r_i(x)$, such that $d_h = \deg r_h(x) - \deg r_i(x)$, for $h = 0, 1$, and $d_0 \geq d_1 > 0$, (2.8.9) is satisfied by a unique pair of polynomials $s_i(x)$, $t_i(x)$ such that $\deg s_i(x) < d_1$, $\deg t_i(x) < d_0$. Three polynomials $r_0(x)$, $r_1(x)$, and $r_l(x) = \gcd(r_0(x), r_1(x))$ define a unique pair of polynomials $(s_l(x), t_l(x))$ satisfying (2.8.9) for $i = l$, such that $\deg s_l(x) < \deg(r_1(x)/r_l(x))$, $\deg t_l(x) < \deg(r_0(x)/r_l(x))$. If $\gcd(r_0(x), r_1(x)) = 1$, then (2.8.9) defines the polynomial $t_l(x) = (1/r_1(x)) \bmod r_0(x)$, that is, the reciprocal of $r_1(x)$ modulo $r_0(x)$. As we mentioned earlier, the solution of Problem 2.7.4 immediately implies the solution of Problem 2.7.10 concerning division in the algebra A_u. In both cases, the solution involves $O(M(n) \log n)$ ops (for $M(n)$ in (2.4.1), (2.4.2)), due to Theorem 2.8.10.   ☐

Equations (2.8.9) for $i = 1, 2, \ldots, l - 1$ have an interesting interpretation: they define rational functions $r_i(x)/t_i(x)$ with bounded degrees of their numerators $r_i(x)$ and denominators $t_i(x)$ (that is, $\deg r_i(x) < \deg r_1(x)$, $\deg t_i(x) < \deg r_0(x) - \deg r_i(x)$ by Theorem 2.8.9) which represent the polynomial $r_1(x)$ modulo $r_0(x)$. Formalizing this observation, we employ the EEA to solve the following important problem.

Problem 2.9.1. Rational function reconstruction. Given non-negative integers m, n, and $N = m + n$ and two polynomials, $u(x)$ of degree $N + 1$ and $v(x)$ of degree of at most N, find two polynomials, $r(x)$ of degree of at most m and $t(x)$ of degree of at most n, such that (2.7.3) holds under these bounds on the degree (generally distinct from the bounds stated in Problem 2.7.10), that is,

$$r(x) = s(x)u(x) + t(x)v(x) = t(x)v(x) \bmod u(x),$$
$$\deg r(x) \leq m, \quad \deg t(x) \leq n,$$
(2.9.1)

for some polynomial $s(x)$.

Problem 2.7.4 on computing the reciprocal of a polynomial modulo a polynomial is a special case of Problem 2.9.1, where $m = 0$, $N = n$, and $r(x) = 1$.

Here is another important special case.

Problem 2.9.2. (m, n) Padé approximation. Solve Problems 2.9.1 where $u(x) = x^{N+1}$.

The solutions to Problem 2.9.1 define a (unique) rational function $r(x)/t(x)$ satisfying (2.9.1). The function is reconstructed modulo $u(x)$ from the integers m and n and the polynomials $u(x)$ and $v(x)$. Unlike the case of Problem 2.7.10, polynomial $r(x)$ is unknown, but the bounds on the degrees of $r(x)$ and $t(x)$ are stronger in Problem 2.9.1 than in Problem 2.7.10, provided that $m > 0$ in Problem 2.9.1.

The solutions to Problem 2.9.1 for all pairs (m, n) of non-negative integers such that $m + n \leq N$ and for fixed polynomials $u(x) = r_0(x)$ of degree $N + 1$ and $v(x) = r_1(x)$ of degree of at most N form the (triangular) *rational function table*, associated with the pair of polynomials $u(x)$ and $v(x)$. In the special case of Problem 2.9.2 where $u(x) = x^{N+1}$, $N = 0, 1, \ldots$, the table is called the *Padé approximation table* associated with a fixed (polynomial or) formal power series $v(x)$. By increasing N, we arrive at an unbounded table. The next theorem states the *computational cost estimates* for the solution of Problem 2.9.2.

Theorem 2.9.3. *For a fixed natural N and for $h = 0, 1, \ldots, N$, all the $(N - h, h)$-th antidiagonal entries of the rational function table associated with two polynomials $u(x)$ and $v(x)$ in Problem 2.9.1 of rational function reconstruction can be computed in $O(N^2)$ ops, whereas for any fixed pair (m, n), $m + n = N$, the solution $(r(x), s(x))$ to Problem 2.9.1 (and in particular to Problems 2.7.4 and/or 2.9.2) can be computed in $O(M(N) \log N)$ ops.*

Proof. A comparison of (2.8.9) and (2.9.1) shows that the entries of the antidiagonal of this table are related to each other via (2.8.6)–(2.8.9) and, consequently, via the EEA. Therefore, we arrive at the following algorithm.

Algorithm 2.9.4. Antidiagonal of a rational function table.

INPUT: A natural number N and two polynomials, $u(x)$ of degree $N + 1$ and $v(x)$ of degree of at most N, as in Problem 2.9.1.

OUTPUT: The $(N - h, h)$ entries for $h = 0, 1, \ldots, N - l$ of the rational function table associated with the polynomials $u(x)$ and $v(x)$, where $l = \deg \gcd(u, v)$.

COMPUTATION: Write $r_0(x) = u(x)$, $r_1(x) = v(x)$, $t_0(x) = 0$, $t_1(x) = 1$ and compute the coefficients of the polynomials $r_i(x)$, $t_i(x)$ for $i = 2, 3, \ldots, l + 1$ defined by (2.8.6) and (2.8.8), that is, apply the EEA to the pair of polynomials $u(x) = r_0(x)$, $v(x) = r_1(x)$. Output the pairs $(r_i(x), t_i(x))$ of the successive entries of the EEA. They successively fill the entries $(N - h, h)$, $h = 0, 1, \ldots, N - l$, of the antidiagonal of the rational function table, associated with the polynomials $u(x)$ and $v(x)$ and having the pair $(r_1(x), t_1(x)) = (v(x), 1)$ as the entry $(N, 0)$. In particular, the entry $(N - h, h)$ of the table is filled with the pair $(r_i(x), t_i(x))$ where $\deg r_i(x) \leq N - h$ and is maximal over all i under the latter bound. (An equivalent rule is that each pair $(r_i(x), t_i(x))$ fills exactly $\deg q_{i-1}(x)$ successive entries of this antidiagonal, where $q_{i-1}(x)$ is a polynomial in (2.8.6)–(2.8.8).)

Theorem 2.8.10 and Algorithm 2.9.4 together support Theorem 2.9.3.   □

Uniqueness of the solution: One may verify that for a fixed input m, n, $u(x)$, and $v(x)$ the resulting solution $r(x)$, $t(x)$ to Problem 2.9.1 is unique up to scaling, and therefore we can make it unique by the requirement that $\gcd(r, t) = 1$ and the polynomial $t(x)$ be monic [GG99, page 108].

Remark 2.9.5. To yield a specified entry $(N - i, i)$ of the table, superfast modification of Algorithm 2.9.4 can be applied (see Algorithm 2.8.6). In the special case where $u(x) = x^{N+1}$ for a fixed natural N, Algorithm 2.9.4 computes the antidiagonal entries $(N - h, h)$, $h = 0, 1, \ldots, N$, of the Padé approximation table associated with the polynomial (or formal power series) $v(x)$. An alternative Main Diagonal Algorithm, also running in $O(M(n) \log N)$ time for a specified i, $i \leq N$ and in $O(N^2)$ time for all $i \leq N$, was presented in [BGY80] for computing the main diagonal, that is, the (i, i)-th entries of the latter table for $i = 1, 2, \ldots, N$, provided that all its entries (i, j) are distinct for $i, j = 0, 1, \ldots, N$. This provision holds for generic input polynomial or formal power series $v(x)$.

The Explicit Reconstruction Condition: The explicit expression

$$v(x) = (r(x)/t(x)) \bmod u(x)) \tag{2.9.2}$$

immediately follows from equations (2.9.1) if

$$\gcd(u, t) = 1. \tag{2.9.3}$$

It can be proved that under the degree bounds (2.9.1), the latter requirement can be fulfilled if and only if

$$\gcd(r_i, t_i) = 1 \tag{2.9.4}$$

for the output pair $r_i(x)$, $t_i(x)$ in Algorithm 2.9.4.

In Section 3.5 we recall some well-known extensions of the solution of Problem 2.9.1 where explicit formulation (2.9.2) and, consequently, conditions (2.9.3) and (2.9.4) are most meaningful.   □

Let us next extend the computation of Padé approximation to the computation of the gcd of two polynomials (at the cost $O(M(N) \log N)$, N being the maximal degree of the input polynomials).

Algorithm 2.9.6. Gcd of two polynomials via Padé approximation.

INPUT: Natural m and n and the coefficients of two polynomials $u(x)$ and $v(x)$, $v(0) \neq 0$, of degrees m and n, respectively.

OUTPUT: The coefficients of $\gcd(u, v)$.

COMPUTATION:

1. Compute the first $N + 1 = m + n + 1$ Taylor's coefficients $a_0, \ldots, a_N$ of the power series

$$a(x) = \frac{u(x)}{v(x)} = \sum_{i=0}^{+\infty} a_i x^i$$

(see Problems 2.5.2 and 2.4.1).

2. Compute the (m, n) Padé approximation $(r(x), t(x))$ to the power series $a(x)$, which outputs the $(m - \rho, n - \rho)$ Padé approximation for the maximum integer ρ, $\rho \leq \min\{m, n\}$, such that $r(x)/t(x) = u(x)/v(x)$.

3. Compute and output $\gcd(u, v) = u(x)/r(x) = v(x)/t(x)$.

The correctness of the algorithm follows from uniqueness of Padé approximation.

Instead of using the ratio $a(x) = u(x)/v(x)$ in the above algorithm, we may use the ratio $a(x) = r_j(x)/r_k(x)$ for any fixed pair $(r_j(x), r_k(x))$ of distinct polynomials, in the polynomial remainder sequence $\{r_i(x)\}$, $i = 0, 1, \ldots, l$ in (2.8.6) provided that $j \leq l, k \leq l$.

Here is another important application of Padé approximation.

Problem 2.9.7. Computation of the minimum span for a linear recurrence (Berlekamp–Massey Problem). Given $2s$ numbers $a_0, \ldots, a_{2s-1}$, compute the minimum natural $n \leq s$ and n numbers $c_0, \ldots, c_{n-1}$ such that $a_i = c_{n-1}a_{i-1} + \cdots + c_0 a_{i-n}$, for $i = n, n + 1, \ldots, 2s - 1$.

Solution: The solution $n, c_0, \ldots, c_{n-1}$ is unique and is given by the degree n and the coefficients $c_0, \ldots, c_{n-1}$ of the minimum span polynomial $t(x) = x^n - \sum_{i=0}^{n-1} c_i x^i$, such that the pair of polynomials $(r(x), t(x))$ is an (m, n) Padé approximation to the polynomial $a(x) = \sum_{i=0}^{2s-1} a_i x^i$, where $\deg r(x) < s$, $N = 2s - 1$, $m = s - 1$, $n = s$. $\square$

Finally, let us extend Problem 2.9.1 and Algorithm 2.9.4 to their integer analogs.

Problem 2.9.8. Rational number reconstruction. Given three integers m, u, and v, $1 \leq m \leq u, 0 \leq v < u$, compute a pair of integers r and t satisfying the relations

$$r = tv \bmod u, \quad |r| < m, \quad 0 \leq t \leq u/m. \tag{2.9.5}$$

Solution: Apply the EEA with $r_0 = u$, $r_1 = v$, $t_0 = 0$, $t_1 = 1$. Select the pair r_i, t_i satisfying the inequality $|r| < m$ in (2.9.5) for the minimal i. Output $r = r_i, t = t_i$ if $t_i > 0$. Otherwise output $r = -r_i, t = -t_i$. $\square$

The Explicit Reconstruction Condition: If (2.9.4) holds, then we have

$$r/t = v \bmod u. \tag{2.9.6}$$

Otherwise (2.9.5), (2.9.6) has no solution.

Figure 2.2: Correlation among polynomial and matrix computations II. ("A" and "P" are abbreviation for "Algorithm" and "Problem", respectively; the arrows show reductions of problems to each other.)

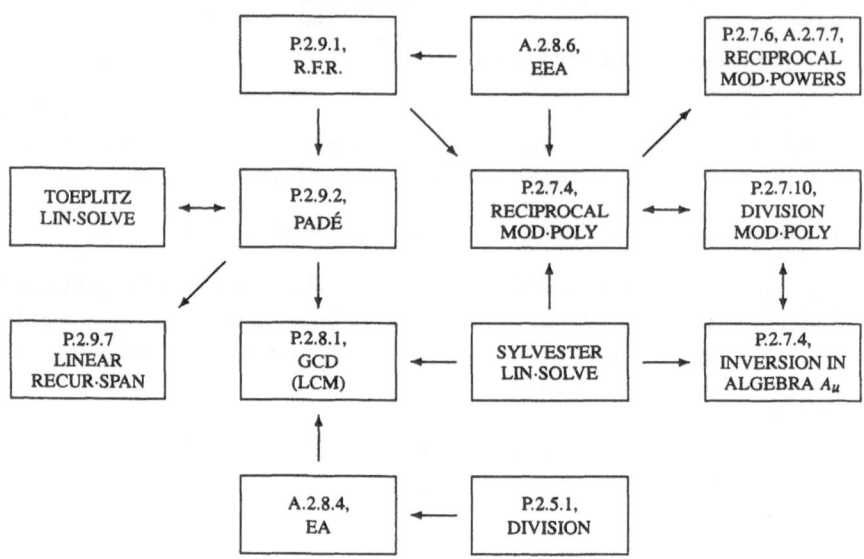

Computational Cost: By applying the superfast version of the integer EA, we obtain the solution by using

$$b = O(\mu(d) \log d) \qquad (2.9.7)$$

bit-operations where $d = \lceil \log_2 u \rceil$ and $\mu(d)$ bit-operations are sufficient to multiply two integers modulo $2^d + 1$ (see Exercise 2.9).

Uniqueness of the solution: If (2.9.4), (2.9.6) hold, $t > 0$ and $|r| < m/2$ then the solution is unique. In particular, if the unknown integers r and t satisfy (2.9.4)–(2.9.6), $t > 0$, and $2r^2 < u$, we apply the above algorithm for

$$m = \lceil \sqrt{2u} \rceil \qquad (2.9.8)$$

and recover a unique pair of r and t.

2.10   Matrix algorithms for rational interpolation and the EEA

It is well known that the EEA and, consequently, the superfast algorithms of the previous section based on the EEA are very unstable numerically. The problem can be cured (at least partly) by using structured matrix algorithms.

In Section 2.8, we already reduced testing co-primality of two polynomials to some structured matrix problems. Let us next *extend this approach to the rational function reconstruction (Problem 2.9.1)*.

In this extension, unlike the previous section, we first compute the polynomials $s(x)$ and $t(x)$ in (2.9.1) and then immediately recover the polynomial $r(x)$ from the equation in (2.9.1), that is,

$$r(x) = s(x)u(x) + t(x)v(x).$$

Let us assume that $\deg u(x) = m$, $\deg v(x) = n$, $\deg r(x) = k$, bound the degrees of the polynomials $s(x)$ and $t(x)$ based on Theorem 2.8.8, and write $r(x) = \sum_{i=0}^{k} r_i x^i$, $s(x) = \sum_{i=0}^{n-k-1} s_i x^i$, $t(x) = \sum_{i=0}^{m-k-1} t_i x^i$, $u(x) = \sum_{i=0}^{m} u_i x^i$, $v(x) = \sum_{i=0}^{n} v_i x^i$, $r_k = 1$. We recall that in Problem 2.9.1, the polynomials $u(x)$ and $v(x)$ and the degree k are fixed, and the polynomials $r(x)$, $s(x)$, and $t(x)$ are unknown. Let the latter polynomials be normalized by scaling, which makes the polynomial $r(x)$ monic. By equating the coefficients of x^i for $i = k, k+1, \ldots, m+n-k-1$ on both sides of (2.7.3), we arrive at the vector equation

$$(U_k, V_k) \begin{pmatrix} \mathbf{s} \\ \mathbf{t} \end{pmatrix} = \mathbf{e}_0 \qquad (2.10.1)$$

where $\mathbf{s} = (s_i)_{i=0}^{n-k-1}$, $\mathbf{t} = (t_i)_{i=0}^{m-k-1}$, and the matrices U_k and V_k are the two submatrices obtained by the deletion of the last k columns and the first $2k$ rows of the matrices of U and V in (2.7.2), respectively. (If we deleted only the first k rows and the last k columns, we would have had the vector $\mathbf{r} = (r_0, \ldots, r_k, 0, \ldots, 0)^T$ instead of $\mathbf{e}_0$ on the right-hand side of (2.10.1). The deletion of the next k rows and, consequently, the first k components of the vector $\mathbf{r}$ turns $\mathbf{r}$ into the vector $\mathbf{e}_0$ since $r_k = 1$.) The $(m+n-2k) \times (m+n-2k)$ submatrix $S_k = (U_k, V_k)$ of the $(m+n) \times (m+n)$ resultant matrix S is called a *subresultant matrix*.

We reduced Problem 2.9.1 directly to a linear system of $m+n-2k$ equations with the subresultant matrix $S_k = (U_k, V_k)$ having structure of Toeplitz type. The algorithms in Chapter 5 yield a superfast solution to such systems. Compared to the polynomial approach of the previous section, the latter matrix approach avoids the application of the EEA and its inherent most severe problems of numerical stability. One may similarly compute all other output polynomials of the EEA as well, even though this involves a little more ops than in the direct computation via the EEA. The latter extension of the matrix approach relies on the following basic result.

Theorem 2.10.1. *The matrix S_k is non-singular if and only if one of the remainder (subresultant) polynomials $r_i(x)$ computed by the EEA has degree k. In this case there exists a scalar c_i such that*

$$s_i(x) = c_i \sum_{j=0}^{n-k-1} y_j x^j,$$

$$t_i(x) = c_i \sum_{j=0}^{m-k-1} z_j x^j,$$

$$S_k \begin{pmatrix} \mathbf{y} \\ \mathbf{z} \end{pmatrix} = \mathbf{e}_0,$$

$$\mathbf{y} = (y_j)_{j=0}^{n-k-1},$$

$$\mathbf{z} = (z_j)_{j=0}^{m-k-1}.$$

By this theorem, all monic associates of the remainders $r_i(x)$ (for all i) can be computed by the EEA which solves the linear systems (2.10.1) for $k = 0, 1, \ldots, l$; $l = \deg \gcd(u, v) \leq \min\{m, n\}$. Simple rules for the subsequent computation of the scalars c_i and, consequently, the remainders $r_i(x)$ can be found in Exercise 2.18. Then the polynomials $q_i(x)$, $s_i(x)$, and $t_i(x)$ for all i can be easily obtained from (2.8.6)–(2.8.9). The overall arithmetic cost of all these computations is $O(nM(n))$ ops [P89, page 1478]. The version of [BG95] yields the same arithmetic cost bound but stronger numerical stability due to the application of *univariate Bezoutians*, which is an important class of structured matrices. (Univariate Bezoutians can be defined in various equivalent ways, e.g., as the *inverses of non-singular Toeplitz or Hankel matrices*. See a generalized definition in Exercise 4.11 of Chapter 4.)

2.11   Matrix approach to Padé approximation

Let us next extend the matrix approach of the previous section to the computation of $\gcd(u, v)$ (Problem 2.8.1). We keep denoting by m, n, and k the degrees of the polynomials $u(x)$, $v(x)$, and $\gcd(u, v)$. Recall that $\gcd(u, v)$ is one of the remainders computed by the EEA and suppose that we know its degree k. Then the algorithm of the previous section computes the co-factors $s(x)$ and $t(x)$ by solving linear system (2.10.1). Then we may immediately compute the gcd by substituting the co-factors into (2.7.3). For degree k, we have (2.8.2), that is, $2k = 2 \deg(\gcd(u, v)) = m + n - \mathrm{rank}(S)$ where S is the matrix in (2.7.2). Our superfast algorithms of Chapter 5 apply to the computation of both the rank and solution of the linear system.

Let us next simplify the matrix computations for the Padé approximation Problem 2.9.2, where the polynomial equation (2.7.3) takes the following form:

$$t(x)v(x) \bmod x^{N+1} = r(x).$$

Let us write $t(x) = \sum_{i=0}^{n} t_i x^i$, $v(x) = \sum_{i=0}^{N} v_i x^i$, $r(x) = \sum_{i=0}^{m} y_i x^i$, where $N = m + n$. By equating the coefficients of x^i, $i = 0, 1, \ldots, m + n$, on both sides of the polynomial equations, we arrive at the following linear system:

$$\begin{pmatrix} v_0 & & & & & O \\ v_1 & \ddots & & & & \\ \vdots & \ddots & \ddots & & & \\ v_{m-n-1} & & \ddots & \ddots & & \\ v_{m-n} & \ddots & & \ddots & v_0 & \\ \vdots & \ddots & \ddots & & v_1 & \\ \vdots & & \ddots & \ddots & \vdots & \\ v_{m-1} & & & \ddots & v_{m-n-1} \\ v_m & \ddots & & & v_{m-n} \\ \vdots & \ddots & \ddots & & \vdots \\ \vdots & & \ddots & \ddots & \vdots \\ \vdots & & & \ddots & \vdots \\ v_{m+n} & \cdots & \cdots & \cdots & v_m \end{pmatrix} \begin{pmatrix} t_0 \\ \vdots \\ \vdots \\ t_n \end{pmatrix} = \begin{pmatrix} y_0 \\ \vdots \\ \vdots \\ \vdots \\ \vdots \\ \vdots \\ \vdots \\ \vdots \\ y_m \\ 0 \\ \vdots \\ \vdots \\ 0 \end{pmatrix}. \qquad (2.11.1)$$

Next, move the right-hand sides of these equations to the left-hand side, by appending the vector $(y_0, \ldots, y_m)^T$ at the bottom of the vector $(t_0, \ldots, t_n)^T$ and by appending the matrix $\begin{pmatrix} -I_{m+1} \\ O \end{pmatrix}$ on the right of the coefficient matrix. Finally, multiply the first column $(v_0, \ldots, v_{m+n})^T$ by t_0 and move the product on the right-hand side of the linear system. This gives us the system of equations

$$\begin{pmatrix} U & -I_{m+1} \\ V & O \end{pmatrix} \begin{pmatrix} t_1 \\ \vdots \\ t_n \\ y_0 \\ \vdots \\ y_m \end{pmatrix} = -t_0 \begin{pmatrix} v_0 \\ \vdots \\ v_{m+n} \end{pmatrix} \qquad (2.11.2)$$

where

$$V = \begin{pmatrix} v_m & \cdots & v_{m-n+1} \\ \vdots & \ddots & \vdots \\ v_{m+n-1} & \cdots & v_m \end{pmatrix}, \quad U = \begin{pmatrix} 0 & & & O \\ v_0 & 0 & & \\ \vdots & \ddots & \ddots & \\ v_{m-1} & \cdots & & \end{pmatrix}.$$

The latter system is equivalent to system (2.11.1) and to the original problem of Padé approximation. The solution is immediately defined by the solution to the Toeplitz subsystem formed by the last n equations with n unknowns (for $t_0 = 1$ or for $t_0 = 0$),

$$\begin{pmatrix} v_m & \cdots & v_{m-n+1} \\ \vdots & \ddots & \vdots \\ v_{m+n-1} & \cdots & v_m \end{pmatrix} \begin{pmatrix} t_1 \\ \vdots \\ t_n \end{pmatrix} = -t_0 \begin{pmatrix} v_{m+1} \\ \vdots \\ v_{m+n} \end{pmatrix}. \qquad (2.11.3)$$

This system (and thus the original problem of Padé approximation as well) either is non-singular and has a solution $\mathbf{t} = (t_1, \ldots, t_n)^T$ where $t_0 = 1$ or is singular and has infinitely many solutions for $t_0 = 0$. (The resulting ratio of the polynomials $\sum_i y_i x^i / \sum_j t_j x^j$ is always unique, however.)

Furthermore, we have non-singularity of the $r \times r$ leading principal submatrix of the Hankel matrix JV for the reflection matrix J of Definition 2.1.1 and for the Toeplitz coefficient matrix V in the linear system (2.11.3), where $r = \operatorname{rank}(V)$. We arrive at the following algorithm.

Algorithm 2.11.1. Padé approximation.

INPUT: Two natural numbers m and n, a polynomial $v(x) = \sum_{i=0}^{m+n} v_i x^i$, and black box algorithms for the computation of the rank of a Toeplitz matrix and the solution of a non-singular Toeplitz linear system of equations (see Chapter 5).

OUTPUT: The (m, n) Padé approximation to the polynomial $v(x)$ by two polynomials $y(x) = \sum_{i=0}^{m} y_i x^i$ and $t(x) = \sum_{i=0}^{n} t_i x^i$.

COMPUTATION:

1. Compute the value $r = \operatorname{rank}(V)$, $V = (v_{m+i-j})$ being the coefficient matrix of the Toeplitz linear system (2.11.3), where $r \times r$ is the maximum size of a non-singular leading principal submatrix of the matrix JV.

2. If $r = n$, let $t_0 = 1$, and compute the solution $t_n, \ldots, t_1$ to the non-singular Toeplitz linear system (2.11.3). If $r < n$, let $t_0 = 0$. Stage 1 has given us an $r \times r$ non-singular Toeplitz submatrix of the matrix V, and we solve the system (2.11.3) by applying the given black box Toeplitz solver.

3. Compute the values $y_0, \ldots, y_m$ from the linear system (2.11.2).

The converse reduction of a Toeplitz linear system to Padé approximation and, therefore, to the EEA yielded *the first ever superfast solution of a general non-singular Toeplitz linear system* [BGY80]. Let us recall this reduction because its techniques are of independent interest even though our superfast algorithms of Chapters 5–7 cover a much more general class of matrices and have superior numerical stablity. The reduction from a Toeplitz linear system to Padé approximation relies on the *celebrated formula by Gohberg and Semencul* for the inverse $X = T^{-1} = (x_{ij})_{i,j=0}^{n-1}$ of a non-singular $n \times n$ Toeplitz matrix V, that is,

$$x_{0,0} X = Z(\mathbf{x}) Z^T (J\mathbf{y}) - Z(Z\mathbf{y}) Z^T (ZJ\mathbf{x}) \qquad (2.11.4)$$

provided that $\mathbf{x} = X\mathbf{e}_0$, $\mathbf{y} = X\mathbf{e}_{n-1}$, and $x_{0,0} \neq 0$. The formula reduces the solution of a non-singular Toeplitz linear system $T\mathbf{z} = \mathbf{b}$ to computing the vectors $X\mathbf{e}_0$ and $X\mathbf{e}_{n-1}$, that is, vectors of the first and the last columns of the matrix $X = T^{-1}$, provided that $x_{0,0} \neq 0$. To reduce computing the vector $X\mathbf{e}_0$ to Problem 2.9.2 of Padé approximation, it is sufficient to remove the upper part of linear system (2.11.1) where $y_m \neq 0$, $m = n$ to arrive at the following linear system of $n + 1$ equations:

$$
\begin{pmatrix}
v_n & v_{n-1} & \cdots & & v_0 \\
v_{n+1} & \ddots & \ddots & & \vdots \\
\vdots & & \ddots & \ddots & v_{n-1} \\
v_{2n} & \cdots & & v_{n+1} & v_n
\end{pmatrix}
\begin{pmatrix}
t_0 \\
t_1 \\
\vdots \\
t_n
\end{pmatrix}
=
\begin{pmatrix}
y_n \\
0 \\
\vdots \\
0
\end{pmatrix}.
$$

The system has general Toeplitz matrix V of coefficients. If this is a non-singular linear system, the algorithm computes its solution vector $\mathbf{t}$ (and then the first column $\mathbf{t}/y_n$ of the matrix V^{-1}) by applying the EEA to the input polynomials $r_0(x) = u(x) = x^{2n+1}$, $r_1(x) = v(x) = \sum_{i=0}^{2n} v_i x^i$. The Padé approximation problem does not degenerate in this case because the matrix V is non-singular by assumption. To compute the vector $\mathbf{t}/y_n$, given by a specific entry of the Padé table, we have a superfast algorithm by Theorem 2.8.10. Similarly, we may compute the last column of the matrix V^{-1}, which is given by the first column of the inverse of the Toeplitz matrix $J V J$. We arrive at an algorithm that solves a non-singular linear system with an $(n + 1) \times (n + 1)$ Toeplitz matrix V by using $O(M(n) \log n)$ ops provided that $x_{0,0} = (V^{-1})_{0,0} \neq 0$. The latter assumption has been relaxed in [BGY80], where the matrix V was embedded into an $(n + 2) \times (n + 2)$ Toeplitz matrix T and the problem was solved for the matrix T replacing the matrix V if $x_{0,0} = 0$.

Due to the algorithm for the solution of Problem 2.9.7 and Algorithm 2.9.6, the superfast algorithms for the Padé approximation are immediately extended to computing the minimum span of a linear recurrence and a polynomial gcd.

2.12 Conclusion

Various fundamental polynomial and rational computations have a close correlation to matrices with structures of Toeplitz/Hankel types. The correlation and the Toeplitz/Hankel structure enable superfast computations, which can be further accelerated if some additional (e.g., circulant) structure is available. In this chapter we explored and exploited this correlation (see Figures 2.1 and 2.2) and covered some fundamental problems of polynomial computations and various related topics. We showed superfast solution algorithms for these problems and for multiplying Toeplitz and Hankel matrices and their inverses by vectors. The study gave us better insight into the subjects and was applied to both numerical FFT based implementation of the algorithms and computations in the fields that do not support FFT.

2.13   Notes

Sections 2.2-2.3   An historical account and some pointers to huge literature on FFT can be found in [C87], [C90]. See also [VL92] on the implementation of FFT. Theorem 2.3.1 and Corollary 2.3.2 are Proposition 4.1 and Corollary 4.1 of Chapter 3 of [BP94].

Section 2.4   The evaluation/interpolation method is due to [T63]. Bound (2.4.1) is due to [CK91] (also see [S77]). Computing convolution by binary segmentation was introduced in [FP74], the nomenclature of binary segmentation in [BP86]. On the correlation between computations with integers and polynomials with integer coefficients, see [T63], [SS71], [AHU74], [FP74], [HW79], [S82], [CGG84], [S85], [BP86], [S88], [BP94], [K98], and [GG99]. On matrix operations via binary segmentation, see [P84, Section 40]. On fast multivariate polynomial multiplication, see [CKL89] and [P94]. In [ASU75] and [K98], the solution of Problem 2.4.2 is attributed to L. I. Bluestein 1970 and L. R. Rabiner, R. W. Schaffer and C. M. Rader 1969. The solution of Problem 2.4.3 and Theorem 2.4.4 are due to [ASU75]. On parallel algorithms, computaional models, and complexity estimates, see [L92], [Q94], [BP94] and [LP01].

Section 2.5   Superfast solutions of Problem 2.5.3 are from [S72], [Ku74] (polynomial versions), and [BP86] (matrix version). The duality of the two versions as well as of several other pairs of polynomial and matrix superfast algorithms was observed in [BP86]. On solving Problem 2.5.4 over any field, see [P97a], [P00a], and the references therein. (The algorithm of [P00a] does not solve the more general problem of the recovery of the linear recurrence coefficients; the title and a claim in the introduction are erroneous.) The solution of Problem 2.5.5 is from [GKL87].

Section 2.6   On circulant matrices, see [D79]. Theorem 2.6.4 is from [CPW74].

Section 2.7   Theorems 2.7.2 and 2.7.3 and the nomenclature of the Horner basis are from [C96]. For earlier bibliography on the algorithms of this section, see, e.g., [BP94], [GG99]. On application of the computations in the quotient algebra A_u to univariate polynomial rootfinding, see [C96], [BP96]; on further non-trivial extensions to solving a system of multivariate polynomials, see [MP98], [MP00], and the bibliography therein.

Section 2.8   The Euclidean algorithm (for integers) was published in the Euclidean *Elements* about 300 B.C. but traced back about two centuries earlier than that in [K98]. Theorem 2.8.3 can be found, e.g., in [Z93]. The superfast algorithm for Problem 2.8.1 is due to [M73], [AHU74], and [BGY80] and extends the algorithm of [L38], [K70], and [S71] for computing the gcd of a pair of positive integers, both less than 2^d, using $O((d \log d) \log \log d)$ bit-operations. On further extensive study of the EA and EEA, see [K98], [GG99]. On Theorem 2.8.9, see [G84]. On numerical instability of the EAA, the definitions and the main properties of approximate gcds, and algorithms for their computation, see [S85], [EGL96], [EGL97], [CGTW95], [P98], [Pa], and the references therein. On various applications of approximate gcds, see [PB73], [K80],

[B81], [Kar87], [SC93]. On continued fraction expansion, see [Z93], [K98], [GG99], and the bibliography therein.

Section 2.9 Further information and bibliography on Padé approximation can be found in [G72], [BGY80], [BG-M96], on rational function reconstruction in [GY79], [G84], [GG99]. On rational number reconstruction, see [HW79], [Wan81], [WGD82], [UP83], [S86], [A89], [KR89], [GG99], and the bibliography therein. Algorithm 2.9.6 is due to [K88]. On Problem 2.9.7 and its solution, see [Be68], [BGY80]. The solution has applications to coding theory [Be68], to sparse multivariate polynomial interpolation (see Section 3.10), and to parallel matrix computations [KP91], [KP92], [BP94, Chapter 4], [P96].

Section 2.10 Theorem 2.10.1 is taken from [G84]. On univariate Bezoutians, see [HR84], [LT85], [W90], [BP94, Chapter 2, Section 9], [MP00], and the references therein. On multivariate Bezoutians, see [MP00] and the references therein.

Section 2.11 The section follows [BGY80]. Non-singularity of the $r \times r$ northeastern submatrix of the matrix in (2.11.3) was proved in [G72] (see Theorem 2 in [BGY80]). The Gohberg–Semencul formula (2.11.4) is from [GS72].

2.14 Exercises

2.1 Complete all omitted proofs of the results in this chapter (see Exercises 2.22–2.24).

2.2 Specify the constants hidden in the "O" notation in the arithmetic computational complexity estimates of this chapter.

2.3 Write programs to implement the algorithms of this chapter (see some sample pseudocodes in the next section). Then apply the algorithms to some specific input matrices and/or polynomials (as in Exercises 2.13 and 2.19–2.21).

2.4 The work space is the memory space required for computation in addition to the input and output space. Estimate the work space required for the algorithms of this chapter.

2.5 (a) Extend Figures 2.1 and 2.2 as much as you can.

   (b) Examine the extended figures and the algorithms that support the problem reductions shown by arrows. Estimate how many ops are required for each reduction.

2.6 Reduce multiplication of a $g \times h$ block matrix by a vector to multiplication of gh matrix blocks by vectors and performing addition of the gh resulting vectors. Apply this reduction to multiply the matrix $\Omega_{m,n} = (\omega_n^{ij})_{i=0,j=0}^{m-1,n-1}$ by a vector using $O(M(l)r/l)$ ops for $l = \min(m,n)$, $r = \max(m,n)$, and $M(n)$ in (2.4.1), (2.4.2).

2.7 (A. Toom, A. Schönhage) Use the equations $2ab = (a + b)^2 - a^2 - b^2$ and $a^2 = (1/a - 1/(a + 1))^{-1} - a$ to reduce polynomial (integer) multiplication to squaring a polynomial (an integer) and to reduce squaring to the computation of a reciprocal. In a similar reduction of multiplication to squaring for matrices, rely on the matrix equation

$$\begin{pmatrix} A_3 - R & B_1 \\ A_1 & R \end{pmatrix}^2 - \begin{pmatrix} A_4 - R & -B_2 \\ A_2 & R \end{pmatrix} = \begin{pmatrix} X_1 & X_3 \\ X_2 & X_4 \end{pmatrix}$$

where $R = B_1 + B_2$, $X_3 = A_3 B_1 + A_4 B_2$, $X_4 = A_1 B_1 + A_2 B_2$. Relate the respective estimates for the computational cost of the three problems.

2.8 Compare the ops count for the three convolution algorithms:

(a) the classical algorithm (using $9(n + 1)^2 - 2n - 1$ ops),

(b) the divide-and-conquer algorithm using $O(n^{\log_2 3})$ ops, $\log_2 3 = 1.584962...$, obtained by recursive application of the following equation:

$$(p_0 + x^k p_1)(q_0 + x^k q_1)$$
$$= p_0 q_0 (1 - x^k) + (p_1 + p_0)(q_1 + q_0) x^k + p_1 q_1 (x^{2k} - x^k)$$

where p_0, p_1, q_0, and q_1 are polynomials in x of degree of at most $k - 1$ (see the next exercise),

(c) the one based on Toom's evaluation-interpolation techniques, leading to cost bounds (2.4.1), (2.4.2),

(d) for each n from 1 to 10^6, specify the fastest of the three algorithms.

(e) Replace x^k by $\sqrt{-1}$ in the equation of part (b) and deduce an algorithm that multiplies two complex numbers by performing real multiplication three times and real addition and subtraction five times.

(f) Can you modify this algorithm to use fewer operations of real multiplication? Real addition/subtraction?

2.9 For the three convolution algorithms of the previous exercise, there are three counterparts for the problem of integer multiplication modulo $2^d + 1$, that is, the classical algorithm, the Karatsuba algorithm from [KO63] (similar to the one of Exercise 2.8(b)), and Schönhage–Strassen's of [SS71], [AHU74]). They involve $O(d^3)$, $O(d^{\log_2 3})$, and $O((d \log d) \log \log d)$ bit-operations, respectively. (Schönhage–Strassen's advanced algorithm improved the earlier algorithm of [T63] using $O(d^{1+\epsilon})$ bit-operations for any positive ϵ.)

(a) Deduce the latter bound, $O((d \log d) \log \log d)$.

(b) Specify the constants in the "O" notation above [Ba].

(c) For each of these algorithms specify the range for the precision d where the algorithm uses fewer bit-operations than the two others.

(d) [B76], [BB84], [A85]. Let $f(x)$ be any one of the elementary functions $\exp x$, $\sin x$, $\cos x$ or $\tan x$ or the inverse of this function. For a complex x and $d > 1$, approximate $f(x)$ within 2^{-d} by using $O(\mu(d)\log d)$ bit-operations where $\mu(d)$ is the selected estimate of parts (a)–(c).

2.10 Prove that 3 is a primitive 16-th root of 1 in $\mathbb{Z}_{17}$. Use this property to compute the coefficients of the product of the two polynomials $p(x) = \sum_{i=0}^{7}(i+1)x^i$,

$q(x) = \sum_{i=0}^{7}(i+2)x^i$ in $\mathbb{Z}_{17}$.

2.11 (a) Apply the evaluation-interpolation techniques to multiply several polynomials.

(b) Find the correlation between ϵ and N for which any precision approximation algorithms of [B84] and [PLS92] for polynomial division become equivalent to each other (see the end of Section 2.6).

2.12 (a) Prove that the $n \times n$ matrices $\omega_{2n} Z_1$ and Z_{-1} are similar to each other, that is, $\omega_{2n} Z_1 A = A Z_{-1}$ for some non-singular matrix A.

(b) Prove that the $(2n) \times (2n)$ matrices $Z_1^{(2n)}$ and $\operatorname{diag}(Z_1, Z_{-1})$ are similar to each other where $Z_1^{(2n)}$ denotes the $(2n) \times (2n)$ matrix Z_1. Interpret this property as the equation $x^{2n} - 1 = (x^n - 1)(x^n + 1)$.

(c) [AB91] Show that the Moore–Penrose generalized inverse A^+ (cf. Definition 4.6.1 in Chapter 4) is a circulant matrix if and only if A is a circulant matrix.

2.13 Compute a polynomial $w(z)$ such that $w(z)(1 + 2z - 3z^2) = 1 \bmod z^3$.

2.14 Show that the power sums of polynomial roots may non-uniquely define the coefficients of a polynomial over a finite field. Consider

(a) the polynomials of degree 3 over $\mathbb{Z}_2$ and

(b) the polynomials x^m and $(x+1)^m$ over $\mathbb{Z}_k$ where k divides m.

2.15 (a) Elaborate on the binary segmentation techniques for basic operations with polynomials and general matrices where all entries are integers.

(b) Estimate the bit-operation cost of these operations and compare with the cost of the same operations performed without using the binary segmentation.

(c) Try to extend the techniques to operations with Toeplitz matrices filled with bounded integers.

2.16 (a) Under the notation of Theorem 2.7.1, assume that $u(0) \neq 0$ and represent every matrix of the algebra A_u as the product $Z(\mathbf{u})T$ where T is a Toeplitz matrix (see Theorem 2.7.3).

(b) Represent every Toeplitz matrix as the product $Z^{-1}(\mathbf{u}) F_u(\mathbf{f})$ for appropriate vectors $\mathbf{u}$ and $\mathbf{f}$.

2.17 Specify extensions of Algorithms 2.8.4, 2.8.6, and 2.8.7 to the computation of integer gcds and to the continued fraction approximation of a rational number (see Remark 2.8.11).

2.18 [G84] Prove the following expressions for the scalars c_i of Theorem 2.10.1. Let $r_i^*(x)$ denote the monic associates of the polynomials $r_i(x)$ (that is, the monic polynomials obtained from the polynomials $r_i(x)$ by scaling) and let $b_i r_i^*(x) = r_{i-2}^*(x) - q_{i-1}^*(x) r_{i-1}^*(x)$ (see (2.8.6)). Then

$$c_i = c_0 b_2 b_4 b_6 \ldots b_i \text{ if } i \text{ is even,}$$
$$c_i = c_1 b_3 b_5 b_7 \ldots b_i \text{ if } i \text{ is odd,}$$
$$r_i(x) = c_i r_i^*(x) \text{ for all } i.$$

2.19 Given the polynomials $m_0(x) = x^2 + 1$ and $m_1(x) = 2x + 5$, compute the polynomials $w_0(x) \bmod m_0(x)$ and $w_1(x) \bmod m_1(x)$ such that $w_0(x) m_1(x) = 1 \bmod m_0(x)$, $w_1(x) m_0(x) = 1 \bmod m_1(x)$.

2.20 Find the polynomial $p(x)$ of degree 2 such that $p(x) \bmod (x^2 + 1) = 5x - 16$, $p(x) \bmod (2x + 5) = 20$.

2.21 Compute the (2,2) Padé approximation to the polynomial $1 + 2x - 3x^2 + x^3 - 2x^4 + x^5$.

2.22 (a) Let Problem 2.9.7 have input $a_0, \ldots, a_{2s-1}$ and output $n, c_0, \ldots, c_{n-1}$. Prove that the linear system of k equations $a_i = c_{k-1} a_{i-1} + \ldots + c_0 a_{i-k}$ in $c_0, \ldots, c_{k-1}$, $i = 0, 1, \ldots, k-1$, is non-singular for $k = n$ and is singular for $k < n$.

(b) Prove equation (2.8.2).

(c) Prove non-singularity of the $r \times r$ leading principal submatrix of the matrix JV for the matrices J of Definition 2.1.1 and $V = (v_{m+i-j})_{i,j=0}^{n-1}$ in (2.11.2), (2.11.3), and for $r = \text{rank}(V)$.

(d) Prove uniqueness of the solutions to Problems 2.7.10 and 2.9.1 (up to scaling).

2.23 Complete the algorithm of [BGY80] for a Toeplitz linear system of equations $V\mathbf{z} = \mathbf{b}$. That is, define an embedding of the $(n+1) \times (n+1)$ non-singular Toeplitz matrix V into an $(n+2) \times (n+2)$ non-singular Toeplitz matrix $T = (t_{ij})$ with $t_{00} \neq 0$ and the recovery of the inverse V^{-1} from T^{-1}, as suggested in [BGY80] (see Section 2.12).

2.24 (a) Prove Gohberg–Semencul's formula (2.11.4).

(b) Prove its following modification [H79], [T90]. Let $T = (t_{i-j})_{i,j=0}^{n-1}$ be an $n \times n$ non-singular Toeplitz matrix, let t_{-n} be any scalar (e.g., $t_{-n} = 0$), and write $p_n = -1, \tilde{\mathbf{t}} = (t_{i-n})_{i=0}^{n-1}, \mathbf{p} = (p_i)_{i=0}^{n-1} = T^{-1}\tilde{\mathbf{t}}, \mathbf{q} = (p_{n-i})_{i=0}^{n-1}$, $\mathbf{x} = T^{-1}\mathbf{e}_0, \mathbf{u} = ZJ\mathbf{x}$. Then $T^{-1} = Z(\mathbf{p})Z^T(\mathbf{u}) - Z(\mathbf{x})Z^T(\mathbf{q})$.

(c) [GO94]. Define matrix T and vectors $\tilde{\mathbf{t}}$ and $\mathbf{x}$ as in part (b). Let a, b and f be three scalars such that $bf \neq 1$, $b \neq 0$. Write $\mathbf{t} = (t_i)_{i=0}^{n-1}$, $\mathbf{v} = T^{-1}(\mathbf{t} - b\tilde{\mathbf{t}} + a\mathbf{e}_0)$. Then prove the following matrix equation:

$$T^{-1} = \frac{1}{1 - bf}\left(Z_f(\mathbf{x})Z_{1/b}(\mathbf{v}) - Z_f(\mathbf{v} - (1 - bf)\mathbf{e}_0)Z_{1/b}(\mathbf{x})\right).$$

(d) How many parameters are involved in each of the above expressions for the inverse matrix T^{-1} and in the one of [GS72] in the case where T is a symmetric Toeplitz matrix?

2.15 Appendix. Pseudocodes

In this section, we show sample pseudocodes for the following algorithms:

- Convolve Two Vectors.

- Multiply a Lower Triangular Toeplitz Matrix by a Vector.

- Multiply an $N \times N$ Circulant Matrix by a Vector.

- Multiply an $N \times N$ Skew-Circulant Matrix by a Vector.

- Multiply a General Toeplitz Matrix by a Vector (two algorithms).

- Invert a Lower Triangular Toeplitz Matrix.

- Invert a Lower Triangular Toeplitz Matrix Having the Dimension of a Power of 2 (this algorithm is used as a subroutine in the previous one).

Note: In this section, lg is logarithm base 2 and N is an integral power of 2.

Pseudocode 2.15.1. Convolve Two Vectors.

   This algorithm uses Fast Fourier Transform (FFT) and Inverse Fast Fourier Transform (IFFT) to multiply two univariate polynomials; it is assumed that subroutines for FFT and IFFT are available and that all polynomials are represented by their coefficient vectors.

INPUT:

>   vector **U** (coefficients of a polynomial),
>
>   vector **V** (coefficients of a polynomial)

OUTPUT:

>   vector **W** (coefficients of the polynomial product)

PROCEDURES:

1. Calculate H, the dimension of the return vector

2. Calculate N, the smallest integral power of 2 which is not less than H ($H \leq N < 2H$; since N is an integral power of 2, FFT and IFFT can be applied to N-dimensional vectors using the N-th roots of 1)

3. Perform FFT on each input vector embeded into a vector of dimension N by appending zero components; obtain vectors **fU** and **fV** of dimension N

4. Pairwise multiply components of **fU** and **fV**; obtain vector **fP**

5. Perform IFFT on vector **fP**; return vector **P** of dimension N

6. Return the vector **W** of dimension H, obtained by extracting the subvector of the H leading components of **P**

```
vector Convolve( U , V )
{
    H = dimension( U ) + dimension( V ) - 1

    N = 2^ceil( lg( H ) )

    fU = FFT( N , U )
    fV = FFT( N , V )

    fW[i] = fU[i] * fV[i]   for i from 0 to N-1

    P = IFFT( N , fW )

    W[i] = P[i] for i from 0 to H-1
    return W
}
```

Pseudocode 2.15.2. Multiply a Lower Triangular Toeplitz Matrix by a Vector.
This algorithm multiplies a square lower triangular Toeplitz matrix by a vector using FFT and IFFT to convolve two vectors or equivalently to multiply two polynomials with these coefficient vectors. The lower triangular Toeplitz matrix is represented by its first column vector **L**.

Example.   The lower triangular Toeplitz matrix

$$\begin{pmatrix} 1 & 0 & 0 \\ -2 & 1 & 0 \\ 3 & -2 & 1 \end{pmatrix}$$

is represented by the vectors $\mathbf{L} = (1\ -2\ 3)^T$.

INPUT:

   vector **L** (representing an $n \times n$ lower triangular Toeplitz matrix L),

   vector **V** (of dimension n)

OUTPUT:

   vector $L\mathbf{V}$

PROCEDURES:

   1. Convolve the two input vectors using FFT and IFFT

   2. Extract the leading subvector of the convolution vector

```
vector LTTbyVectorMultiply( L , V )
{
    n = dimension( L )
    U = Convolve( L , V )

    LV[i] = U[(n-1)+i] for i from 0 to n-1
    return LV
}
```

Pseudocode 2.15.3. Multiply an $N \times N$ Circulant Matrix by a Vector

This algorithm multiplies a circulant matrix by a vector using FFT and IFFT to compute the positive wrapped convolution of two vectors or equivalently to multiply modulo $x^N - 1$ two polynomials in x with these coefficient vectors. The circulant matrix is represented by its first column vector $\mathbf{A}$; N is an integral power of 2.

INPUT:

        vector $\mathbf{A}$ (of dimension N representing an $N \times N$ circulant matrix $Z_1(\mathbf{A})$)

        vector $\mathbf{U}$ (of dimension N)

OUTPUT:

        vector $\mathbf{X}$, the product $Z_1(\mathbf{A})\mathbf{U}$

```
vector CirculantByVectorMultiply( A , U )
{
    N = dimension( A )

    fA = FFT( N , A )
    fU = FFT( N , U )

    fX[i] = fA[i] * fU[i] for i from 0 to N-1
    X = IFFT(fX)
    return X
}
```

Pseudocode 2.15.4. Multiply an $N \times N$ Skew-Circulant Matrix by a Vector

This algorithm multiplies a skew-circulant matrix by a vector using FFT and IFFT to compute the negative wrapped convolution of two vectors or equivalently to multiply modulo $x^N + 1$ two polynomials in x with these coefficient vectors. The skew-circulant matrix is represented by its first column vector **B**. The procedure (like FFT and IFFT) uses precomputed values of the roots of 1 which make up the input vector Ω. N is an integral power of 2.

INPUT:

      vector **B** (of dimension N representing a circulant matrix $Z_{-1}(\mathbf{B})$)

      vector **U** (of dimension N)

      vector $\Omega = (\omega_i)$, $(\omega_i = w_{2N}^i)$, $w_{2N} = \exp(\pi\sqrt{-1}/N)$ (a primitive $2N$-th root of 1)

OUTPUT:

      vector **Y**, the product $Z_{-1}(\mathbf{B})\mathbf{U}$

```
vector SkewCirculantByVectorMultiply( B , U )
{
    N = dimension( B )
    B_[i] = B[i] * Omega[i] for i from 0 to N-1

    fB_ = FFT( N , B_ )
    fU = FFT( N , U )

    fV[i] = fB_[i] * fU[i]   for i from 0 to N-1

    V = IFFT( N , fV )

    Y[i] = V[i] / Omega[i] for i from 0 to N-1
    return Y
}
```

Pseudocode 2.15.5. Multiply a General Toeplitz Matrix by a Vector via Convolution

The algorithm multiplies a square Toeplitz matrix T by a vector $\mathbf{V}$ using FFT and IFFT to convolve two vectors or equivalently to multiply two univariate polynomials with these coefficient vectors. The Toeplitz matrix is represented by the vector made up of the reversed first row and the first column of the matrix.

Example. The Toeplitz matrix

$$\begin{pmatrix} -3 & 4 & 6 \\ 1 & -3 & 4 \\ 2 & 1 & -3 \end{pmatrix}$$

is represented by the vector $\mathbf{T} = (6 \ 4 \ -3 \ 1 \ 2)^T$.

INPUT:

        vector $\mathbf{T}$ (of dimension $2n - 1$ representing an $n \times n$ Toeplitz matrix T),

        vector $\mathbf{V}$ (of dimension n)

OUTPUT:

        vector $T\mathbf{V}$ (of dimension n)

PROCEDURES:

1. Calculate the dimension of the vector $\mathbf{T}$

2. Convolve the two vectors using FFT and IFFT

3. Extract the middle segment of the vector returned by part 2

```
vector TbyVectorMultiply( T , V )
{
    n = ( dimension( T ) + 1 ) / 2
    U = Convolve( T , V )

    TV[i] = U[(n-1)+i] for i from 0 to n-1
    return TV
}
```

Pseudocode 2.15.6. Multiply a General Toeplitz Matrix by a Vector via Splitting

The algorithm first splits the matrix into a sum of a circulant and a skew-circulant matrices, then multiplies each of the summands by the input vectors and sums the product vectors. A Toeplitz matrix is represented by the vector made up of the reversed first row and the first column of the matrix.

INPUT:

vector **T** (of dimension $2n - 1$),

vector **V** (of dimension n)

OUTPUT:

vector T**V** (of dimension n)

PROCEDURES:

1. Embed T into a Toeplitz matrix S of dimension $N = 2^k$ where k is integral

2. Embed **V** into a vector **U** of dimension N

3. Split S into a sum of a circulant matrix $Z_1(\mathbf{A})$ represented by its first column **A** and a skew-circulant matrix $Z_{-1}(\mathbf{B})$ represented by its first column **B**

4. Multiply $Z_1(\mathbf{A})$ by **U** by calling the circulant-by-vector subroutine

5. Multiply $Z_{-1}(\mathbf{B})$ by **U** by calling the skew-circulant-by-vector subroutine

6. Calculate the leading subvector of the vector $(Z_1(\mathbf{A}) + Z_{-1}(\mathbf{B}))\mathbf{U}$ of dimension n

Note: T is a leading principal (northwestern) submatrix of S; all entries of S not defined by T are zeros. See (2.6.2) for part 3, (2.6.4) for part 4, and (2.6.5), (2.6.6) for part 5.

```
vector TbyVectorMultiplyBySplittting( T , V )
{
    H = ( dimension( T ) + 1 ) / 2

    N = 2^ceil( lg( H ) )

//  embed matrix T into a Toeplitz matrix S
    S[i] = 0           for i from 0     to N-H-1
    S[N-H+i] = T[i] for i from 0        to 2H-2
    S[i] = 0           for i from N+H-1 to 2N-2

//  embed vector V into a vector U
    U[i] = V[i] for i from 0 to H-1
    U[i] = 0    for i from H to N-1

//  split matrix T into the sum of a pair of
//  circulant and skew-circulant matrices
    A[0] = S[N-1]
    B[0] = 0
    A[i] = (S[N-1+i]+S[i-1])/2 for i from 1 to N-1
    B[i] = (S[N-1+i]-S[i-1])/2 for i from 1 to N-1

    X = CirculantByVectorMultiply( A , U )
    Y = SkewCirculantByVectorMultiply( B , U )

//  TV = X + Y
    TV[i] = X[i] + Y[i] for i from 0 to n-1
    return TV
}
```

Pseudocode 2.15.7. Invert a Lower Triangular Toeplitz Matrix.

This algorithm inverts a lower triangular Toeplitz matrix represented by its first column vector. Note that the inverse is also a lower triangular Toeplitz matrix, so we only need to compute its first column. The input matrix and all other matrices involved in the algorithms are identified with the vectors representing them.

INPUT:

   vector **L** (the first column of a lower triangular Toeplitz matrix L)

OUTPUT:

   vector **V** (the first column of the inverse matrix)

PROCEDURES:

1. Calculate N, the smallest integral power of 2 which is not less than the dimension of the vector **L**

2. Extend the vector **L** to represent an $N \times N$ lower triangular Toeplitz matrix W

3. Call recursive function to invert the matrix W

4. Return the leading subvector of the vector returned by recursive function of part 3

Note: If the matrix is singular, the return vector is the null vector (filled with zeros).

```
vector InvertLTTMatrix( L )
{
    H = dimension( L )

    if ( L[0] == 0 ) then
         return ( the null vector of dimension H )

    N = 2^ceil( lg( H ) )

    W[i] = L[i] for i from 0 to H-1
    W[i] = 0    for i from H to N-1

    Z = InvertLTTMatrixDimensionPower2( N , W )

    V[i] = Z[i] for i from 0 to H-1
    return V
}
```

Pseudocode 2.15.8. Invert a Non-singular Lower Triangular Toeplitz Matrix Having the Dimension of a Power of 2.

This recursive algorithm inverts a non-singular lower triangular Toeplitz matrix L of dimension $N = 2^h$ where h is integral. The input matrix and all other matrices involved in the algorithms are identified with the vectors representing them.

INPUT:

integer N (a power of 2)

vector $\mathbf{L}$ of dimension N (the first column of a non-singular lower triangular Toeplitz matrix L)

OUTPUT:

vector $\mathbf{Z}$ (the first column of the inverse matrix)

PROCEDURES:

1. Return the answer in the trivial case where the matrix has dimension 1

2. Invert the leading principal (northwestern) block L_0 of L having half-size $N/2$-by-$N/2$ by recursively calling the function for the half size input; the function returns a vector $\mathbf{Z_0}$ of dimension $N/2$ representing the inverse L_0^{-1} of the block L_0

3. Calculate $Y = (-1) * (L_0^{-1}) * (T_1) * (L_0^{-1})$ where T_1 is the southwestern block of L having half-size

4. Merge the vector representations of L_0^{-1} and Y

5. Return the result of part 4, the vector representation of the inverse of the input matrix

Notes:

- T_1 is an $N/2 \times N/2$ Toeplitz matrix found in the southwestern quadrant of the input matrix L.

- L_0 is an $N/2 \times N/2$ lower triangular Toeplitz matrix found in the northwestern quadrant of the input matrix L.

- L_0 is also found in the southeastern quadrant of L.

- L_0 is represented by its first column $\mathbf{L_0}$.

```
vector InvertLTTMatrixDimensionPower2( N , L )
{
    if ( N = 1 ) then return ( 1 / L[0] )

    Z_0 = InvertLTTMatrixDimensionPower2( N/2 , L_0 )

//  X is the product of T_1 by the first column of the
//  inverse of the northwestern input block L_0

    X = TbyVectorMultiply( T_1 , Z_0 )

    Z_1 = LTTbyVectorMultiply( -Z_0 , X )

    Z[i] = Z_0[i+1]      for i from  0  to N-1
    Z[i] = Z_1[i-(N/2)] for i from N/2 to N-1
    return Z
}
```

Chapter 3

Matrix Structures of Vandermonde and Cauchy Types and Polynomial and Rational Computations

Summary   In this chapter, we complement our subjects of the previous chapter by studying fundamental polynomial and rational computations (this includes the Nevanlinna–Pick and Nehari problems of rational interpolation and approximation as well as sparse multivariate polynomial interpolation) together with the related computations with matrices of Vandermonde and Cauchy types. We also show applications to loss-resilient encoding/decoding and to diagonalization of matrix algebras, which leads us to the discrete sine and cosine transforms. Figures 3.3–3.6 summarize correlation among computational problems in these areas.

3.1   Multipoint polynomial evaluation

Problem 3.1.1. Multipoint polynomial evaluation. Given the coefficients of a polynomial $v(x) = \sum_{i=0}^{n-1} v_i x^i$ and a set of node points $t_0, \ldots, t_{n-1}$, compute the values $r_0 = v(t_0), \ldots, r_{n-1} = v(t_{n-1})$.

Problem 3.1.1 can be stated equivalently as follows.

Problem 3.1.2.   Vandermonde-by-vector product.   Given a Vandermonde matrix $V = V(\mathbf{t}) = (t_i^j)_{i,j=0}^{n-1}$ and a vector $\mathbf{v} = (v_j)_{j=0}^{n-1}$, compute the vector $\mathbf{r} = (r_i)_{i=0}^{n-1} = V\mathbf{v}$.

In the case where the t_i are the n-th roots of 1, that is, where $V = \Omega$, $\mathbf{t} = \mathbf{w}$, Problems 3.1.1 and 3.1.2 turn into the problem of the DFT($\mathbf{v}$) computation.

Graph for selective reading

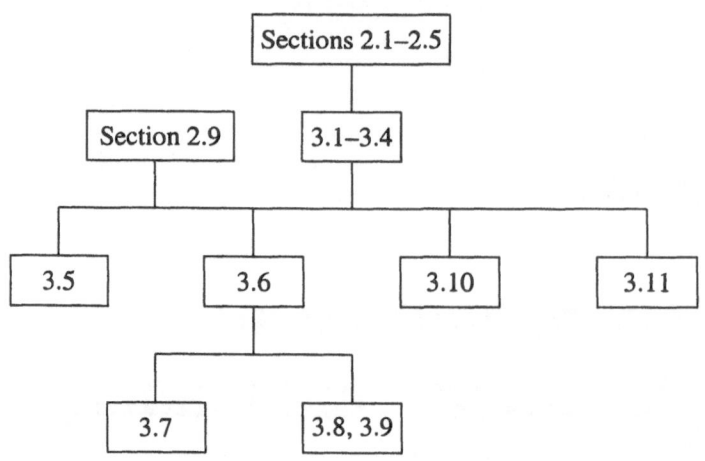

Solution: The Moenck–Borodin algorithm solves Problem 3.1.1 in $O(M(n) \log n)$ ops for $M(n)$ in (2.4.1), (2.4.2) based on the two following simple observations.

Fact 3.1.3. $v(a) = v(x) \bmod (x - a)$ for any polynomial $v(x)$ and any scalar a.

Fact 3.1.4. $w(x) \bmod v(x) = (w(x) \bmod (u(x)v(x))) \bmod v(x)$ for any triple of polynomials $u(x)$, $v(x)$, and $w(x)$.

Algorithm 3.1.5. Superfast multipoint polynomial evaluation.

INITIALIZATION: Write $k = \lceil \log_2 n \rceil$, $m_j^{(0)} = x - x_j$, $j = 0, 1, \ldots, n - 1$; $m_j^{(0)} = 1$ for $j = n, \ldots, 2^k - 1$ (that is, pad the set of the moduli $m_j^{(0)} = x - x_j$ with ones, to make up a total of 2^k moduli). Write $r_0^{(k)} = v(x)$.

COMPUTATION:

1. *Fan-in process* (see Figure 3.1).   Compute recursively the "supermoduli" $m_j^{(h+1)} = m_{2j}^{(h)} m_{2j+1}^{(h)}$, $j = 0, 1, \ldots, 2^{k-h} - 1$; $h = 0, 1, \ldots, k - 2$.

2. *Fan-out process* (see Figure 3.2).   Compute recursively the remainders $r_j^{(h)} = r_{\lfloor j/2 \rfloor}^{(h+1)} \bmod m_j^{(h)}$, $j = 0, 1, \ldots, \min\{n, \lceil n/2^h \rceil - 1\}$; $h = k - 1, k - 2, \ldots, 0$.

OUTPUT: $p(x_i) = r_i^{(0)}$, $i = 0, 1, \ldots, n - 1$.

To prove *correctness* of the algorithm, first apply Fact 3.1.4 recursively to obtain that $r_j^{(h)} = v(x) \bmod m_j^{(h)}$ for all j and h. Now, correctness of the output $v(x_i) = r_i^{(0)}$ follows from Fact 3.1.3.

To estimate the *computational cost* of the algorithm, represent its two stages by the same binary tree (see Figures 3.1 and 3.2), whose nodes are the "supermoduli" $m_j^{(h)}$ at the *fan-in* stage 1, but turn into the remainders $r_j^{(h)}$ at the *fan-out* stage 2.

Figure 3.1: Fan-in process, $m_j^{(h+1)} = m_{2j}^{(h)} m_{2j+1}^{(h)}$

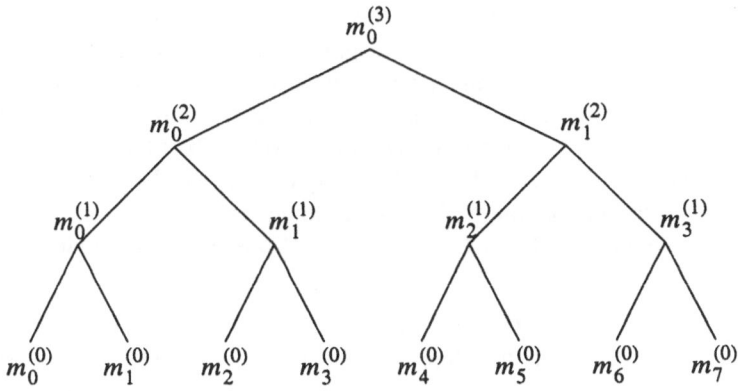

Figure 3.2: Fan-out process, $r_j^{(h)} = r_{\lfloor j/2 \rfloor}^{(h+1)} \bmod m_j^{(h)} = v(x) \bmod m_j^{(h)}$

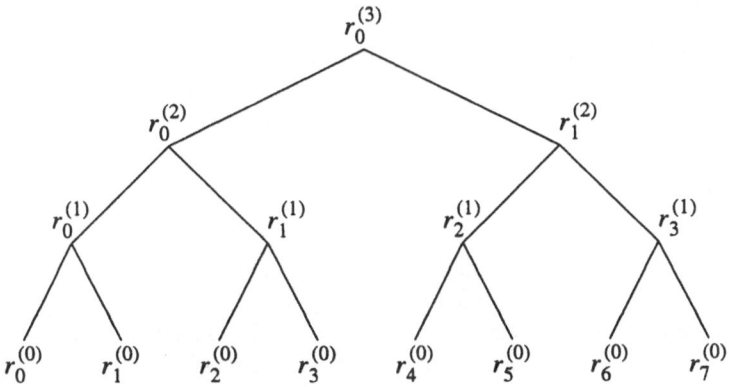

At each level h of the tree, the algorithm computes 2^{k-h} products of pairs of polynomials of degree 2^h at stage 1 and 2^{k-h} remainders of the division of polynomials of degree of at most 2^{h+1} by "supermoduli" of degree 2^h. Each time multiplication/division uses $O(M(2^h))$ ops for $M(n)$ in (2.4.1), (2.4.2), which makes up a total of $O(2^{k-h}M(2^h))$ ops at the h-th level, and $O(\sum_{h=0}^{k-1} 2^{k-h}M(2^h)) = O(M(2^k)k)$ ops at all levels. Recall that $n \le 2^k < 2n$ and obtain the claimed bound of $O(M(n)\log n)$ ops. $\qquad\square$

Remark 3.1.6. The fan-in computation at stage 1 depends only on the set $\{t_0, \ldots, t_{n-1}\}$ and can be viewed as (cost-free) *preprocessing* if the node set is fixed and only the polynomial $v(x)$ varies. Similar observations hold for the solution of many other problems in this chapter.

Remark 3.1.7. The Moenck–Borodin algorithm applies over an arbitrary field. In the case of computations with rounding errors, the algorithm generally leads to numerical stability problems, particularly due to the ill conditioned operation of polynomial division, performed recursively at the fan-out stage 2. In the case where the node points are the n-th roots of 1, for $n = 2^k$, weak numerical stability and simultaneously the cost bound of $O(n \log n)$ ops can be achieved, however, because the algorithm can be turned into the FFT algorithm, as we show in our next example. Besides the set of the 2^k-th roots of 1, there are some other good node sets. The FFT bound of $O(n \log n)$ ops as well as numerical stability of the computations can be also achieved for the evaluation of a polynomial on the Chebyshev node sets. Furthermore, Theorem 2.4.4 implies the bound of $O(M(n))$ on the number of ops where the node sets are of the form $\{at^{2i} + bt^i + c\}_i$.

Example 3.1.8. Moenck–Borodin algorithm as FFT. Let $n = K = 2^k$. Let $\delta_j(i)$ denote the i-th trailing bit in the binary representation of j, that is, $\delta_j(i) = 0$ if (j mod 2^i) $< 2^{i-1}$, $\delta_j(i) = 1$ otherwise. Write $d(h, j) = \sum_{i=1}^{h} \delta_j(i) H / 2^i$ for $H = 2^h$, so $d(h, 0) = 0, d(h, 1) = H/2, d(h, 2) = H/4, d(h, 3) = H/2 + H/4, \ldots$. Order the K-th roots of 1 so that $t_j = \omega_n^{d(k,j)}$, $j = 0, 1, \ldots, K - 1$, for ω_K of Definition 2.1.2. In this case, all "supermoduli" $m_j^{(h)}$ turn into the binomials $x^{2^h} - \omega_{K/2^h}^{d(k-h,j)}$, which we assume to be available cost-free, and the fan-out process of recursive division requires at most $1.5K$ ops at each of the $k = \log_2 K$ levels.

Remark 3.1.9. As a by-product of the fan-in process of stage 1 of Algorithm 3.1.5, which uses $O(M(n) \log n)$ ops, we compute the coefficients of a polynomial $u(x) = \prod_{j=1}^{n}(x - x_j)$ given its roots $x_1, \ldots, x_n$. The coefficients are the *elementary symmetric functions* of the roots (up to the factor -1), and the algorithm is sometimes called **the symmetric function algorithm**.

Remark 3.1.10. Problem 3.1.1 and its solution algorithms are immediately extended to the case where we have m points $t_0, \ldots, t_{m-1}$ for $m > n$ or $m < n$. The solution requires $O(E(l)r/l)$ ops provided $l = \min\{m, n\}$, $r = \max\{m, n\}$, and $E(l)$ ops are sufficient for the solution where $n = l$. $E(l) = O(M(l) \log l)$ for a general set $\{t_i\}$ but decreases to $O(M(l))$, where $t_i = at^{2i} + bt^i + c$ for fixed scalars a, b, c, and t and for all i (see Remark 3.1.7). Similarly, Problem 3.1.2 is extended.

Remark 3.1.11. Problem 3.1.1 and its solution as well as the problems of Lagrange interpolation and the Chinese Remainder computation of Sections 3.3 and 3.5 have natural extensions to integers (see the Notes to Section 2.4).

3.2 Modular reduction and other extensions

Problem 3.1.1 has the following natural extension.

Problem 3.2.1. Modular reduction of a polynomial. Given the coefficients of polynomials $v(x)$, $m_1(x), \ldots, m_k(x)$, where $\deg m_i(x) = d_i$, $i = 1, \ldots, k$; $n = \sum_{i=1}^{k} d_i > \deg v(x)$, compute the polynomials $r_i(x) = v(x) \bmod m_i(x)$ for all i.

Solution: Algorithm 3.1.5 is extended immediately, except that the best order of the enumeration of the leaves of the tree depends on the set of degrees $\{d_i = \deg m_i(x), \ i = 1, \ldots, k\}$.

To decrease the overall complexity of the solution, it is most effective to use *Huffman's trees* where, for a pair of the smallest degrees d_i and d_j (ties are broken arbitrarily), the leaves-moduli $m_i(x)$ and $m_j(x)$ are identified as two children of the "supermodulus" $m_i(x)m_j(x)$ of degree $d_i + d_j$. Then the process is recursively repeated for the same tree but with the leaves $m_i(x)$ and $m_j(x)$ assumed deleted. The deletion process stops where only a single node, $\prod_{i=1}^{k} m_i$, is left, and then all deleted links and nodes are reinstated.

Let us estimate the number $a = a(v, m_1, \ldots, m_k)$ of ops involved in such a computation of the polynomials $r_1(x), \ldots, r_k(x)$ (together with all "supermoduli", including the polynomial

$$u(x) = \prod_{i=1}^{k} m_i(x), \tag{3.2.1}$$

and all remainders represented by the nodes of the tree). Let $H(\mathbf{d})$ denote the entropy of the vector $\mathbf{d}/n$, that is, let

$$H(\mathbf{d}) = -\sum_i (d_i/n) \log(d_i/n) \le \log n. \tag{3.2.2}$$

Then one may verify that

$$a = O(M(n)(H(\mathbf{d}) + 1)) \tag{3.2.3}$$

for $M(n)$ in (2.4.1), (2.4.2); moreover, this bound is optimal within the factor of $O(M(n)/n)$. □

An important special case is the computation of the values $v^{(i)}(t)/i!$, $i = 0, 1, \ldots, k - 1$, of the first k Taylor's coefficients of a degree n polynomial $v(x)$ at a point t, that is, the case where $m_i(x) = (x - t)^i$, $i = 1, \ldots, k$, for a fixed t. In this case we yield the better bound of $O(M(k)n/k)$ if we first compute the polynomial $r(x) = v(x) \bmod (x - t)^k$ as the remainder of the division of the polynomials $r(x)$ by $(x - t)^k$ (this involves $O(M(k)n/k)$ ops) and then solve Problem 2.4.3 (Taylor's shift of the variable) by using $M(k)$ ops. Generalizing, we arrive at the following:

Problem 3.2.2. Hermite evaluation of a polynomial on a fixed set of points. Given a polynomials $v(x)$ of degree n, a set of field elements $t_1, \ldots, t_k$ (node points), and positive integers $d_1, \ldots, d_k$, compute the polynomials $r_{i,j}(x) = v(x) \bmod (x - t_i)^j$, $j = 1, \ldots, d_i$; $d_i \le n$; $i = 1, \ldots, k$.

Solution: By applying the above algorithm, we solve Problem 3.2.2 using

$$a = O\left(\sum_{i=1}^{k} M(d_i)n/d_i\right) \tag{3.2.4}$$

ops. The latter bound implies that

$$a = O(M(n)) \tag{3.2.5}$$

in the case where $k = 1$, $d_1 = O(n)$ and, more generally, where $\prod_i d_i = O(n)$. $\quad\square$

Problem 3.2.2 can be expressed as the problem of multiplication by a vector of the *confluent Vandermonde matrix*, which is a block column vector

$$V(\mathbf{t}, d) = (V_{t_i,d})_{i=1}^{k} \tag{3.2.6}$$

with the block entries $V_{t_i,d}$ of the form $\left(\frac{1}{j!}(\frac{d^j(x^l)}{dx^j})_{x=t_i}\right)_{j,l=0}^{d,n}$, $d = \max_i d_i$.

Generalizing Vandermonde matrices, we also define an $m \times n$ *polynomial Vandermonde matrix*,

$$V_{\mathbf{P}}(\mathbf{t}) = (P_j(t_i))_{i,j} \tag{3.2.7}$$

where $\mathbf{P} = (P_j(x))_{j=0}^{n-1}$ is a fixed basis in the linear space of polynomials of degree less than n, and $\mathbf{t} = (t_i)_{i=0}^{m-1}$ is a fixed vector of the node points. Multiplication of this matrix by a vector $\mathbf{c} = (c_j)_j$ amounts to evaluating the polynomial $\sum_j c_j P_j(x)$ on the node set $\{t_i\}_{i=0}^{m-1}$. In the special case of the monomial basis where $P_j(x) = x^j$, $j = 0, 1, \ldots, n - 1$, we come back to ill conditioned Vandermonde matrices. An appropriate choice of the pair $\mathbf{P}$, $\mathbf{t}$ may improve the condition of the matrix $V_{\mathbf{P}}(\mathbf{t})$. This is the case for a real vector $\mathbf{t}$ with components in the range $[-1, 1]$ and for the bases made up of the Chebyshev polynomials of the first or second kind.

3.3 Lagrange polynomial interpolation

Here are the equivalent (polynomial and matrix) versions of the converses to the equivalent Problems 3.1.1 and 3.1.2.

Problem 3.3.1. Lagrange polynomial interpolation. Given the node set (or vector) $\{t_i\}_{i=0}^{n-1}$ of n distinct points $t_0, \ldots, t_{n-1}$ and the set (or vector) of values $\{r_i\}_{i=0}^{n-1}$, compute a set (or vector) $\{v_j\}_{j=0}^{n-1}$ such that $\sum_{j=0}^{n-1} v_j t_i^j = r_i$, $i = 0, 1, \ldots, n - 1$, that is, recover the coefficients of a polynomial $v(x) = \sum_{j=0}^{n-1} v_j x^j$ from its values at n distinct points $t_0, \ldots, t_{n-1}$.

Problem 3.3.2. Solution of a Vandermonde linear system of equations. Given a Vandermonde matrix $V = V(\mathbf{t}) = (t_i^j)_{i,j=0}^{n-1}$ (for n distinct values t_i) and a vector $r = (r_i)_{i=0}^{n-1}$, solve the linear system $V\mathbf{v} = \mathbf{r}$, that is, compute the vector $\mathbf{v} = V^{-1}\mathbf{r}$. ($\det V = \prod_{i<k}(t_i - t_k) \neq 0$, so V is a non-singular matrix.)

The problem is ill conditioned unless $|t_i| = 1$ for all i.

Solution: To yield a superfast solution algorithm, recall the *Lagrange interpolation formula*

$$v(x) = u(x) \sum_{i=0}^{n-1} r_i w_i / (x - t_i), \qquad (3.3.1)$$

where r_i and t_i are the input values,

$$u(x) = u_t(x) = \prod_{i=0}^{n-1} (x - t_i) = x^n + \sum_{i=0}^{n-1} u_i x^i, \qquad (3.3.2)$$

$$w_i = 1 / \prod_{k=0, k \neq i}^{n-1} (t_i - t_k)) = 1/u'(t_i), \quad i = 0, 1, \ldots, n, \qquad (3.3.3)$$

and $u'(x)$ is the formal derivative of the polynomial $u(x)$. Then compute the polynomial $v(x)$ by means of *manipulation with partial fractions*.

Algorithm 3.3.3. Polynomial interpolation.

COMPUTATION:

1. Compute the coefficients of the polynomials $u(x)$ and $u'(x)$.

2. Compute the values of the polynomial $u'(x)$ at the points t_i for all i; compute the values $w_i = 1/u'(t_i)$ for all i.

3. Recursively sum all the partial fractions $r_i w_i / (x - t_i)$ together and output the numerator of the resulting partial fraction $v(x)/u(x)$.

To keep the cost of the computation lower, order the summation of the partial fractions according to the fan-in method of Algorithm 3.1.5 and Figure 3.1, that is, first sum pairwise the original partial fractions and then recursively sum the pairs of the resulting sums, which are partial fractions themselves. At stage 1, we multiply the "moduli" $m_i(x) = x - t_i$ together by using $O(M(n) \log n)$ ops in the fan-in method. Stage 2 is reduced to multipoint evaluation of the polynomial $u'(x)$, so its cost is still $O(M(n) \log n)$ ops. Computation at stage 1 by the fan-in method also gives us all the denominators required at stage 3 (at no additional cost). Computation of the numerator of the sum of each pair of partial fractions in stage 3 is reduced to performing twice multiplication of polynomials of bounded degree, and the overall cost of performing stage 3 is $O(M(n) \log n)$. Remark 3.1.6 applies here as well, that is, the computation of the polynomials $u(x)$ and $u'(x)$ at stage 1 and the denominators at stage 3 can be considered preprocessing if the node set is fixed and only the polynomial $v(x)$ varies.

Example 3.3.4. Interpolation at the 2^k-th roots of 1 (see Example 3.1.8). If the input set is the Fourier set, $\{t_j = \omega^j, \ j = 0, 1, \ldots, K - 1\}$, $K = 2^k \geq n$, $\omega = \omega_K$ being a primitive K-th root of 1, then interpolation amounts to the inverse DFT, and the above algorithm can be reduced to an FFT-like process. Forming the "supermoduli" in this case, we should group the nodes $t_j = \omega^j$ as in Example 3.1.8. Then all the "supermoduli" remain binomials of the form $x^H - \omega^{jH}$, $H = 2^h$, and we assume that we have their coefficients $-\omega^{jH}$ cost-free. Furthermore, the summation of partial fractions takes the form

$$\frac{g(x)}{x^H - \omega^{jH}} + \frac{g^*(x)}{x^H + \omega^{jH}} = \frac{(g(x) + g^*(x))x^H + (g(x) - g^*(x))\omega^{jH}}{x^{2H} - \omega^{2jH}},$$

where $g(x)$ and $g^*(x)$ are polynomials of degree less than H, and computation of the numerators amounts to the inverse FFT. As in the proof of Theorem 2.4.4, we may extend the algorithm to the case where the input set is a scaled and/or shifted Fourier set $\{t_j = a\omega^j + b, \ j = 0, 1, \ldots, K - 1\}$, $K = 2^k \geq n, a \neq 0$.

3.4 Polynomial interpolation (matrix method), transposed Vandermonde matrices, and composition of polynomials

The matrix approach enables some improvement of polynomial interpolation algorithms as well as effective superfast multiplication by a vector of a transposed Vandermonde matrix and its inverse, which are problems of substantial independent interest.

Superfast solution of Problems 3.3.1 and 3.3.2 by matrix methods: Recall the following factorization formula:

$$V^{-1} = J Z_f(\mathbf{u_t} + f e_0) V^T D^{-1}((u'_\mathbf{t}(t_i)(f - t_i^n))_{i=0}^{n-1}), \qquad (3.4.1)$$

where $V = V(\mathbf{t}) = (t_i^j)_{i,j=0}^{n-1}$, the node points $t_0, \ldots, t_{n-1}$ are distinct, $\mathbf{t} = (t_i)_{i=0}^{n-1}$, the vector $\mathbf{u_t} = (u_i)_{i=0}^{n-1}$ and the polynomial $u_\mathbf{t}(x)$ are defined by (3.3.2), and the matrices J, $D(\mathbf{v})$, and $Z_f(\mathbf{v})$ are defined in Definition 2.1.1. (See Exercise 3.10(a) for an alternative factorization formula.)

Let us denote by $e(n)$, $i(n)$, and $f(n)$ the numbers of ops required for the solution of Problems 3.1.1 (evaluation) and 3.3.1 (interpolation), and for the fan-in computation of the polynomial $u_\mathbf{t}(x)$, respectively (see Remark 3.1.6). Due to (3.4.1), $i(n)$ is bounded by $f(n)+e(n)+m(n)+O(n \log n)$ where the term $f(n)+e(n)$ covers the ops involved at the stage of the computation of the vector $\mathbf{u_t}$ and the values $u'_\mathbf{t}(t_i)$ for all i, the term $O(n \log n)$ covers the ops involved in the computation of the values $u'_\mathbf{t}(t_i)(f - t_i^n)$ for all i and the vector $\mathbf{u_t} + f e_0$ for a given vector $\mathbf{u_t}$, and the term $m(n)$ is the overall cost of multiplication by vectors of the three matrix factors on the right-hand side of (3.4.1). The first and third times multiplication involves $O(M(n))$ and n ops, respectively. For the multiplication of a transposed Vandermonde matrix V^T by a vector, we combine the algorithms of Section 3.1 and the next theorem, due to Tellegen.

Theorem 3.4.1. *Let $a(M_\mathbf{v})$ be the minimum number of ops sufficient to multiply a fixed $k \times l$ matrix M by a vector. Let $a(M_\mathbf{v}^T)$ denote the similar bound for the transpose M^T. Then $a(M_\mathbf{v}^T)+l = a(M_\mathbf{v})+k$ provided that the matrix M has no zero rows or columns.*

Applying Tellegen's Theorem, we perform the multiplication by using $e(n)$ ops. Summarizing, we obtain

$$i(n) \leq 2e(n) + f(n) + O(M(n) + n \log n). \qquad (3.4.2)$$

The stages of matrix-by-vector multiplication replace stage 3 of Algorithm 3.3.3. With Tellegen's Theorem 3.4.1, the matrix approach gives us a little faster solution because the cost of the fan-in stage is already covered by the term $f(n)$. If the vector $\mathbf{u}_t$ is computed as the cost-free preprocessing, then the term $f(n)$ can be deleted from the right-hand side of (3.4.2). As a by-product of (3.4.1), we obtain the factorization

$$V^{-T} = (V^T)^{-1} = D^{-1}((u_t'(t_i)(f - t_i^n))_{i=0}^{n-1})VJZ_f^T(\mathbf{u}_t + f\mathbf{e}_0), \qquad (3.4.3)$$

which *enables superfast solution of a transposed Vandermonde linear system of equations. Another option is to reduce the latter problem to Problem 3.3.2 by applying Tellegen's Theorem 3.4.1. Both ways we arrive at cost bound (3.4.2).* □

The results of this section and Section 3.1 support various further applications of the Toom evaluation/interpolation method (see Section 2.5). For instance, here is an application to the solution of the following problem.

Problem 3.4.2. Composition of polynomials. Given the coefficients of two polynomials $s(x)$ (of degree m) and $t(x)$ (of degree n), compute the coefficients of the polynomial $w(x) = s(t(x))$.

Solution: Let $mn + 1 \le K = 2^k < 2mn + 2$, let ω_K be a given primitive K-th root of 1, compute the values $s(t(\omega_K^i))$, $i = 0, 1, \ldots, K - 1$, and then interpolate to the polynomial $w(x) = s(t(x))$ at the overall cost of performing $O(mn \log(mn) + nM(m) \log m)$ ops. If no primitive K-th roots of 1 are available, we replace the vector $(\omega_K^i)_{i=0}^{K-1}$ by any vector $\mathbf{v} = (v^i)_{i=0}^{K-1}$ with K distinct components v^i, and yield the polynomial $w(x)$ in $O(M(mn) \log(mn))$ ops. □

Problem 3.4.3. Composition of power series. Given a natural n and two formal power series $s(t)$ and $t(x)$ such that $t(0) = 0$, compute the polynomial $w(x) = s(t(x))$ mod x^{n+1}. (Note that we may replace the power series $t(x)$ and $s(t)$ by polynomials $t(x)$ modulo x^{n+1} and $s(t)$ modulo t^{n+1}, respectively, due to the assumption that $t(0) = 0$.)

Solution: The problem is a special case of Problem 3.4.2, whose solution involves $O(n^2 \log^2 n)$ ops. (Here, we assume that FFT can be applied.) Computing modulo x^{n+1}, we decrease the latter bound to $O((n \log n)^{3/2})$. Fix an integer h of the order of $\sqrt{n/\log n}$. Represent $t(x)$ as the sum $t_0(x) + x^h t_1(x)$ where $t_0(x) = t(x)$ mod x^h. Evaluate the coefficients of the polynomial $s(t_0(x))$ by solving Problem 3.4.2 at the cost $O(nh \log^2 n) = O((n \log n)^{3/2})$. Then write the generalized Taylor expansion

$$s(t(x)) = s(t_0(x)) + s'(t_0(x))x^h t_1(x)$$
$$+ s''(t_0(x))(x^h t_1(x))^2/2! + \cdots + s^{(g)}(t_0(x))(x^h t_1(x))^g/g!$$

where $g = \lceil n/h \rceil$ and $s^{(g)}(t)$ denotes the derivative of $s(t)$ of order g. To compute the polynomials

$$(1/j!)s^{(j)}(t_0(x)) \text{ mod } x^{n-jh+1} \text{ for } j = 1, 2, \ldots, g,$$

recursively differentiate in x the polynomial $s(t_0(x))$ mod x^{n-jh+1}, with the j-th differentiation step followed by the division of the resulting derivative by $j t_0'(x)$ for

$j = 1, 2, \ldots, g$. Then multiply modulo x^{n+1} the results by the precomputed powers of $x^h t_1(x)$ and sum all the products. This amounts to performing $O(n/h) = O(\sqrt{n \log n})$ times multiplication of polynomials modulo x^{n+1}, so the overall cost is $O((n \log n)^{3/2})$. $\qquad\qquad\qquad\qquad\qquad\qquad\qquad\qquad\qquad\qquad\qquad\qquad\qquad$ $\square$

3.5   Chinese remainder algorithm, polynomial and rational Hermite interpolation, and decomposition into partial fractions

Polynomial interpolation can be generalized based on the following *Chinese Remainder Theorem.*

Theorem 3.5.1. *Given polynomials $m_i(x)$ and $r_i(x)$, $i = 0, 1, \ldots, k$, where $\deg r_i(x) < \deg m_i(x)$ and the polynomials $m_i(x)$ are pairwise co-prime, there exists a unique polynomial $v(x)$ such that*

$$r_i(x) = v(x) \bmod m_i(x), \quad i = 1, \ldots, k, \tag{3.5.1}$$

$$\deg v(x) < N = \sum_{i=1}^{k} \deg m_i(x). \tag{3.5.2}$$

Based on Theorem 3.5.1, we reverse Problem 3.2.1 of modular reduction of a polynomial.

Problem 3.5.2. Modular reconstruction of a polynomial. Given polynomials $m_i(x)$, $r_i(x)$, $i = 0, 1, \ldots, k$, satisfying the assumptions of Theorem 3.5.1, compute the polynomial $v(x)$ satisfying (3.5.1), (3.5.2).

Solution: The *Chinese Remainder Algorithm* solves Problem 3.5.2 by extending Algorithm 3.3.3 as follows.

Algorithm 3.5.3. Chinese Remainder Algorithm (modular reconstruction of a polynomial).

COMPUTATION:

1. Compute the polynomial $u(x) = \prod_{i=1}^{k} m_i(x)$ of (3.2.1).

2. Compute the polynomials
$$u_i(x) = (u(x)/m_i(x)) \bmod m_i(x), \quad i = 1, \ldots, k.$$

3. Compute the polynomials $w_i(x), i = 1, \ldots, k$, satisfying the following relations:
$$w_i(x)u_i(x) = 1 \bmod m_i(x), \tag{3.5.3}$$
$$\deg w_i(x) < \deg m_i(x). \tag{3.5.4}$$

Figure 3.3: Correlation among polynomial and matrix computations III. ("A" and "P" are abbreviation for "Algorithm" and "Problem", respectively; the arrows show reductions of problems to one another.)

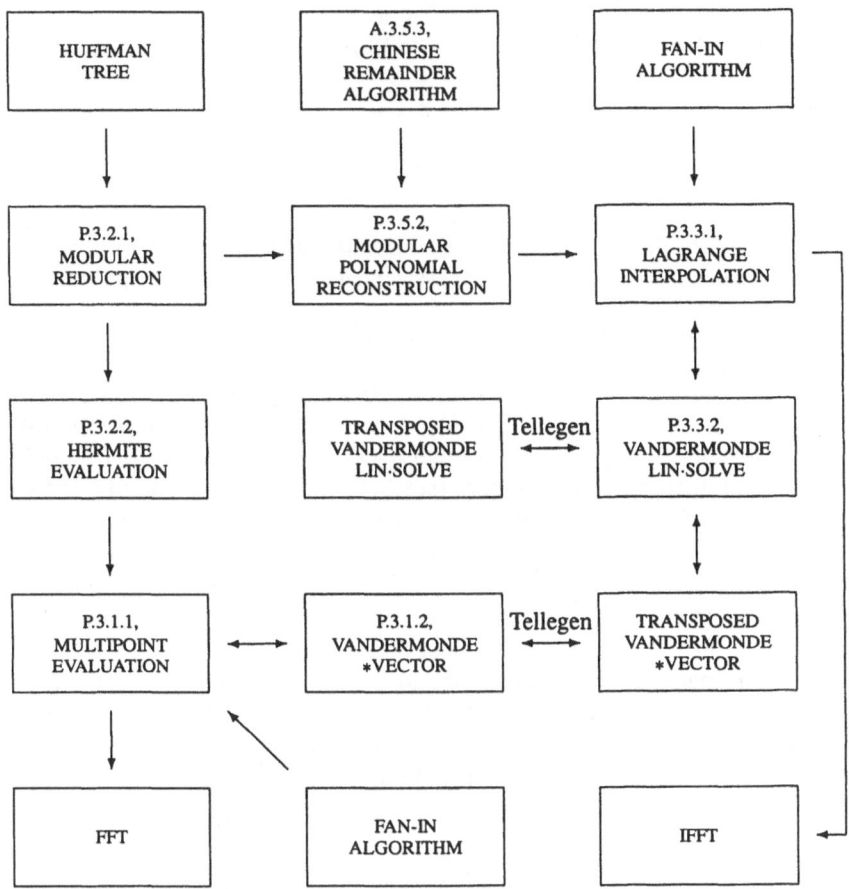

(For every linear polynomial $m_i(x) = x - t_i$, we have $w_i(x) = 1/u'(t_i)$ as in Section 3.3.)

4. Compute and output the polynomial

$$v(x) = \sum_{i=1}^{k}(r_i(x)w_i(x) \bmod m_i(x))u(x)/m_i(x),   \qquad (3.5.5)$$

by summing all partial fractions $(r_i(x)w_i(x) \bmod m_i(x))/m_i(x)$.

The correctness of the algorithm follows immediately from (3.5.3)–(3.5.5) and Theorem 3.5.1.

To estimate the computational cost, we write

$$n = \sum_{i=1}^{k} d_i, \quad d_i = \deg m_i(x). \tag{3.5.6}$$

Theorem 3.5.4. *Problem 3.5.2 of modular reconstruction of a polynomial can be solved by using*

$$a = O(M(n)(H(\mathbf{d}) + 1)) = O(M(n) \log n) \tag{3.5.7}$$

ops for n in (3.5.6), $M(n)$ in (2.4.1), (2.4.2), $H(\mathbf{d}) = -\sum_i (d_i/n) \log(d_i/n)$ in (3.2.2), and $\mathbf{d} = (d_i)_{i=1}^{k}$.

Proof. We apply the fan-in process with Huffman's tree at stage 1 to compute the polynomial $u(x)$, and we reverse this process at stage 2 to compute the polynomials

$$u_i(x)m_i(x) = u(x) \bmod m_i^2(x), \quad i = 1, \ldots, k.$$

In both of these processes we use $O(M(n)(H(\mathbf{d}) + 1))$ ops because $\deg m_i(x) = d_i$ and $\deg m_i^2(x) = 2d_i$ (see (3.2.3)). At stage 2 we also perform division of the polynomials $u_i(x)m_i(x)$ by $m_i(x)$ to output $u_i(x)$ for $i = 1, \ldots, k$. We apply the Toom evaluation/interpolation techniques (see Section 2.4 and the end of Section 2.6) because the remainders vanish in this case. Since $\deg(u_i(x)m_i(x)) < 2d_i$, the overall cost of division is $\sum_i O(M(d_i)) = O((H(\mathbf{d}) + 1)n)$ ops, well within bound (3.5.7). At stage 3, we solve k Problems 2.7.4 of the computation modulo the polynomials $m_i(x)$ of the reciprocals $w_i(x)$ in (3.5.3) and (3.5.4) for $i = 1, \ldots, k$. This involves a total of $O(\sum_{i=1}^{k} M(d_i) \log d_i)$ ops, which is also within bound (3.5.7). At stage 4, we recursively sum together all the partial fractions $(r_i(x)w_i(x) \bmod m_i(x))/m_i(x)$ and output the numerator of the resulting partial fraction. (Only this stage depends on the input remainders $r_i(x)$, that is, stages 1–3 can be assumed to be cost-free preprocessing if the "moduli" $m_i(x)$ are fixed and only the remainders $r_i(x)$ vary.) By applying Huffman's tree to arrange this fan-in process, we also keep its cost within bound (3.5.7). $\square$

In the particular case where the "moduli" $m_i(x)$ are of the form $(x - x_i)^{d_i}$, we arrive at the **Hermite polynomial interpolation problem** for a polynomial $v(x)$, that is, the problem of the evaluation of the polynomial $v(x)$ and its first $d_i - 1$ higher order derivatives $v^{(l)}(x)$, $l = 1, \ldots, d_i - 1$ at $x = x_i$ for $i = 1, \ldots, k$. (Formally, we may require only the computation of the values $v^{(d_i-1)}(x_i)/d_i!$ for $i = 1, \ldots, k$, but the values $v^{(h)}(x_i)/d_i!$ for $h < d_i - 1$ are computed as a by-product.) In this case, the computation at stage 3 of Algorithm 3.5.3 is simplified because we solve simpler Problem 2.7.8 instead of Problem 2.7.4. Bound (3.5.7) on the overall cost of performing Algorithm 3.5.3, however, comes from other stages. Thus (3.5.7) *bounds the complexity of the solution of a confluent Vandermonde linear system*, $V(\mathbf{t}, d)\mathbf{x} = \mathbf{r}$ with the matrix $V(\mathbf{t}, d)$ of equation (3.2.6).

Algorithm 3.5.3 can be extended to the solution of the following problem.

Problem 3.5.5. Partial fraction decomposition. Given polynomials $m_0(x), \ldots,$ $m_k(x)$ (pairwise co-prime), and $v(x)$ such that $\deg v(x) < \deg u(x)$ where $u(x) =$

$\prod_{i=1}^{k} m_i(x)$, (the latter degree bound can be relaxed by means of the division of the polynomials $v(x)$ by $u(x)$ with a remainder), compute the coefficients of all the polynomials of the unique set $\{s_1(x), \ldots, s_k(x)\}$ such that

$$\frac{v(x)}{u(x)} = \sum_{i=1}^{k} \frac{s_i(x)}{m_i(x)},$$

$\deg s_i(x) < \deg m_i(x)$ for all i.

Solution: The solution follows the pattern of the Chinese Remainder Algorithm 3.5.3 for the solution of Problem 3.5.2 and has the same asymptotic cost bound. First compute the polynomials $r_i(x) = v(x) \bmod m_i(x)$, for $i = 0, \ldots, k$, such that

$$r_i(x) = (u(x)s_i(x)/m_i(x)) \bmod m_i(x).$$

Then successively compute the polynomials

$$u(x), \quad u_i(x) = (u(x)/m_i(x)) \bmod m_i(x), \quad w_i(x) \bmod m_i(x)$$

such that $w_i(x)u_i(x) = 1 \bmod m_i(x)$ (by solving Problem 2.7.4 of the inversion of a polynomial modulo a polynomial), and finally compute the polynomials $s_i(x) = w_i(x)r_i(x) \bmod m_i(x)$.

As well as for Problem 3.5.2, simplifications are possible where $m_i(x) = (g_i(x))^{b_i}$ and further where $g_i(x) = x - t_i$. If $t_i = a\omega_n^{i-1}$, $i = 1, 2, \ldots, k$, $n = 2^h \geq k$, for a scalar a and ω_n being the n-th root of 1, then Problem 3.5.5 can be reduced to FFT's and solved by using $O(n \log n)$ ops. □

Let us next combine Algorithms 3.5.3 and 2.9.4 to state and solve several problems of *rational interpolation*, that is, the variations of Problem 2.9.1 of rational function reconstruction where the input polynomial $v(x)$ is given implicitly by its residues modulo the factors $m_i(x)$ of the polynomial $u(x)$ in (3.2.1). We specify several problems of rational inetrpolation depending on the choice of $m_i(x)$ and the expressions relating $r_i(x)$, $m_i(x)$ and $v(x)$.

Due to (3.2.1) and (3.5.2), we have

$$\deg v(x) < \deg u(x), \quad v(x) = v(x) \bmod u(x),$$

which means that the pair $(r(x), t(x)) = (v(x), 1)$ is the $(N, 0)$-th entry of the rational function table for the input pair $(u(x), v(x))$ (cf. Section 2.9). Therefore, application of Algorithm 2.9.4 yields all $(N - i, i)$-th entries of this table for $i = 1, 2, \ldots, N$. Let $(r(x), t(x))$ be such a pair, so

$$t(x)v(x) = r(x) \bmod u(x).$$

By combining this equation with (3.2.1) and (3.5.1), we obtain that

$$t(x)r_i(x) = r(x) \bmod m_i(x), \quad i = 1, \ldots, k. \tag{3.5.8}$$

That is, by applying Algorithms 3.5.3 and 2.9.4, we solve the *problem of rational function reconstruction where the polynomial $v(x)$ is given only implicitly — by its* residues modulo the divisors $m_1(x), \ldots, m_k(x)$ of the polynomial $u(x)$ in (3.2.1). Let us state the latter problem in two forms, that is, based on (3.5.8), where the rational function $r(x)/t(x)$ is involved implicitly, and based on the more explicit variations,

$$r_i(x) = (r(x)/t(x)) \bmod m_i(x), \quad i = 1, \ldots, k \qquad (3.5.9)$$

(compare the Explicit Reconstruction Conditions of Section 2.9).

Problem 3.5.6.

a) **The modified (m, n) rational interpolation problem.** Given natural $k, m, n, d_1, \ldots, d_k$, and N, such that $\sum_i d_i = N + 1 = m + n + 1$; pairwise co-prime polynomials $m_i(x)$, and polynomials $r_i(x), i = 1, \ldots, k$, satisfying the relations

$$\deg r_i(x) < \deg m_i(x) = d_i, \quad i = 1, 2, \ldots, k,$$

compute two polynomials $r(x)$ and $t(x)$ of degrees of at most m and n, respectively, such that the polynomial $t(x)$ is monic and (3.5.8) hold for the polynomial

$$u(x) = \prod_{i=1}^{k} m_i(x) \text{ in (3.2.1).}$$

b) **The (m, n) rational interpolation problem.** The same problem as Problem 3.5.5 but with the requirement of satisfying (3.5.9).

Problem 3.5.6 a) always has a solution, which we may make unique if we minimize the degrees of the polynomials $r(x)$ and $t(x)$ removing all their common non-constant divisors, so that $\gcd(r, t) = 1$. Problem 3.5.6 b) has a well defined solution if and only if $\gcd(u, t) = 1$ (see (2.9.3)), and if so, this is also a solution to Problem 3.5.5 a).

In light of the above comments, we may view Problem 2.9.2 as the *modified (m, n) Padé approximation problem*, and we may view the special case of Problem 3.5.6 b) where $k = 1$ and $d_1 = N$ as the (m, n) *Padé approximation problem*, with the Explicit Reconstruction Condition (2.9.3), (2.9.4) for the associated Problem 2.9.2. In the latter special case, $u(x) = x^{N+1}$, and the condition (2.9.3) turns into the inequality $t(0) \neq 0$.

Two other special cases have been intensively studied independently of Problem 3.5.6 b). Let us state them explicitly starting with the case where

$$m_i(x) = (x - x_i)^{d_i} \text{ for all } i. \qquad (3.5.10)$$

Problem 3.5.7. (m, n) **Hermite rational interpolation.** Given a set of h distinct points x_i, their multiplicities d_i, and the values $v^{(g-1)}(x_i), g = 1, \ldots, d_i; i = 1, \ldots, h$, where $v(x) = v^{(0)}(x)$ is an unknown analytic function or formal power series, find two polynomials, $r(x)$ of degree of at most m and $t(x)$ of degree of at most n, such that $m + n + 1 = N + 1 = \sum_i d_i$, and $\left(\frac{d^{g-1}}{dx^{g-1}} (v(x) - r(x)/t(x)) \right)_{x=x_i} = 0, g = 1, \ldots, d_i;$ $i = 1, \ldots, h$. (These requirements are fulfilled if $v(x) - r(x)/t(x) = a(x)u(x)$ for $u(x)$ in (3.2.1), $m_i(x)$ in (3.5.10), and some analytic function (or formal power series) $a(x)$.)

In the special case where

$$m_i(x) = x - x_i, \quad i = 1, \ldots, k, \tag{3.5.11}$$

that is, where $d_i = 1$ for all i in (3.5.10), we arrive at the following:

Problem 3.5.8. (m, n) **Cauchy rational interpolation problem.** Given two sets of $N + 1$ distinct node points $x_0, \ldots, x_N$ and $N + 1$ values $v_0, \ldots, v_{N+1}$, compute the coefficients of two polynomials $r(x) = \sum_{j=0}^m r_j x^j$ and $t(x) = \sum_{j=0}^n t_j x^j$ such that $r(x_i)/t(x_i) = v_i$ for $i = 0, 1, \ldots, N$, where $N = m + n$.

Remark 3.5.9. The latter problem can be rewritten as the following homogeneous structured system of linear equations:

$$(V_{N,m}(\mathbf{x}), D(\mathbf{v})V_{N,n}(\mathbf{x})) \begin{pmatrix} \mathbf{r} \\ -\mathbf{t} \end{pmatrix} = \mathbf{0}$$

where $V_{N,k}(\mathbf{x}) = (x_i^j)_{i=0, j=0}^{N,k}$ for $k = m$ and $k = n$, $\mathbf{v} = (v_i)_{i=0}^N$, $\mathbf{r} = (r_j)_{j=0}^m$, $\mathbf{t} = (t_j)_{j=0}^n$. The algorithms of Chapter 5 solve this linear system by using a total of $O(M(N) \log N)$ ops, although the solution polynomial $t(x)$ may vanish for $x = x_i$ for some i. In this section we solve the more general Problems 3.5.6 at the same computational cost by combining Theorem 2.9.3 and 3.5.4.

Solution of Problems 3.5.6–3.5.8:   The solution is given by the following algorithm:

Algorithm 3.5.10. The modified (m, n) rational interpolation.

COMPUTATION: First apply Algorithm 3.5.3 to compute the polynomial $v(x)$ of degree of at most N satisfying equations

$$v(x) \bmod m_i(x) = r_i(x), \quad i = 1, \ldots, k.$$

This is the solution of Problem 3.5.2 of modular reconstruction of a polynomial, which is the modified $(N, 0)$ rational interpolation Problem 3.5.6 a) for polynomials $m_i(x)$ and $r_i(x)$ satisfying (3.5.8). Now apply Algorithm 2.9.4 to compute the antidiagonal entries $(N - i, i)$ of the rational function reconstruction table for $i = 1, 2, \ldots, n$. Output the (m, n)-th entry, which is the desired solution of Problem 3.5.6 a).

If $\gcd(u, t) = 1$ for the input polynomial $u(x)$ and the output polynomial $t(x)$, then the computed pair of polynomials $(r(x), t(x))$ also defines the solution of Problem 3.5.6 b). In the special cases of Hermite and Cauchy Problems 3.5.7 and 3.5.8, the assumption that $\gcd(u, t) = 1$ means that $t(x_i) \neq 0$ for all i, which is necessary and sufficient for the existence of the solution of these two problems. Even if this condition does not hold, one may always solve the *modified problems of (m, n) Hermite and/or Cauchy rational interpolation*, where the expression $t(x)v(x) - r(x)$ replaces the expression $v(x) - r(x)/t(x)$.

The solution of Problems 3.5.7 and 3.5.8 involves the computation of the polynomial $u(x)$, but this polynomial is computed by Algorithm 3.5.3 as a by-product.   □

By applying Algorithm 3.5.10 and combining Theorems 2.9.3 and 3.5.4, we obtain the following result.

Theorem 3.5.11. *Problem 3.5.6 a) of the modified (m, n) rational interpolation (covering the modified (m, n) Hermite and Cauchy rational interpolation problems and Problem 2.9.2 of (m, n) Padé approximation) can be solved by using $O(M(N) \log N)$ ops for $N = m + n$ and for $M(N)$ in (2.4.1), (2.4.2). This also solves Problem 3.5.6 b) of (m, n) rational interpolation (and, consequently, the problems of (m, n) Hermite and/or Cauchy rational interpolation) if and only if the output polynomial $t(x)$ of the solution algorithm for Problem 3.5.5 satisfies (2.9.3).*

Remark 3.5.12. The problems, theorems, and algorithms of this section can be easily extended from polynomials to integers. Theorem 3.5.1, Problem 3.5.2, and the Chinese Remainder Algorithm 3.5.3 can be immediately extended if we replace the polynomials $m_i(x)$, $r_i(x)$, $u(x)$, $v(x)$, and $w_i(x)$ by non-negative integers $m_i > 1$, r_i, u, v, and w_i where all integers m_i are co-prime and if we replace (3.5.2) and (3.5.4) by the following relations:

$$v < N = \prod_{i=1}^{k} m_i,$$

$$w_i < m_i, \quad i = 1, \ldots, k.$$

In this case Algorithm 3.5.3 recovers a non-negative integer v from the smaller integers $r_i = v \bmod m_i$, $i = 1, \ldots, k$. This technique enables lower precision computation of large integers. For instance, $\det M$ for an $n \times n$ matrix M filled with smaller integers ranging, say from -100 to 100, can be as large as $(100\sqrt{n})^n$, which is already $\approx 10^{40}$ for $n = 16$. The exact computation of $\det M$, however, can be performed modulo several smaller primes m_i with a lower precision. Exception is the final stage where the Chinese Remainder Algorithm for integers is applied to recover $\det M$ from the values $(\det M) \bmod m_i$ for all i, using only a small number of ops. A similar approach turned out to be effective in various important geometric computations where only the sign of the determinant was sought.

3.6 Cauchy matrix computations. Polynomial and rational interpolation and multipoint evaluation

Problem 3.6.1. Cauchy-by-vector product. Multiply a Cauchy matrix $C(\mathbf{s}, \mathbf{t}) = \left(\frac{1}{s_i - t_j}\right)_{i,k=0}^{n-1}$ by a vector $\mathbf{v} = (v_k)_{k=0}^{n-1}$, that is, compute the values $\sum_{k=0}^{n-1} v_k / (s_i - t_k)$ for $i = 0, \ldots, n - 1$.

Solution I: Rewrite the linear combination of the reciprocals, $\sum_{k=0}^{n-1} v_k / (x - t_k)$, as a ratio of two polynomials, $p(x) = p_{\mathbf{t}, \mathbf{v}}(x)$ and $u(x) = u_{\mathbf{t}}(x)$, by applying the fan-in method for recursive summation of the partial fractions $v_k / (x - t_k)$, as in Section 3.3. Then apply Moenck–Borodin's Algorithm 3.1.5 for polynomial evaluation to compute the values $p(s_i)$, $u(s_i)$ and finally compute and output the ratios $p(s_i)/u(s_i)$ for $i = 0, 1, \ldots, n - 1$. The solution requires $O(M(n) \log n)$ ops. $\qquad \square$

Figure 3.4: Correlation among polynomial and matrix computations IV. ("A" and "P" are abbreviation for "Algorithm" and "Problem", respectively; the arrows show reductions of problems to one another.)

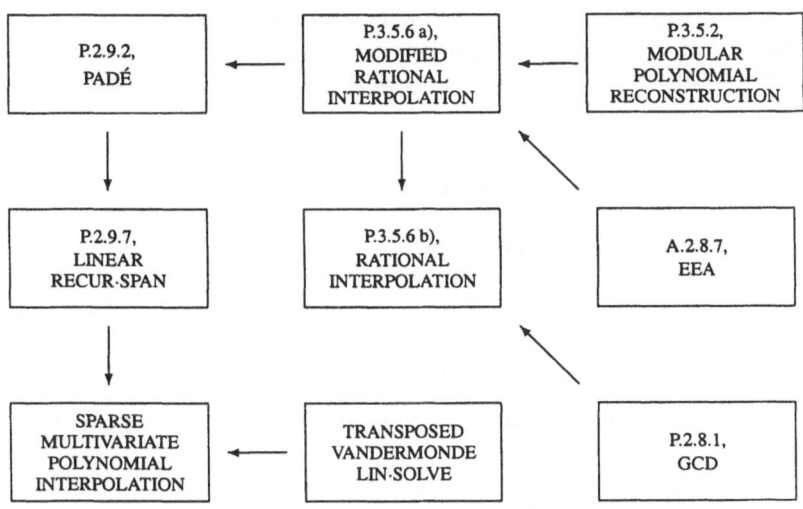

Remark 3.6.2. Any rational function $p(x)/u(x)$, $\deg p(x) < n = \deg u(x)$, whose poles $t_0, \ldots, t_{n-1}$ are simple and available, can be represented as a linear combination of the reciprocals $\sum_k v_k/(x - t_k)$. Given two polynomials $p(x)$ and $u(x)$ and the n distinct zeros $t_0, \ldots, t_{n-1}$ of the polynomial $u(x)$, we may recover the values v_k for all k by applying the solution algorithm for Problem 3.5.5 of partial fraction decomposition.

The above solution of Problem 3.6.1 can be extended to solve the following celebrated problem.

Problem 3.6.3. Trummer's problem. Multiply by a vector $\mathbf{v} = (v_k)_{k=0}^{n-1}$ the matrix $C = C_0(\mathbf{s}, \mathbf{s})$, obtained from the Cauchy matrix $C(\mathbf{s}, \mathbf{s})$ by setting its diagonal entries to zeros.

Solution: First compute the polynomials $p(x) = p_{\mathbf{s},\mathbf{v}}(x)$ and $u(x) = u_{\mathbf{s}}(x)$ as in the solution of Problem 3.6.1 and then evaluate the vector $(C\mathbf{v})_{i=0}^{n-1} = \left(\frac{2p'(s_i) - v_i u''(s_i)}{2u'(s_i)} \right)_{i=0}^{n-1}$ by using $O(M(n) \log n)$ ops. $\qquad \square$

Both $O(M(n) \log n)$ time algorithms for Problems 3.6.1 and 3.6.3 apply over any field but are prone to numerical stablility problems when they are performed numerically with finite precision. *Approximate solutions* to Problem 3.6.1 as well as Trummer's Problem 3.6.3 can be computed by the numerically stable **Fast Multipole Algorithm** of [GR87]. The latter advanced technique is not required, however, in some important cases where the input node sets $\{s_i\}$ and $\{t_j\}$ are special.

Example 3.6.4. Problems 3.6.1 and 3.6.3 on the set of roots of 1. For the input set $\{t_j = \omega^j, \; j = 0, 1, \ldots, K-1\}, n \le K = 2^k < 2n$, we have $u(x) = x^K - 1$, and the same algorithm as in Example 3.3.4 reduces the recursive summation of the fractions $r_j w_j/(x - t_j)$ to the FFT-like process, which yields the polynomial $p(x)$ in $O(n \log n)$ ops. Then the values $u(s_i), u'(s_i),$ and $u''(s_i)$ for any set $\{s_i, i = 0, \ldots, n-1\}$ can be computed in $O(n \log n)$ ops. For special node sets $\{s_i\}$ such as the shifted and scaled K-th roots of 1, the values $p(s_i), i = 0, 1, \ldots, n-1$, can be also computed in $O(n \log n)$ or $O(M(n))$ ops (see Example 3.1.8).

Solution II of Problem 3.6.1 (matrix version):   Let us write

$$D(s, t) = \operatorname{diag}(u_t(s_i))_{i=0}^{n-1}, \quad D'(t) = \operatorname{diag}(u'_t(t_i))_{i=0}^{n-1} \tag{3.6.1}$$

for the polynomial $u_t(x) = \prod_{i=0}^{n-1}(x - t_i)$ of equation (3.3.2). Then we have the vector equation

$$D(s, t)C(s, t)v = (p(s_i))_{i=0}^{n-1} \tag{3.6.2}$$

for any vector v and the associated numerator polynomial $p(x) = p_{t,v}(x)$. Recall that

$$p(x) = \sum_j v_j u_j(x) \tag{3.6.3}$$

where $u_j(x) = u_t(x)/(x - x_j)$. Recall also that $u_j(t_j) = u'_t(t_j)$ for all j and deduce that

$$(u_j(s_i))_{i,j} = V(s)V^{-1}(t)\operatorname{diag}(u'_t(t_j)). \tag{3.6.4}$$

Combine equations (3.6.1)–(3.6.4), to obtain the following vector equation:

$$C(s, t)v = D^{-1}(s, t)V(s)V^{-1}(t)D'(t)v.$$

We have deduced this equation for any vector v and, therefore, have arrived at the following important *factorization of a Cauchy matrix*:

$$C(s, t) = D^{-1}(s, t)V(s)V^{-1}(t)D'(t). \tag{3.6.5}$$

(3.6.5) reduces the solution of Problem 3.6.1 to application of the fan-in method for the computation of the polynomial $u_t(x)$ and to polynomial evaluation and interpolation. This involves more ops than in our previous solution though asympotically still $O(M(n) \log n)$ ops.   □

Factorization (3.6.5) has several other applications. It immediately implies the expression

$$\det C(s, t) = \prod_{i<j}(s_i - s_j)(t_i - t_j)/\prod_{i,j}(s_i - t_j)$$

and consequently the following *strongest non-singularity* result.

Corollary 3.6.5. *An $n \times n$ Cauchy matrix $C(s, t)$ is defined and non-singular in any field if and only if the $2n$ values $s_0, \ldots, s_{n-1}, t_0, \ldots, t_{n-1}$ are distinct. If so, all submatrices of the matrix $C(s, t)$ are also non-singular.*

Another application of (3.6.5) is the **reduction of polynomial evaluation and interpolation to Problem 3.6.1** of Cauchy-by-vector multiplication, with the motivation of exploiting the power of the Fast Multipole Algorithm. Indeed, by (3.6.5), we have the following **factorizations of a Vandermonde matrix and its inverse:**

$$V(s) = D(s, t)C(s, t)(D'(t))^{-1}V(t), \tag{3.6.6}$$

$$V^{-1}(t) = V^{-1}(s)D(s, t)C(s, t)(D'(t))^{-1}, \tag{3.6.7}$$

where we may choose any vector t for a fixed s and vice-versa. By choosing the vectors $s = a\mathbf{w}$ and/or $t = b\mathbf{w}$ where $\mathbf{w} = (\omega_n^i)_{i=0}^{n-1}$ is the vector of the n-th roots of 1 from Definition 2.1.2 and a and b are two scalars, we turn the Vandermonde matrices $V(t)$ and $V(s)$, respectively, into the matrices $\Omega^{(a)} = (a\omega^{ij})_{i,j}$ and $\Omega^{(b)} = (b\omega^{i,j})_{i,j}$ of scaled DFT_n. This reduces each of the problems of polynomial interpolation and multipoint evaluation essentially to an FFT and one of the two Problems 3.6.1 — for Cauchy matrices $C(a\mathbf{w}, t)$ and $C(s, b\mathbf{w})$, respectively, where we may choose parameters a and b at our convenience, to facilitate the application of the Fast Multipole Algorithm.

(3.6.5) allows various *extensions and modifications.* Let us show some related **Cauchy factorizations** (see Exercise 3.10 for more examples). We may rewrite (3.6.5) replacing the Vandermonde matrices $V(s)$ and $V(t)$ by polynomial Vandermonde matrices $V_P(s)$ and $V_P(t)$ (see (3.2.7)) for any fixed basis $\mathbf{P} = (P_0, P_1, \ldots, P_{n-1})^T$ in the linear space of polynomials of degree of at most $n - 1$. The same derivation of (3.6.5) also applies to the generalized factorization

$$C(s, t) = D^{-1}(s, t)V_P(s)V_\mathbf{P}^{-1}(t)D'(t), \tag{3.6.8}$$

which turns into (3.6.5) where $\mathbf{P}$ is the monomial basis, $\mathbf{P} = (1, x, x^2, \ldots, x^{n-1})^T$.

Furthermore, a Cauchy matrix $C(s, t)$ can be expressed via two Cauchy matrices $C(s, q)$ and $C(q, t)$ for any vector q, for instance, as follows:

$$C(s, t) = \operatorname{diag}\left(\frac{u_q(s_i)}{u_t(s_i)}\right)_{i=0}^{n-1} C(s, q) \operatorname{diag}\left(\frac{u_t(q_j)}{u_q'(q_j)}\right)_{j=0}^{n-1} C(q, t).$$

To prove the latter equation, rewrite (3.6.5) and (3.6.1) twice — for the matrices $C(s, q)$ and $C(q, t)$ replacing the matrix $C(s, t)$, then substitute the resulting expressions on the right-hand side of the above equation, and compare the resulting equation with (3.6.5).

The vector q can be chosen to facilitate the solution of Problem 3.6.1 for the input matrix $C(s, t)$. For instance, one may choose $q = a\mathbf{w}$ for a scalar a, thus reducing a large part of the computations to the application of FFT and inverse FFT. On the other hand, we observe that our solution of Problem 3.6.1 for $C(s, t)$ in the case where $t = a\mathbf{w}$ (see Example 3.6.4) is a little simpler than the solution we have shown so far in the case where $s = a\mathbf{w}$. We may, however, reduce one case to another by just combining Tellegen's Theorem 3.4.1 and the following simple equation:

$$C^T(s, t) = -C(t, s). \tag{3.6.9}$$

Let us next consider the converse of Problem 3.6.1.

Problem 3.6.6. Solution of a Cauchy linear system of equations. Solve a nonsingular Cauchy linear system of n equations, $C(\mathbf{s}, \mathbf{t})\mathbf{v} = \mathbf{r}$ for an unknown vector $\mathbf{v}$ and three given vectors $\mathbf{r}, \mathbf{s},$ and $\mathbf{t}$.

Solution: This is the problem of interpolation to a rational function represented in the polarized form, $r(x) = \sum_{k=0}^{n-1} v_k/(x - t_k)$ given the sets $\{t_k\}_{k=0}^{n-1}$, $\{s_i\}_{i=0}^{n-1}$, and $\{r(s_i)\}_{i=0}^{n-1}$. The problem can be solved in $O(M(n) \log n)$ ops, based on the inversion of factorizations (3.6.5) or (3.6.8).

It is instructive to derive this inverse factorization in terms of polynomial evaluation and interpolation. First compute the coefficients of the polynomial $u_\mathbf{t}(x) = \prod_{k=0}^{n-1}(x - t_k)$ by means of the fan-in method (using a total of $O(M(n) \log n)$ ops). Then apply Moenck–Borodin's Algorithm 3.1.5 to compute the values $u_\mathbf{t}(s_i)$ by using $O(M(n) \log n)$ ops. Then compute the values $p(s_i) = r(s_i)u_\mathbf{t}(s_i)$, $i = 0, 1, \ldots, n$, by using $n + 1$ ops, apply the algorithms of Sections 3.3 or 3.4 to interpolate to the polynomial $p(x)$ by using $O(M(n) \log n)$ ops, and finally compute the values $v_k = p(t_k)/u'_\mathbf{t}(t_k)$, $k = 0, 1, \ldots, n - 1$ (here again, we use $O(M(n) \log n)$ ops). The matrix version of this algorithm is given by the inverse of equations (3.6.5) and (3.6.8), that is, by the equation

$$C^{-1}(\mathbf{s}, \mathbf{t}) = (D'(\mathbf{t}))^{-1} V_P(\mathbf{t}) V_P^{-1}(\mathbf{s}) D(\mathbf{s}, \mathbf{t}). \tag{3.6.10}$$

Alternatively, one may combine the inverse equation (3.6.10) with (3.6.8) where $\mathbf{s}$ and $\mathbf{t}$ are interchanged and obtain that

$$C^{-1}(\mathbf{s}, \mathbf{t}) = \mathrm{diag}(u_\mathbf{s}(t_i)/u'_\mathbf{t}(t_i))_{i=0}^{n-1} C(\mathbf{t}, \mathbf{s}) \, \mathrm{diag}(u_\mathbf{t}(s_i)/u'_\mathbf{s}(s_i))_{i=0}^{n-1}. \tag{3.6.11}$$

(3.6.11) reduces polarized rational interpolation to computing polynomials $u_\mathbf{s}(x)$ and $u_\mathbf{t}(x)$, multipoint polynomial evaluation, and Problem 3.6.1, for whose solution one may utilize the Fast Multipole Algorithm. □

Various problems of rational interpolation are associated with matrices having more general structures of Cauchy type, such as Loewner matrices $\left(\frac{u_i - v_j}{s_i - t_j} \right)_{i,j}$ [F85] or Pick matrices (see Section 3.8).

Let us conclude this section with *polynomial* versions of the **reduction of Problems 3.1.1 and 3.3.1 to Problems 3.6.1 and 3.6.6,** respectively. This is done simply by replacing the polynomial $v(x) = \sum_{i=0}^{n-1} v_i x^i$ of Problems 3.1.1 and 3.3.1 by the rational function $\frac{v(x)}{x^N - a^N} = \sum_{i=0}^{N-1} \frac{s_i}{x - t_i}$ where a is a fixed non-zero scalar, $t_i = a\omega_N^{i-1}$, $i = 1, \ldots, N$, $s_i = 0$ for $i \geq n$, $N = n_+ = 2^{\lceil \log_2 n \rceil}$. The transition from the coefficients $v_0, \ldots, v_{n-1}$ to the numerator values $s_0, \ldots, s_{n-1}$ is the simplest special case of Problem 3.5.5 of partial fraction decomposition. The problem can be reduced to scaling (n ops) and a small number of FFT's (see Problem 3.5.5) for $m_i(x) = x - a\omega_N^i$, $i = 0, 1, \ldots, n - 1$, that is, the transition requires $O(n \log n)$ ops.

Then again we may solve Problems 3.6.1 and 3.6.6 efficiently by the Fast Multipole Algorithm. As in the case of similar reduction based on (3.6.6) and (3.6.7), for appropriate choices of the scalar a, this approach implies superior numerical algorithms for Problems 3.1.1 and 3.3.1 versus their direct solution via Vandermonde matrix computations.

Figure 3.5: Correlation among polynomial and matrix computations V. ("A" and "P" are abbreviation for "Algorithm" and "Problem", respectively; the arrows show reductions of problems to one another.)

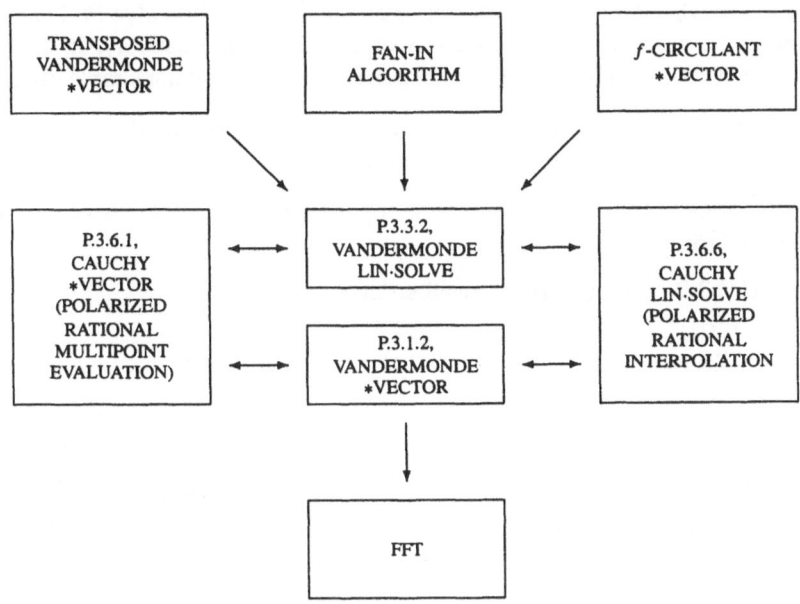

3.7 Loss (erasure)-resilient encoding/decoding and structured matrices

For demonstration, let us next apply Vandermonde matrices to encoding/decoding based on Corollary 3.6.5. (We do not use the results of this section elsewhere in the book.) Suppose that a string of packets of information must be sent via a channel where some packets can be lost and only m of n packets surely reach the receiver, $n > m$. Then fix an $n \times m$ Vandermonde matrix $V_{n,m} = (s^{ij})_{i=0,j=0}^{n-1,m-1}$ where all values $1, s, s^2, \ldots, s^{n-1}$ are distinct and use this matrix as a *generator matrix for the loss(erasure)-resilient encoding/decoding*.

At the encoding stage, we compute the vector $V_{n,m}\mathbf{h}$ where $\mathbf{h} = (h_j)_{j=0}^{m-1}$ is the vector of m information packets. With $n - m$ packets lost, we arrive at a subvector $\mathbf{v}$ of dimension m of the vector $V_{n,m}\mathbf{h}$ and at the respective $m \times m$ submatrix V of the matrix $V_{n,m}$. At the decoding stage, we recover the vector $\mathbf{h} = V^{-1}\mathbf{v}$. V is a square Vandermonde matrix, $V = (s_i^j)_{i,j=0}^{m-1}$, where $s_i = s^{h(i)}$, $0 \le h(i) < n$, and all exponents $h(i)$ and, consequently, all values s_i are distinct for $i = 0, 1, \ldots, m - 1$. Therefore, $\det V = \prod_{i<j}(s_i - s_j) \ne 0$, that is, the matrix V is non-singular. The encoding and decoding of the information is reduced to the solution of Problems 3.1.2 (see Remark 3.1.10) and 3.3.2 with the matrices $V_{n,m}$ and V, respectively, where we may choose any value of the scalar parameter s.

Furthermore, Problem 3.1.2 where $t_i = s^i$ for all i is immediately reduced to $h = \lceil n/m \rceil$ Problems 2.4.2 of Generalized DFT_m and thus can be solved by using $O(M(m)n/m)$ ops (see Exercise 2.6 and Remark 3.1.10). If the computation is in the field containing the K-th roots of 1 where $n \le K = 2^k < 2n$, then we may choose the value $s = \omega_K$ and apply FFT to speed up the solution. To solve Problem 3.3.2 of computing the vector $\mathbf{h} = V^{-1}\mathbf{v}$, we rely on (3.4.1) (or alternatively, on the equation in Exercise 3.10(a)), Tellegen's Theorem 3.4.1, and the following simple observation.

Theorem 3.7.1. *Let $a(M)$ ops be sufficient to multiply a matrix M by a vector. Let X be a submatrix of a matrix Y. Then $a(X) \le a(Y)$.*

With the cited tools we reduce the computation of the vector $\mathbf{h} = V^{-1}\mathbf{v}$ to the following steps.

Algorithm 3.7.2. Loss (erasure)-resilient decoding.

INITIALIZATION: Fix a scalar f such that $f \ne s^{im}$ for $i = 0, 1, \ldots, n-1$.

COMPUTATION:

1. Compute the coefficient vector $\mathbf{u_s}$, $\mathbf{s} = (s_i)_{i=0}^{m-1}$ of the polynomial $u_s(x) - x^m = \prod_i (x - s_i) - x^m$ (see (3.3.2)) by means of the fan-in process in Figure 3.1, by using $O(M(m) \log m)$ ops.

2. Recall that $s_i = s^{h(i)}$, $0 \le h_i < n$, for $i = 0, 1, \ldots, m-1$, and compute the values $u_s'(s_i)$ and $u_s'(s_i)(f - s_i^m)$ for $i = 0, 1, \ldots, n-1$ by using $O(M(m)n/m)$ ops (see Problem 3.1.1 and Remark 3.1.9).

3. Compute the vector $\mathbf{y} = D^{-1}((u_s'(s_i)(f - s_i^m))_{i=0}^{m-1})\mathbf{v}$.

4. Apply Tellegen's Theorem 3.4.1 and then Theorem 3.7.1 to reduce the computation of the vector $\mathbf{z} = V^T\mathbf{y}$ to multiplication by a vector of the matrix $V_{n,m}$. (By Remark 3.1.10, the latter multiplication requires $O(M(m)n/m)$ ops.)

5. In $O(M(m))$ additional ops, compute and output the vector $\mathbf{h} = JZ_f(\mathbf{u_s} + f\mathbf{e}_0)\mathbf{z}$ (see Sections 2.4 and 2.6).

Summarizing, we have the following result.

Theorem 3.7.3. *Loss (erasure)-resilient encoding/decoding with $n \times m$ generator matrix, $m < n$, uses $O(M(m)n/m)$ ops at the encoding stage and $f(m) + O(M(m)n/m)$ ops at the decoding stage, for $M(m)$ in (2.4.1), (2.4.2) provided $f(m) = O(M(m) \log m)$ ops are sufficient in the fan-in process that computes the vector $\mathbf{u_s}$ for a given vector $\mathbf{s} = (s_i)_{i=0}^{m-1}$.*

Remark 3.7.4. There is a minor computational advantage in using (3.4.1) with $f = 1$. This choice of the parameter f in Algorithm 3.7.2 is possible where $s_i \ne 1$ for $i = 0, 1, \ldots, m-1$. To exclude equation $s_i = 1$ for all i, we replace the first row of the matrix $V_{n,m}$, say by a row $(t^j, \ j = 1, \ldots, m-1)$ for a scalar t distinct from $1, s, s^2, \ldots, s^{n-1}$.

Remark 3.7.5. The presented encoding/decoding algorithm is different from the more customary algorithm with the generator matrix of the form $\begin{pmatrix} I \\ C(\mathbf{s}, \mathbf{t}) \end{pmatrix}$ for two vectors $\mathbf{s}$ and $\mathbf{t}$ of dimensions $n - m$ and m, respectively. In the latter version, encoding consists in multiplying the $(n - m) \times m$ Cauchy matrix $C(\mathbf{s}, \mathbf{t})$ by a vector. Then at the decoding stage, we multiply by vectors the matrices $C(\mathbf{r}, \mathbf{q})$ and $C^{-1}(\mathbf{r}, \mathbf{p})$ where $\mathbf{r}$ is a subvector of the vector $\mathbf{s}$ and where the vector $\mathbf{t}$ is made up of two subvectors $\mathbf{p}$ and $\mathbf{q}$ associated with the two sets of the lost and received packets of messages, respectively. The decoding is simplified because the dimension l of vectors $\mathbf{p}$ and $\mathbf{r}$ is less than $\min\{n - m, m\}$. On the other hand, the process becomes more complicated because partition of the vector $\mathbf{t}$ into the subvectors $\mathbf{p}$ and $\mathbf{q}$ may change when new packets of messages are sent, and this restricts the possibility of preprocessing at the decoding stage. Then two *node polynomials* $u_\mathbf{p}(x)$ and $u_\mathbf{q}(x)$ instead of a single polynomial $u_\mathbf{s}(x)$ must be computed, at the double time-cost. Furthermore, the Vandermonde structure of the matrices $V_{n,m}$ and V is easier to exploit than the Cauchy structure of the matrices $C(\mathbf{s}, \mathbf{t})$, $C(\mathbf{r}, \mathbf{q})$, and $C^{-1}(\mathbf{r}, \mathbf{p})$.

Remark 3.7.6. Both Vandermonde and Cauchy type codes of this section belong to the important class of generalized Reed–Solomon codes. Some other loss-resilient codes perform even slightly faster, in linear time, at the price of allowing randomization.

Figure 3.6: Correlation of loss-resilient encoding decoding to structured matrix computations. ("A" and "P" are abbreviation for "Algorithm" and "Problem", respectively; the arrows show reductions of problems to one another.)

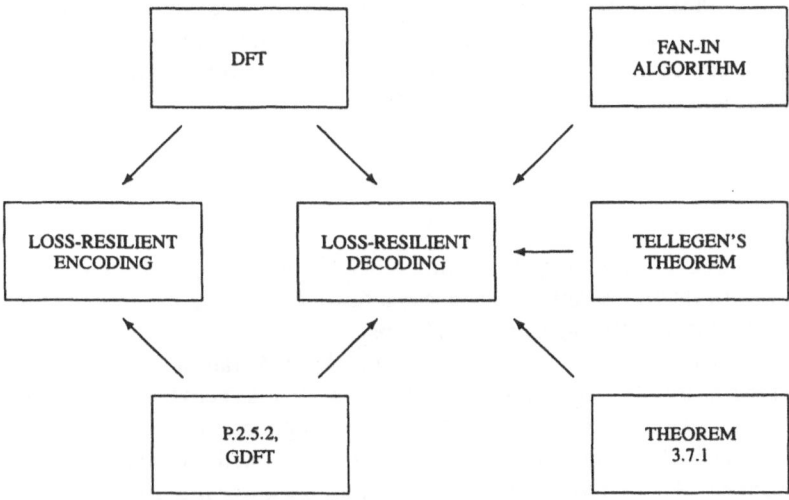

3.8 Nevanlinna–Pick interpolation problems

The Nevanlinna–Pick celebrated problems of rational interpolation have highly important applications in systems, control and circuit theory. The problems arise in the study
of *transfer functions*, which are analytic and bounded in a fixed region such as a half-
plane. We state these problems in various forms and recall their known formulations
as structured matrix problems. We begin with the simplest scalar form.

Problem 3.8.1. Scalar Nevanlinna–Pick problem.

INPUT: n distinct points $\{z_k\}_{k=1}^{n}$ in the open right half-plane Π^{+}, such that $\operatorname{Re} z_k > 0$
for all k; n points $\{f_k\}_{k=1}^{n}$ inside the unit circle centered at the origin, that is, $|f_k| < 1$
for all k.

OUTPUT: A rational scalar function $f(z)$ such that

1. $f(z)$ is analytic in the open half-plane Π^{+}.

2. $f(z)$ is passive, that is, $|f(z)|$ does not exceed 1 in the closed right half-plane,
$$\sup_{z \in \Pi^{+} \cup \mathbb{R}\sqrt{-1}} |f(z)| \leq 1.$$

3. $f(z)$ meets the interpolation conditions $f(z_k) = f_k$, for $k = 1, 2, \ldots, n$.

Passivity condition 2 is dictated by the energy conservation requirement. That is,
function $f(z)$ should not exceed the input energy. Because of the latter condition as
well as restriction 1 on the poles of the output function $f(z)$, *the problem is very different from those in Section 3.5*. Subsequent variations differ from the problems of
Section 3.5 even more, in particular, their outputs are matrix functions.

The simplest version of the Nevanlinna–Pick rational interpolation problem has a
solution if and only if the corresponding Pick matrix,

$$R = \left[\frac{1 - f_i f_j^*}{z_i + z_j^*} \right]_{i,j}$$

is positive definite [P16] (see Definition 4.6.3 of Chapter 4 on positive definiteness of
a matrix). In [N29] Nevanlinna proposed a fast recursive algorithm that computed the
solution to this problem by using $O(n^2)$ ops.

In a generalization of Problem 3.8.1, the $M \times N$ rational matrix function $F(z)$
(called the *null-pole coupling function* in [BGR90]) is recovered not from its matrix
values at selected complex points z_k but from the application of $F(z)$ to some selected
vectors, that is, from the action of $F(z_k)$ into some fixed direction for each z_k. This
motivated M. G. Krein to introduce the terminology *tangential* for this problem.

Problem 3.8.2. Tangential Nevanlinna–Pick problem.

INPUT: n distinct points z_k in the open right half-plane Π^{+}, n non-zero row vectors
$\mathbf{x}_k$ of dimension M, and n row vectors $\mathbf{y}_k$ of dimension N, $k = 1, \ldots, n$.

OUTPUT: A rational $M \times N$ matrix function $F(z)$ satisfying the following conditions:

1. $F(z)$ is analytic in the open half-plane Π^+.

2. $F(z)$ is passive, that is, $\sup\limits_{z \in \Pi^+ \cup \mathbb{R}\sqrt{-1}} \|F(z)\| \le 1$ for a fixed matrix norm (see Definition 6.1.1 in Chapter 6 on matrix norms).

3. $F(z)$ meets the tangential interpolation conditions: $\mathbf{x}_k F(z_k) = \mathbf{y}_k$ for $k = 1, 2, \ldots, n$.

Again, the problem has a solution if and only if the generalized Pick matrix

$$R = \left[\frac{\mathbf{x}_i \mathbf{x}_j^* - \mathbf{y}_i \mathbf{y}_j^*}{z_i + z_j^*} \right]_{i,j} \tag{3.8.1}$$

is positive definite. Solutions of this tangential variant of the problem by fast $O(n^2)$ algorithms can be found in several places.

Clearly, Problem 3.8.1 is a special case of Problem 3.8.2 where $M = N = 1$, the vectors $\mathbf{x}_k$ and $\mathbf{y}_k$ turn into scalars, $\mathbf{x}_k = 1$, $\mathbf{y}_k = f_k$, $k = 1, \ldots, n$, and we write $f(z)$ instead of $F(z)$. Here is an important extension.

Problem 3.8.3. Tangential confluent Nevanlinna–Pick problem.

INPUT: n distinct points $\{z_k\}$ in the open right half-plane Π^+ with their multiplicities $\{m_k\}$, $k = 1, \ldots, n$; n row vectors $\{\mathbf{x}_{k,j}\}_{j=1}^{m_k}$ where $\mathbf{x}_{k,j}$ are row vectors of dimension M, and n row vectors $\{\mathbf{y}_{k,j}\}_{j=1}^{m_k}$ where $\mathbf{y}_{k,j}$ are row vectors of dimension N, $k = 1, \ldots, n$.

OUTPUT: A rational $M \times N$ matrix function $F(z)$ satisfying the following conditions:

1. $F(z)$ is analytic in the open half-plane Π^+.

2. $F(z)$ is passive.

3. $F(z)$ meets the tangential confluent interpolation conditions:

$$(\mathbf{x}_{k,j})_{j=1}^{m_k} \begin{bmatrix} F(z_k) & F'(z_k) & \cdots & \frac{F^{(m_k-1)}(z_k)}{(m_k-1)!} \\ 0 & F(z_k) & \ddots & \vdots \\ \vdots & \ddots & \ddots & F'(z_k) \\ 0 & \cdots & 0 & F(z_k) \end{bmatrix} = (\mathbf{y}_{k,j})_{j=1}^{m_k},$$

for $k = 1, 2, \ldots, n$.

Let us next display an important special case of Problem 3.8.3, where $(\mathbf{x}_{k,j})_{j=1}^{m_k} = \mathbf{e}_0$ for all k.

Problem 3.8.4. Carathéodory–Fejér problem.

INPUT: As for Problem 3.8.3 where $(\mathbf{x}_{k,j})_{j=1}^{m_k}$ is the first coordinate row vector $\mathbf{e}_0^T$ of dimension M for $k = 1, \dots, n$.

OUTPUT: As for Problem 3.8.3 where condition 3 is simplified to the form $F(z_k) = y_{k,1}, F'(z_k) = y_{k,2}, \dots, F^{(m_k-1)}(z_k)/(m_k - 1)! = y_{k,m_k}, k = 1, \dots, n$.

Let us next extend Problem 3.8.2 by adding *boundary conditions*.

Problem 3.8.5. Tangential boundary Nevanlinna–Pick problem. This is Problem 3.8.2 where

1. the points $z_1, \dots, z_l$ lie on the imaginary line $\mathbb{R}\sqrt{-1}, l < n$;

2. $\|\mathbf{x}_k\| = \|\mathbf{y}_k\|, k = 1, \dots, l$;

3. $\mathbf{x}_k F'(z_k)\mathbf{y}_k^* = -\rho_k, k = 1, \dots, l$,

for some fixed positive scalars ρ_k, called *coupling numbers*.

The necessary and sufficient condition that this problem has a solution is positive definiteness of the following generalized Pick matrix:

$$R = (r_{i,j})_{i,j}, \quad r_{i,j} = \begin{cases} \rho_i & \text{for } i = j \leq l, \\ \dfrac{\mathbf{x}_i\mathbf{x}_j^* - \mathbf{y}_i\mathbf{y}_j^*}{z_j + z_j^*} & \text{otherwise.} \end{cases} \tag{3.8.2}$$

It is also possible to extend Problems 3.8.3 and 3.8.5 to **the tangential confluent boundary Nevanlinna–Pick problem**.

By the *Möbius transformation* of the variable, $w = \frac{z-1}{z+1}$, all these problems can be extended from the functions $F(z)$ on the right half-plane to those on the unit disc $\{w : |w| \leq 1\}$.

An effective practical approach to the solution of Problems 3.8.1–3.8.5 is given by the **State-Space Method**, due to R. E. Kalman. The method was introduced for linear time-invariant dynamical systems but turned out to be fundamental for many other engineering and mathematical problems as well. The method relies on *realization* of the transfer matrix function in the form

$$W(z) = E + C(zI_m + A^*)^{-1}K$$

where A, K, C, and E are four fixed matrices of sizes $m \times m, m \times (M+N), (M+N) \times m$, and $(M + N) \times (M + N)$, respectively, and m exceeds $M + N$ but is minimized to yield *minimal* realization,

$$\boxed{W(z)} = \boxed{E} + \boxed{C} \boxed{(zI_m + A^*)^{-1}} \boxed{K}$$

In this way nonlinearity is confined to the inversion of the matrix $zI_m + A^*$. The matrix function $W(z)$ can be described via its *global State-Space representation*

$$W(z) = I_{M+N} - B^*(zI_m + A^*)^{-1}R^{-1}BD, \qquad (3.8.3)$$

which involves the inverse of a structured matrix R. The complete class of the transfer functions $F(z)$ is now recovered from the realization matrix function

$$W(z) = \begin{pmatrix} W_{00}(z) & W_{01}(z) \\ W_{10}(z) & W_{11}(z) \end{pmatrix} \qquad (3.8.4)$$

as follows:

$$F(z) = (W_{00}(z)G(z) + W_{01}(z))(W_{10}(z)G(z) + W_{11}(z))^{-1} \qquad (3.8.5)$$

where $G(z)$ is an arbitrary rational matrix function satisfying requirements 1 and 2 of the respective interpolation problem, for instance, $G(z) = 0$.

Thus, the entire problem of obtaining the transfer function $F(z)$ (in factorized form) has been reduced to the inversion of a structured matrix R. For Problems 3.8.2–3.8.5 this matrix satisfies a Sylvester type equation,

$$AR + RA^* = BDB^*, \qquad (3.8.6)$$

where we first specify the matrices A, B, and D for tangential Nevanlinna–Pick Problem 3.8.2:

$$A = \operatorname{diag}(z_k)_{k=1}^n, \qquad (3.8.7)$$

$$B = \begin{pmatrix} \mathbf{x}_1 & \mathbf{y}_1 \\ \vdots & \vdots \\ \mathbf{x}_n & \mathbf{y}_n \end{pmatrix}, \qquad (3.8.8)$$

$$D = \operatorname{diag}(I_M, -I_N) = \begin{pmatrix} I_M & 0 \\ 0 & -I_N \end{pmatrix}, \qquad (3.8.9)$$

D is called a *signature matrix*. Then Sylvester type equation (3.8.6) defines a Cauchy-like matrix R in (3.8.1), whose inversion is covered by the superfast algorithms of Chapter 5. Moreover, *an interesting interpolation interpretation of these algorithms is shown in Section 5.8.*

Recall that in the scalar Nevanlinna–Pick Problem 3.8.1, the vectors $\mathbf{x}_k$ and $\mathbf{y}_k$ turn into scalars, $\mathbf{x}_k = 1$, $\mathbf{y}_k = f_k$, $k = 1, \ldots, n$, that is, the matrix B turns into the matrix $\begin{pmatrix} 1 & \cdots & 1 \\ f_1 & \cdots & f_n \end{pmatrix}^T$. For Problem 3.8.5, equations (3.8.3)–(3.8.9) still apply, and (3.8.6)–(3.8.9) define the matrix R except for its first l diagonal entries (defined by (3.8.2)).

For the tangential confluent Nevanlinna–Pick Problem 3.8.3, equation (3.8.6) and the signature matrix D do not change, the diagonal entries z_k of the matrix A are replaced by Jordan blocks $J_{m_k}(z_k) \in \mathbb{C}^{m_k \times m_k}$, $k = 1, \ldots, n$, and the vectors $\mathbf{x}_k$ and

y_k in the matrix B are replaced by the matrices $(\mathbf{x}_{k,j})_{j=1}^{m_k}$ and $(\mathbf{y}_{k,j})_{j=1}^{m}$, $k = 1, \ldots, n$. That is, for Problem 3.8.3 we have

$$A = \operatorname{diag}(J_{m_k}(z_k))_{k=1}^n \in \mathbb{C}^{m \times m}, \quad m = \sum_{k=1}^n m_k, \qquad (3.8.10)$$

$$J_{m_k}(z_k) = \begin{pmatrix} z_k & & & O \\ 1 & \ddots & & \\ & \ddots & \ddots & \\ O & & 1 & z_k \end{pmatrix} \in \mathbb{C}^{m_k \times m_k}, \qquad (3.8.11)$$

$$B = \begin{pmatrix} \mathbf{x}_{1,1} & \mathbf{y}_{1,1} \\ \vdots & \vdots \\ \mathbf{x}_{1,m_1} & \mathbf{y}_{1,m_1} \\ \mathbf{x}_{2,1} & \mathbf{y}_{2,1} \\ \vdots & \vdots \\ \mathbf{x}_{2,m_2} & \mathbf{y}_{2,m_2} \\ \cdot & \cdot \\ \cdot & \cdot \\ \cdot & \cdot \\ \mathbf{x}_{n,1} & \mathbf{y}_{n,1} \\ \vdots & \vdots \\ \mathbf{x}_{n,m_n} & \mathbf{y}_{n,m_n} \end{pmatrix} \in \mathbb{C}^{m \times (M+N)}. \qquad (3.8.12)$$

The matrices R satisfying (3.8.6), (3.8.10)–(3.8.12) generalize the matrices R in (3.8.1) satisfying (3.8.6)–(3.8.9) (which are the special case where $m_k = 1$ for all k) and are called *confluent Cauchy-like matrices*. The superfast algorithms of Chapter 5 cover the inversion of these matrices as well.

3.9 Matrix Nehari Problem

In this section, we show another important application of the State-Space Method, this time to a *rational approximation problem*.

Problem 3.9.1. Matrix Nehari problem (generic case)

INPUT: n distinct points $\{z_1, \ldots, z_n\}$ in the open left half-plane Π^-, that is, such that $\operatorname{Im} z_k < 0$ for all k; n non-zero column vectors $\mathbf{x}_1, \ldots, \mathbf{x}_n$ of size $M \times 1$, n non-zero column vectors $\mathbf{y}_1, \ldots, \mathbf{y}_n$ of size $N \times 1$, and an $M \times N$ rational matrix function $K(z)$ having only simple poles $z_1 = \cdots = z_n \in \Pi^-$ such that the matrix function $K(z) - (z - z_k)^{-1}\mathbf{x}_k\mathbf{y}_k^*$ is analytic at the points $z = z_k$, $k = 1, \ldots, n$, and $\|K(z)\| = \sup |K(z)| < \infty$ where the supremum is over all points z lying on the imaginary line $\mathbb{R}\sqrt{-1}$.

OUTPUT: An $M \times N$ rational matrix function $F(z)$ with no poles in the open left half-plane Π^- and on the line $\mathbb{R}\sqrt{-1}$ and satisfying $\|K(z) - F(z)\| < 1$ for all $z \in \mathbb{R}\sqrt{-1}$.

The problem has solution if and only if the matrix $I - QP$ is positive definite (see Definition 4.6.3) where

$$-P = \left(\frac{\mathbf{x}_i^* \mathbf{x}_j}{z_i + z_j^*} \right)_{i,j=1}^{n}, \qquad -Q = \left(\frac{\mathbf{y}_i^* \mathbf{y}_j}{z_i^* + z_j} \right)_{i,j=1}^{n} \tag{3.9.1}$$

are positive definite matrices. The solution of Problem 3.9.1 can be reduced to the computation of factorization (3.8.3)–(3.8.5) where $G(z)$ was an arbitrary $M \times N$ matrix function with no poles in the closed left half-plane $\Pi^- \cup \mathbb{R}\sqrt{-1}$ and such that

$$\sup_{z \in \mathbb{R}\sqrt{-1}} \|G(z)\| \leq 1$$

and

$$R = \begin{pmatrix} P & I \\ I & Q \end{pmatrix}, \tag{3.9.2}$$

for P and Q in (3.9.1). We have the factorization

$$R = \begin{pmatrix} P & I \\ I & Q \end{pmatrix} = \begin{pmatrix} I & 0 \\ P^{-1} & I \end{pmatrix} \begin{pmatrix} P & I \\ 0 & Q - P^{-1} \end{pmatrix}, \tag{3.9.3}$$

which reduces the inversion of the matrix R essentially to the inversion of the matrices P and $Q - P^{-1} = -(I - QP)P^{-1}$. Both have Cauchy-like structure (see Example 1.4.1) and can be inverted by the superfast algorithms in Chapter 5. By (3.9.3) the matrix function $I - QP$ is positive definite if and only if the matrix R has exactly n positive and exactly n negative eigenvalues.

Remark 3.9.2. The matrix Nehari–Takagi problem generalizes Problem 3.9.1 by allowing the output approximation function $F(z)$ to have at most k poles in the open left half-plane Π^- (counting the poles with their multiplicities). The reduction of the solution to equations (3.8.3)–(3.8.6), (3.9.1)–(3.9.3) is extended as well, so the matrix R in (3.8.3)–(3.8.6) has the same structure and satisfies (3.9.2). The matrix $I - QP$ is assumed to be non-singular but is now allowed to have up to k negative eigenvalues (counted with their multiplicities), and the problem has a solution if and only if the matrix R in (3.9.2) has at most $n + k$ positive eigenvalues (counted with their multiplicities).

3.10   Sparse multivariate polynomial interpolation

Computations with general multivariate polynomials are closely related to computations with the so-called multilevel structured matrices (in particular with the resultant matrices or Newton polytop matrices), which are represented by fewer parameters and can be multiplied by vectors superfast but do not have known superfast inversion algorithms. In some cases the computations with multivariate polynomials can be reduced to Vandermonde and/or Toeplitz/Hankel type computations. Two examples are given by multivariate polynomial multiplication and sparse multivariate polynomial interpolation. For demonstration, let us next outline the important algorithm of Ben-Or and Tiwari 1988 for the latter problem. We do not use this algorithm elsewhere in the book. In the Notes we cite several alternative algorithms.

In practical computations, *sparse multivariate polynomials* are typically defined by a (black box) subroutine for their evaluation running at a low cost B. An important advantage of this representation is a small amount of data and memory space required. This approach has been effectively applied to several fundamental problems of polynomial computations. A basic step is the solution of the following problem.

Problem 3.10.1. Recovery of a polynomial represented by a black box subroutine for its evaluation: Let

$$h(\mathbf{x}) = \sum_{i=0}^{n-1} c_i m_i(\mathbf{x})$$

be a sparse polynomial where c_i are unknown non-zero coefficients; $m_i(\mathbf{x}) = \prod_{j=1}^{k} x_j^{e_{i,j}}$ are monomials in the components of the vector $\mathbf{x}$ defined by the unknown exponents $e_{i,j}$, $i = 0, \ldots, n-1$; $\mathbf{x} = (x_1, \ldots, x_k)^T$ is the vector of variables; and n is the unknown number of terms of $h(\mathbf{x})$ bounded by a fixed natural $N \geq n$. Of importance practically is the case where $\max_i \sum_{j=1}^{k} e_{i,j}$ is much greater than $\log_k n$, which means that $h(\mathbf{x})$ is a sparse polynomial.

With the input consisting of k, N and a black box subroutine that evaluates $h(\mathbf{x})$ by using B ops, we seek the exponents $e_{i,j}$ and the coefficients c_i for all i and j.

Solution: Let us write $v_i = \prod_{j=1}^{k} p_j^{e_{i,j}}$, $h_s = h(p_1^s, \ldots, p_k^s)$, $i, s = 0, 1, \ldots, n-1$ for k fixed distinct primes $p_1, \ldots, p_k$, such that

$$v_i \neq v_j \text{ for } i \neq j, \quad h_s = \sum_{i=0}^{n-1} c_i v_i^s. \tag{3.10.1}$$

To compute the exponents $e_{i,j}$, we first compute the integers $n, v_0, \ldots, v_{n-1}$ as follows. Consider the auxiliary polynomial

$$u(y) = \prod_{i=0}^{n-1} (y - v_i) = \sum_{q=0}^{n} u_q y^q, \quad u_n = 1, \tag{3.10.2}$$

where the integer n and the values u_q are not known. Then $u(v_i) = 0$ for all i, $\sum_{i=0}^{n-1} c_i v_i^j u(v_i) = 0$ for all j. Substitute (3.10.2) and deduce that

$$\sum_{q=0}^{n} u_q \sum_{i=0}^{n-1} c_i v_i^{q+j} = 0, \quad j = 0, \ldots, n-1.$$

Now substitute (3.10.1) and obtain that

$$\sum_{q=0}^{n} h_{q+j} u_q = 0, \quad j = 0, \ldots, n-1.$$

We first compute the values $h_0, \ldots, h_{2N-1}$ by $2N$ times applying the black box subroutine for the evaluation of the polynomial $h(\mathbf{x})$ and thus using $2NB$ ops. Then we recall that $u_n = 1$ and compute n and u_q for $q = 0, 1, \ldots, n-1$. The latter step amounts to solving Problem 2.9.7, that is, to computing the minimum span for a linear recurrence sequence, which involves $O(M(N) \log N)$ ops.

When the coefficient vector $\mathbf{u}$ of the polynomial $u(y)$ is available, we obtain the integer roots $v_0, \ldots, v_{n-1}$ of this polynomial by applying any root-finding algorithm for a polynomial having only integer roots (see Algorithm 7.2.3 in Section 7.2 and/or [Z90]). Let R denote the number of ops required in this algorithm. (R depends on the algorithm and the magnitudes of the roots v_i; we omit its estimation referring the reader to [L83], [Z90].) Then we compute the exponents $e_{i,j}$ for all i and j by repeatedly dividing the integers v_i by the powers of the primes $p_1, \ldots, p_k$ at the overall cost of performing

$$E = O\left(\sum_{i,j} \log(e_{i,j} + 1)\right)$$

ops.

Next, given a natural n and the exponents $e_{i,j}$, we compute the desired coefficient vector $\mathbf{c} = (c_i)_{i=0}^{n-1}$ by solving the linear system

$$V^T \mathbf{c} = \mathbf{h}, \quad V^T = (v_i^s)_{s,i=0}^{n-1}, \quad \mathbf{h} = (h_s)_{s=0}^{n-1},$$

where $V^T = (v_i^s)_{s,i}$ is an $n \times n$ transposed Vandermonde matrix (see (3.10.1) and (3.4.3)). Summarizing, the entire solution involves $2NB + R + E + O(M(n) \log n)$ ops. □

3.11 Diagonalization of Matrix Algebras. Polynomial Vandermonde Matrices and Discrete Sine and Cosine transforms

In this section we extend Theorem 2.6.4 to the diagonalization of matrix algebras by using Vandermonde and polynomial Vandermonde matrices, which leads us to the important structured matrices of discrete sine and cosine transforms.

Diagonalization of an $n \times n$ matrix M by its similarity transformation is the computation of its *eigendecomposition*

$$M = Y^{-1} D(\mathbf{z}) Y. \tag{3.11.1}$$

Here Y is a non-singular matrix and $\mathbf{z} = (z_j)_{j=0}^{n-1}$ is the vector of the n *eigenvalues* of the matrix M, which are the zeros of the characteristic polynomial $c_M(x) = \det(xI - M)$.

(We assume that the matrix M allows its diagonalization. Generally, it must not; for example, the matrix $\begin{pmatrix} 0 & 1 \\ 0 & 0 \end{pmatrix}$ allows no diagonalization. In this case *eigenproblem* (3.11.1) is generalized: one seeks an $n \times k$ rectangular matrix Y and a $k \times k$ diagonal matrix $D(\mathbf{z})$ satisfying the equation $YM = D(\mathbf{z})Y$ for the largest k.) If all n eigenvalues $z_0, \ldots, z_{n-1}$ are distinct, the k-th column of the $n \times n$ matrix Y^{-1} is the right eigenvector of M associated with the eigenvalue z_k, $k = 0, 1, \ldots, n-1$, and defined uniquely up to scaling by a constant factor.

Equation (3.11.1) actually diagonalizes the entire algebra $A(M)$ of matrix polynomials $v(M)$. Indeed, (3.11.1) implies that

$$v(M) = Y^{-1} \operatorname{diag}(v(z_j))_{j=0}^{n-1} Y. \tag{3.11.2}$$

The next theorem defines diagonalization of the algebra $A(F_u)$ for any fixed polynomial $u(x)$ with n distinct zeros,

$$u(x) = x^n + \sum_{i=0}^{n-1} u_i x^i = \prod_{j=0}^{n-1} (x - z_j). \tag{3.11.3}$$

Theorem 3.11.1. *For two polynomials $u(x)$ of equation (3.11.3) with n distinct zeros $z_0, \ldots, z_{n-1}$ and $v(x)$ and for the vector $\mathbf{z} = (z_j)_{j=0}^{n-1}$, we have*

$$F_u = V^{-1}(\mathbf{z}) D(\mathbf{z}) V(\mathbf{z}), \tag{3.11.4}$$

and consequently

$$v(F_u) = V^{-1}(\mathbf{z}) v(D(\mathbf{z})) V(\mathbf{z}). \tag{3.11.5}$$

Proof. Pre-multiply equation (3.11.4) by the matrix $V(\mathbf{z})$. $\square$

Theorem 2.6.4 is a special case of Theorem 3.11.1 where $u_i = 0$ for $i > 0$, $u_0 = f^n$, and the polynomials $v(F_u)$ form the algebra of f-circulant matrices. Multiplication by a vector of an f-circulant matrix as well as of its inverse (if the matrix is nonsingular) requires $O(M(n))$ ops and is numerically stable. The same operations for a matrix $v(F_u)$ use the order of $M(n) \log n$ ops relying on factorization (3.11.5) but generally are numerically unstable. As a remedy, one may look for special cases (such as where $\mathbf{z} = \mathbf{w}$ is the vector of the n-th roots of 1) and/or for various extensions of Theorem 3.11.1. We are going to combine both directions, but let us first simplify the input matrix M.

We recall that the customary solution algorithms for the eigenproblem begin with the similarity transformation of the input matrix M into an *upper Hessenberg matrix* $H = (h_{i,j})_{i,j=0}^{n-1}$, $h_{i,j} = 0$ for $i > j+1$,

$$H = UMU^{-1}, \quad M = U^{-1}HU, \tag{3.11.6}$$

where one may assume that U is an orthogonal matrix, $U^*U = UU^* = I$.

Factorization (3.11.6) can be computed by direct methods, whereas the subsequent diagonalization relies on iterative processes such as the QR algorithm. These processes become much more effective where the Hessenberg matrix H is real symmetric or Hermitian, in which case it is tridiagonal and can be assumed to be real with no loss of generality.

In our next consideration, however, we follow a distinct path. We seek the *closed form diagonalizations* of the matrices H and M. If the matrix H is *unreduced*, that is, if

$$h_{i,i+1} \neq 0, \quad i = 1, \ldots, n-1,$$

then we have

$$HK = KF_u,$$

where $K = K(H, e_0)$ is the Krylov matrix (see Definition 2.1.1) and $u(x) = c_H(x)$ is the characteristic polynomial of the matrix H. Observe that K is a non-singular triangular matrix. Therefore, under the assumption that the matrix H is unreduced, we have the diagonalizations

$$H = KV^{-1}(\mathbf{z})D(\mathbf{z})V(\mathbf{z})K^{-1},$$
$$M = U^{-1}KV^{-1}(\mathbf{z})D(\mathbf{z})V(\mathbf{z})K^{-1}U. \tag{3.11.7}$$

It is not generally recommended using this approach for numerical computations because of the severe problems of numerical stability involved. At this point, one may look for the classes of interesting special matrices where these problems are avoided, but we consider an alternative direction where we diagonalize a Hessenberg matrix H by a *similarity transformation with polynomial Vandermonde matrices* (compare (3.2.7)).

Theorem 3.11.2. *Let* $\mathbf{Q} = (Q_j(x))_{j=0}^{n-1}$ *be a vector of polynomials, where* $Q_0(x) = 1$, $\deg Q_j(x) = j$; $xQ_j(x) = \sum_{i=0}^{j+1} h_{ij} Q_i(x)$, $j = 0, \ldots, n-1$; $H_{\mathbf{Q}} = (h_{ij})_{i,j=0}^{n-1}$, $h_{ij} = 0$ *for* $i > j+1$, *so that* $\mathbf{Q}^T H_{\mathbf{Q}} = (xQ_j(x))_{j=0}^{n-1}$; $Q_n(x) = h_{nn} \prod_{k=0}^{n-1}(x - z_k)$, $\mathbf{z} = (z_k)_{k=0}^{n-1}$, $z_0, \ldots, z_{n-1}$ *are distinct. Then we have*

$$H_{\mathbf{Q}} = V_{\mathbf{Q}}^{-1}(\mathbf{z})D(\mathbf{z})V_{\mathbf{Q}}(\mathbf{z}), \quad V_{\mathbf{Q}}(\mathbf{z}) = (Q_i(x_j))_{i,j=0}^{n-1}. \tag{3.11.8}$$

Proof. To verify the theorem, multiply the first equation in (3.11.8) by the matrix $V_{\mathbf{Q}}(\mathbf{z})$. $\square$

Theorem 3.11.1 is the special case of the latter theorem where $\mathbf{Q} = (x^j)_{j=0}^{n-1}$ is the monomial basis, in which case $H_{\mathbf{Q}} = F_{Q_n}$ is the Frobenius matrix, and $Q_n(x)$ is its characteristic polynomial.

Combining (3.11.6) and (3.11.8) for $H = H_{\mathbf{Q}}$ defines the similarity transformation (3.11.1) where $M = H_{\mathbf{Q}}$, $Y = V_{\mathbf{Q}}(\mathbf{z})U$. $H_{\mathbf{Q}}$ is called a *confederate* matrix of the vector $\mathbf{Q}$.

For an unreduced upper Hessenberg matrix H, a vector $\mathbf{Q} = (Q_j(x))_{j=0}^{n-1}$ is unique and can be computed based on the equations of Theorem 3.11.2 such that $H = H_{\mathbf{Q}}$. The computation involves $O(n^3)$ ops. Conversely, $O(n^3)$ ops are sufficient to compute the unreduced confederate matrix $H_{\mathbf{Q}}$ from a given vector $\mathbf{Q} = (Q_j(x))_{j=0}^{n-1}$, $\deg Q_j(x) = j$, $j = 0, 1, \ldots, n$. By these estimates, the diagonalization based on

Theorem 3.11.2 is not superfast and is not even fast, but is in closed form for any given pair of a polynomial basis $\mathbf{Q}$ and the associated matrix $H_{\mathbf{Q}}$.

Furthermore, Theorem 3.11.2 enables fast multiplication by a vector of an upper Hessenberg matrix provided that the matrix $H_{\mathbf{Q}}$ is represented as the confederate matrix of the vector $\mathbf{Q}$, the zeros of the polynomial $Q_n(x)$ are known and distinct, and the polynomial Vandermonde matrix $V_{\mathbf{Q}}(\mathbf{z})$ and its inverse can be multiplied by a vector fast. If the latter properties of the matrix $H_{\mathbf{Q}}$ hold, then any matrix from the algebra $A(H_{\mathbf{Q}})$, generated by the matrix $H_{\mathbf{Q}}$ (that is, any matrix polynomial in $H_{\mathbf{Q}}$) can be multiplied by a vector fast.

It remains for us to find interesting special classes of the matrices $V_{\mathbf{Q}}(\mathbf{z})$, for which application of the above approach is efficient. This is the case where the vector $\mathbf{Q}$ is made up of (scaled) Chebyshev polynomials and $V_{\mathbf{Q}}(\mathbf{z})$ are the matrices associated with trigonometric (sine and cosine) transforms. (They are not Vandermonde matrices and cannot be used in diagonalizations (3.11.4) and (3.11.5).)

Example 3.11.3. Discrete Sine transform (DST) I.

$$S = S^{-1} = \sqrt{\tfrac{2}{n+1}} \left(\sin \tfrac{kj\pi}{n+1} \right)_{k,j=1}^{n} \tag{3.11.9}$$

is the matrix of DST I. We have $S = V_{\mathbf{U}}(\mathbf{z})$ where $\mathbf{U}$ is the vector of the Chebyshev polynomials of the second kind. The matrix $S = V_{\mathbf{U}}$ is associated with the tridiagonal matrix $H_{\mathbf{U}}, 2H_{\mathbf{U}} = Z + Z^T = Y_{00}$, and we have the diagonalizations

$$Y_{00} = S \operatorname{diag} \left(2 \cos \tfrac{k\pi}{n+1} \right)_{k=1}^{n} S, \tag{3.11.10}$$

and more generally

$$M = S \operatorname{diag}(SM\mathbf{e}_0) \operatorname{diag}^{-1} \left(\sqrt{\tfrac{2}{n+1}} \sin \tfrac{k\pi}{n+1} \right)_{k=1}^{n} S,$$

provided that M is a matrix from the *algebra* $\tau = A(Y_{00})$ generated by the matrix Y_{00}.

$\square$

Theorem 3.11.4. *For any real vector* $\mathbf{v}$, *the vector* $S\mathbf{v}$ *can be computed in* $O(n \log n)$ *real ops.*

Proof. Combine DFT_{2n+2} with the equation $e^{\sqrt{-1}y} = \cos y + \sqrt{-1} \sin y$. $\square$

Clearly, Theorem 3.11.4 can be immediately extended to the complex vectors $\mathbf{u} + \sqrt{-1}\mathbf{v}$ where $\mathbf{u}$ and $\mathbf{v}$ are real vectors. Together with equation (3.11.2) for $Y = S$, $z_j = 2 \cos \tfrac{(j+1)\pi}{n}, j = 0, \ldots, n-1$, these observations imply the following result:

Corollary 3.11.5. *Given a vector* $\mathbf{v} = (v_i)_{i=0}^{n-1}$ *and a matrix* $M = \sum_{i=0}^{n-1} m_i Y_{00}^i$ *from the algebra* τ *generated by the matrix* $Y_{00} = Z + Z^T$, *the vectors* $M\mathbf{v}$ *and (if* $\det M \neq 0$ *then also)* $M^{-1}\mathbf{v}$ *can be computed in either of the fields of real or complex numbers* $\mathbb{R}$ *or* $\mathbb{C}$ *by using* $O(n \log n)$ *ops.*

In Tables 3.1–3.5 we display the matrices of eight discrete trigonometric transforms, their inverses, the associated polynomials $Q_j(x)$, $j = 1, \ldots, n$, the vectors $\mathbf{z}$ of the zeros of the polynomials $Q_n(x)$, and the real symmetric tridiagonal matrices H_Q having spectra $\mathbf{z}$.

In Section 4.4, we use the following diagonalization result reflected in these tables.

Theorem 3.11.6. *Write* $Y_{00} = Z + Z^T$, $Y_{11} = Y_0 + e_0 e_0^T + e_{n-1} e_{n-1}^T$,

$$ C = C_n^{II} = \sqrt{\tfrac{2}{n}} \left(\mu_j \cos \tfrac{(2j+1)k\pi}{2n} \right)_{k,j=0}^{n-1}, $$

$\mu_0 = \tfrac{1}{\sqrt{2}}$, $\mu_j = 1$ *for* $j > 0$,

$$ S = S_n^I = \sqrt{\tfrac{2}{n+1}} \left(\sin \tfrac{(k+1)(j+1)\pi}{n+1} \right)_{k,j=0}^{n-1} $$

(where C and S denote the matrices of DCT II and DST I in Table 3.1, respectively),

$$ D_C = 2\,\mathrm{diag}\left(\cos \tfrac{j\pi}{n} \right)_{j=0}^{n-1}, $$

$$ D_S = 2\,\mathrm{diag}\left(\cos \tfrac{(j+1)\pi}{n+1} \right)_{j=0}^{n-1}. $$

Then we have

$$ Y_{00} = S D_S S, \quad Y_{11} = C D_C C^T, \quad S^2 = C C^T = I. $$

The eight trigonometric transforms, their variations, and *Hartley transforms* I and II defined by the matrices

$$ H_1 = \left(\cos \tfrac{2kj\pi}{n} + \sin \tfrac{2kj\pi}{n} \right)_{k,j=0}^{n-1} $$

and

$$ H_2 = \left(\cos \tfrac{(2j+1)k\pi}{n} + \sin \tfrac{(2j+1)k\pi}{n} \right)_{k,j=0}^{n-1} $$

serve as real alternatives to complex FFT and have numerous applications.

We conclude this section with some immediate extensions of Theorem 3.11.4 and Corollary 3.11.5.

Theorem 3.11.7. *For any real or complex vector* $\mathbf{v}$ *and the matrix M of any of the eight trigonometric transforms of Table 3.1, the vectors* $M\mathbf{v}$ *and* $M^{-1}\mathbf{v}$ *can be computed by using* $O(n \log n)$ *real ops.*

Corollary 3.11.8. *Given a vector* $\mathbf{v}$ *of dimension n and* $n \times n$ *matrix* $M = \sum_{i=0}^{n-1} m_i H_Q^i$ *from the algebra generated by any matrix H_Q of Table 3.5, the vectors $M\mathbf{v}$ and (if* $\det M \neq 0$ *then also)* $M^{-1}\mathbf{v}$ *can be computed in the fields of real or complex numbers* $\mathbb{R}$ *or* $\mathbb{C}$ *by using* $O(n \log n)$ *ops.*

Table 3.1: The matrices of eight forward and inverse discrete trigonometric transform ($\mu_0 = \mu_n = 1/\sqrt{2}$, $\mu_i = 1$ for $0 < i < n$.)

	T_Q (discrete transform)	T_Q^{-1} (inverse transform)
DCT-I	$C_n^I = \sqrt{\frac{2}{n-1}} \left(\mu_k \mu_{n-1-k} \mu_j \mu_{n-1-j} \cos \frac{kj\pi}{n-1} \right)_{k,j=0}^{n-1}$	$(C_n^I)^T = C_n^I$
DCT-II	$C_n^{II} = \sqrt{\frac{2}{n}} \left(\mu_k \cos \frac{(2j+1)k\pi}{2n} \right)_{k,j=0}^{n-1}$	$(C_n^{II})^T = C_n^{III}$
DCT-III	$C_n^{III} = \sqrt{\frac{2}{n}} \left(\mu_j \cos \frac{(2k+1)j\pi}{2n} \right)_{k,j=0}^{n-1}$	$(C_n^{III})^T = C_n^{II}$
DCT-IV	$C_n^{IV} = \sqrt{\frac{2}{n}} \left(\cos \frac{(2k+1)(2j+1)\pi}{4n} \right)_{k,j=0}^{n-1}$	$(C_n^{IV})^T = C_n^{IV}$
DST-I	$S_n^I = \sqrt{\frac{2}{n+1}} \left(\sin \frac{(k+1)(j+1)\pi}{n+1} \right)_{k,j=0}^{n-1}$	$(S_n^I)^T = S_n^I$
DST-II	$S_n^{II} = \sqrt{\frac{2}{n}} \left(\mu_{k+1} \sin \frac{(k+1)(2j+1)\pi}{2n} \right)_{k,j=0}^{n-1}$	$(S_n^{II})^T = S_n^{III}$
DST-III	$S_n^{III} = \sqrt{\frac{2}{n}} \left(\mu_{j+1} \sin \frac{(2k+1)(j+1)\pi}{2n} \right)_{k,j=0}^{n-1}$	$(S_n^{III})^T = S_n^{II}$
DST-IV	$S_n^{IV} = \sqrt{\frac{2}{n}} \left(\sin \frac{(2k+1)(2j+1)\pi}{4n} \right)_{k,j=0}^{n-1}$	$(S_n^{IV})^T = S_n^{IV}$

3.12 Conclusion

We have examined various problems of polynomial and rational interpolation and multipoint evaluation with extensions and have observed their close correlation to the solution of Vandermonde and Cauchy linear systems of equations (for the interpolation) and multiplication of Vandermonde and Cauchy matrices by vectors (for evaluation). We recall from the previous chapter that non-polarized rational interpolation has a close correlation with linear systems of equations having structures of Toeplitz type. That is, the Toeplitz type structures are correlated to the Cauchy type structures. The correlation and the structure enable superfast computations. Cauchy and Vandermonde structures are also correlated to each other, and this can be exploited for the design of superfast algorithms. Various rational interpolation problems lead to other classes of structured matrices. In Sections 3.8 and 3.9, important problems of numerical rational interpolation and approximation have been reduced to computations with matrices having structures of Cauchy type. In Chapter 5, we show superfast algorithms for these problems. Besides univariate polynomial and rational computations, we have shown applications of structured matrices to loss-resilient encoding/decoding (in Section 3.7) and sparse multivariate polynomial interpolation (in Section 3.10). We have also used Vandermonde and polynomial Vandermonde matrices to extend the diagonalization of f-circulant matrices to any matrix algebra generated by a Frobenius matrix or more generally by an unreduced Hessenberg matrix. This extension has led us to the matrices of the discrete sine and cosine transforms.

Table 3.2: Polynomials $Q_k(y)$, $k = 0, \ldots, n - 1$ (T_i and U_i are the i-th degree Chebyshev polynomials of the first and second kinds.)

	$\{Q_0(x), \quad Q_1(x), \ldots, \quad Q_{n-2}(x), \quad Q_{n-1}(x)\}$
DCT-I	$\{\frac{1}{\sqrt{2}}T_0, \quad T_1, \ldots, \quad T_{n-2}, \quad \frac{1}{\sqrt{2}}T_{n-1}\}$
DCT-II	$\{U_0, \ U_1 - U_0, \ldots, U_{n-2} - U_{n-3}, U_{n-1} - U_{n-2}\}$
DCT-III	$\{\frac{1}{\sqrt{2}}T_0, \quad T_1, \ldots, \quad T_{n-2}, \quad T_{n-1}\}$
DCT-IV	$\{U_0, \ U_1 - U_0, \ldots, U_{n-2} - U_{n-3}, U_{n-1} - U_{n-2}\}$
DST-I	$\{U_0, \quad U_1, \ldots, \quad U_{n-2}, \quad U_{n-1}\}$
DST-II	$\{U_0, \ U_1 + U_0, \ldots, U_{n-2} + U_{n-3}, U_{n-1} + U_{n-2}\}$
DST-III	$\{U_0, \quad U_1, \ldots, \quad U_{n-2}, \quad \frac{1}{\sqrt{2}}U_{n-1}\}$
DST-IV	$\{U_0, \ U_1 + U_0, \ldots, U_{n-2} + U_{n-3}, U_{n-1} + U_{n-2}\}$

3.13 Notes

Sections 3.1–3.3 Algorithm 3.1.5 is from [MB72]; Algorithm 3.3.3 is from [BM74]; [BM71], [F72], and [H72] should be cited as preceding related works. The lower bound of $\lceil n \log_2(n - 1) \rceil$ ops for any straight line algorithm for Problems 3.1.1 and 3.1.2 is due to [S73] (see the definitions and proofs also in [BM75] and/or [B-O83]). This is within the factor of $M(n)/n$ from the record upper bounds. For the evaluation of a polynomial of degree n at a single point, the linear substitution/active operation method of [P66] gives a lower bound of $2n$ ops (which is optimal). On Chebyshev polynomials and Chebyshev node sets, see [R92]. On polynomial evaluation and interpolation on the Chebyshev node sets, see [CHQZ87], [G88], [F95]. On Huffman's tree, see [H52]. Our definition of the confluent Vandermonde matrices is close to those in [BE73], [G62], [GP71], [H88], [H90], [L94], [L95], and [L96]. (3.2.4) and (3.2.5) bound the complexity of multiplication of a confluent Vandermonde matrix by a vector, which improves the record bound of the papers [OS00a] and [OS00b] by the factor of $\log n$. The definition of polynomial Vandermonde matrices is from [KO96], [KO97]. A well-known alternative to Lagrange interpolation is Newton's interpolation (see Exercise 3.8, [A78], [CdB80], [GG99]). On the condition number of Vandermonde matrices, see [G62], [GI88], [Gau90].

Section 3.4 Expression (3.4.1) is from [GO94a] (also see Exercise 3.10(a)). On Tellegen's Theorem 3.4.1, see [DK69], [PSD70]. On composition of polynomials, see, e.g., [GG99]. On composition of power series, see [BK78], [K98].

Table 3.3: Polynomials $Q_n(x)$ and vectors z of their zeros

	$Q_n(x)$	z = { zeros of $Q_n(x)$}
DCT-I	$xT_{n-1} - T_{n-2}$	$(\cos \frac{k\pi}{n-1})_{k=0}^{n-1}$
DCT-II	$U_n - 2U_{n-1} + U_{n-2}$	$(\cos \frac{k\pi}{n})_{k=0}^{n-1}$
DCT-III	T_n	$(\cos \frac{(2k+1)\pi}{2n})_{k=0}^{n-1}$
DCT-IV	$2T_n$	$(\cos \frac{(2k+1)\pi}{2n})_{k=0}^{n-1}$
DST-I	U_n	$(\cos \frac{(k+1)\pi}{n+1})_{k=0}^{n-1}$
DST-II	$U_n + 2U_{n-1} + U_{n-2}$	$(\cos \frac{(k+1)\pi}{n})_{k=0}^{n-1}$
DST-III	T_n	$(\cos \frac{(2k+1)\pi}{2n})_{k=0}^{n-1}$
DST-IV	$2T_n$	$(\cos \frac{(2k+1)\pi}{2n})_{k=0}^{n-1}$

Section 3.5 On the proof and origin of the Chinese Remainder Theorem, see [K98]. Our presentation of Algorithms 3.5.3 and 3.5.5 follows [BM75] and [G86], respectively. On problems of rational function reconstruction and rational interpolation, see [GY79], [GG99], and the references therein. On the extension of the problems, theorems, and algorithms of Section 3.5 to integers, see [AHU74], [BM75], [K98], [GG99]. On the application of the Chinese Remainder Theorem 3.5.1 to the computation of the sign of the determinant of a matrix, see [BEPP99] and the bibliography therein (see [PY99] and [PYa] on an alternative method for the solution of the latter problem).

Section 3.6 The solution of Trummer's Problem 3.6.3 is from [GGS87]. The algorithm uses some observations from [G60]. On earlier derivations of equation (3.6.5), see [K68], [HR84], [FHR93], [GO94a]. Alternative derivation of the expression for det $C(\mathbf{s}, \mathbf{t})$ can be found in [PS72]. Polynomial interpolation and multipoint evaluation were reduced to Problem 3.6.1 and the Fast Multipole Algorithm in [PLST93], [PZHY97], [PACLS98], [PACPS98]. In spite of substantial theoretical development of this approach in the cited papers and successful test results for multipoint polynomial evaluation reported in [PZHY97], we are aware of no public or commercial use of this approach for interpolation or multipoint evaluation.

The resulting approximation algorithms apply to polynomial interpolation and multipoint evaluation for any set of complex node points. There are specialized alternative algorithms for polynomial approximation where the node points lie in a fixed line interval [R88], [P95] or on a circle [OP99, Section 7].

Various extensions and modifications of (3.6.5) can be found in [FHR93], [OP99], [OS00]. (3.6.8) is from [OP99]. Superfast solution of Problem 3.6.6 is due to [G60].

The paper [FHR93] studied interpolation of a rational function made up of the pair of a polynomial and a polarized partial fraction. In this case interpolation is reduced

Table 3.4: Factorization of the matrices T_Q of trigonometric transforms into the product of diagonal matrices W_Q and polynomial Vandermonde matrices $V_Q = V_Q(\mathbf{z})$

DCT-I	$C_n^I = W_Q V_Q$ with $W_Q = \sqrt{\frac{2}{n-1}}\,\mathrm{diag}\left(\frac{1}{\sqrt{2}}, 1, \ldots, 1, \frac{1}{\sqrt{2}}\right)$
DCT-II	$C_n^{II} = W_Q V_Q$ with $W_Q = \sqrt{\frac{2}{n}}\,\mathrm{diag}\left(\frac{1}{\sqrt{2}}, \cos\frac{\pi}{2n}, \ldots, \cos\frac{(n-1)\pi}{2n}\right)$
DCT-III	$C_n^{III} = W_Q V_Q$ with $W_Q = \sqrt{\frac{2}{n}} \cdot I$
DCT-IV	$C_n^{IV} = W_Q V_Q$ with $W_Q = \sqrt{\frac{2}{n}}\,\mathrm{diag}\left(\cos\frac{\pi}{4n}, \cos\frac{3\pi}{4n}, \ldots, \cos\frac{(2n-1)\pi}{2n}\right)$
DST-I	$S_n^I = W_Q V_Q$ with $W_Q = \sqrt{\frac{2}{n+1}}\,\mathrm{diag}\left(\sin\frac{\pi}{n+1}, \ldots, \sin\frac{n\pi}{n+1}\right)$
DST-II	$S_n^{II} = W_Q V_Q$ with $W_Q = \sqrt{\frac{2}{n}}\,\mathrm{diag}\left(\sin\frac{\pi}{2n}, \ldots, \sin\frac{(n-1)\pi}{2n}, \frac{1}{\sqrt{2}}\sin\frac{\pi}{2}\right)$
DST-III	$S_n^{III} = W_Q V_Q$ with $W_Q = \sqrt{\frac{2}{n}}\,\mathrm{diag}\left(\sin\frac{\pi}{2n}, \ldots, \cos\frac{(2n-3)\pi}{2n}, \frac{1}{\sqrt{2}}\sin\frac{(2n-1)\pi}{2n}\right)$
DST-IV	$S_n^{IV} = W_Q V_Q$ with $W_Q = \sqrt{\frac{2}{n}}\,\mathrm{diag}\left(\sin\frac{\pi}{4n}, \sin\frac{3\pi}{4n}, \ldots, \sin\frac{(2n-1)\pi}{4n}\right)$

to a linear system of equations whose coefficient matrix is a 1×2 block matrix with a Vandermonde block and a Cauchy block.

For further studies of the correlation of rational interpolation and structured matrix computation, see our Sections 3.8 and 5.8, the papers [GY79], [BGY80], [BP94], [L94], [L95], [H95], [VK98], [OP98], [OP99], [VHKa], and the bibliography therein.

Section 3.7 Unlike the approach using Vandermonde matrices, the Cauchy matrix approach to loss-resilient encoding/decoding is well studied [MS77], [WB94]. Until the very recent works [P00c] and [Pb], publications on this subject did not exploit modern advanced techniques available for the computations with structured matrices.

Section 3.8 The presentation follows [BGR90], [OP98]. On the first studies of the scalar Nevanlinna–Pick problem, see [P16] and [N29]. On applications of the problems in systems, control and circuit theory, see [DGK81], [K87], [DGKF89], [ABKW90], and the bibliography therein. On various generalizations, see [D89], [D89a], [BGR90], [FF90], [S97], [FFGK98], [S98], and the bibliography therein. On the confluent Problem 3.8.3, see [BGR90], [OS99a], [OS00a], [OS00b]. An extension to the tangential confluent boundary Nevanlinna–Pick problem was announced in [OS00b]. On confluent Cauchy-like matrices, see [OS00a], [OS00b], and [PWa]. The bibliography on Kalman's State-Space Method can be found in [BGR90] and [OP98]. The global State-Space representation (3.8.3)–(3.8.5) and the reduction to the inversion of a structured matrix R in (3.8.6) is due to [BGR90]. On fast algorithms for the tangential Nevanlinna–Pick problem, see, e.g., [KP74], [F75], [DGK79], [D89], [GO93], [SKL-AC94], [GO94b], and [KO97]. On superfast solution algorithms, see the Notes to Section 5.8.

Table 3.5: Associated tridiagonal confederate matrices H_Q

DCT-I	tridiag	$\begin{pmatrix} \frac{1}{\sqrt{2}} & \frac{1}{2} & \cdots & \frac{1}{2} & \frac{1}{\sqrt{2}} \\ 0 & 0 & 0 & \cdots & 0 & 0 & 0 \\ \frac{1}{\sqrt{2}} & \frac{1}{2} & \cdots & \frac{1}{2} & \frac{1}{\sqrt{2}} \end{pmatrix}$
DCT-II	tridiag	$\begin{pmatrix} \frac{1}{2} & \frac{1}{2} & \cdots & \frac{1}{2} & \frac{1}{2} \\ \frac{1}{2} & 0 & 0 & \cdots & 0 & 0 & \frac{1}{2} \\ \frac{1}{2} & \frac{1}{2} & \cdots & \frac{1}{2} & \frac{1}{2} \end{pmatrix}$
DCT-III	tridiag	$\begin{pmatrix} \frac{1}{\sqrt{2}} & \frac{1}{2} & \cdots & \frac{1}{2} & \frac{1}{2} \\ 0 & 0 & 0 & \cdots & 0 & 0 & 0 \\ \frac{1}{\sqrt{2}} & \frac{1}{2} & \cdots & \frac{1}{2} & \frac{1}{2} \end{pmatrix}$
DCT-IV	tridiag	$\begin{pmatrix} \frac{1}{2} & \frac{1}{2} & \cdots & \frac{1}{2} & \frac{1}{2} \\ \frac{1}{2} & 0 & 0 & \cdots & 0 & 0 & -\frac{1}{2} \\ \frac{1}{2} & \frac{1}{2} & \cdots & \frac{1}{2} & \frac{1}{2} \end{pmatrix}$
DST-I	tridiag	$\begin{pmatrix} \frac{1}{2} & \frac{1}{2} & \cdots & \frac{1}{2} & \frac{1}{2} \\ 0 & 0 & 0 & \cdots & 0 & 0 & 0 \\ \frac{1}{2} & \frac{1}{2} & \cdots & \frac{1}{2} & \frac{1}{2} \end{pmatrix}$
DST-II	tridiag	$\begin{pmatrix} \frac{1}{2} & \frac{1}{2} & \cdots & \frac{1}{2} & \frac{1}{2} \\ -\frac{1}{2} & 0 & 0 & \cdots & 0 & 0 & -\frac{1}{2} \\ \frac{1}{2} & \frac{1}{2} & \cdots & \frac{1}{2} & \frac{1}{2} \end{pmatrix}$
DST-III	tridiag	$\begin{pmatrix} \frac{1}{2} & \frac{1}{2} & \cdots & \frac{1}{2} & \frac{1}{\sqrt{2}} \\ 0 & 0 & 0 & \cdots & 0 & 0 & 0 \\ \frac{1}{2} & \frac{1}{2} & \cdots & \frac{1}{2} & \frac{1}{\sqrt{2}} \end{pmatrix}$
DST-IV	tridiag	$\begin{pmatrix} \frac{1}{2} & \frac{1}{2} & \cdots & \frac{1}{2} & \frac{1}{2} \\ -\frac{1}{2} & 0 & 0 & \cdots & 0 & 0 & \frac{1}{2} \\ \frac{1}{2} & \frac{1}{2} & \cdots & \frac{1}{2} & \frac{1}{2} \end{pmatrix}$

Section 3.9 The presentation follows [BGR90], [BGR90a], [GO93], [GO94b], [OP98]. The reduction of the solution to computing factorization (3.8.3)–(3.8.5) is due to [BGR90] and [BGR90a]. Remark 3.9.2 relies on [BGR90], [GO94b].

Section 3.10 The section outlines the algorithm of [B-OT88]. On computations with polynomials represented by a black box subroutine for their evaluation, see [KT90]. On multivariate polynomials and multilevel structured matrices, see, e.g., [MP96], [T96], [MP98], [EP97], [MP00], [EPa]. On fast multivariate polynomial multiplication, see [CKL89], [P94]. On sparse multivariate polynomial interpolation, see [B-OT88], [KL88], [GKS90], [Z90], [KLW90], [ZR99], [KLL00].

Section 3.11 On various aspects of eigendecomposition and diagonalization of general matrices, including factorization (3.11.6), its computation by direct methods, the QR-algorithm, and numerical stability issues, see [GL96] and the bibliography therein.

On the same subjects for real symmetric or Hermitian matrices, see [Par80], [GL96]. Theorem 3.11.2 and the nomenclature of confederate matrices are from [MB79]. On Chebyshev polynomials, see [R92]. Corollary 3.11.5 is from [BC83]. Theorem 3.11.6 can be found in [GKO95]. Tables 3.1–3.5 are from [KO96]. On trigonometric transforms of these tables, their further variations, and Hartley transforms (including numerous applications), see [ANR74], [ER82], [DV87], [BF93], [KO96], [HR98], [S99], and bibliography therein. The paper [KO96] shows specific applications of the eight transforms of Tables 3.1–3.5 and of the matrix algebras generated by the associated tridiagonal matrices H_Q to the computation of preconditioners for the conjugate gradient algorithm applied to Toeplitz matrices.

3.14 Exercises

3.1 Complete all omitted proofs of the results in this chapter. Verify the expressions shown in Tables 3.1–3.5.

3.2 Specify the constants hidden in the "O" notation in the arithmetic computational complexity estimates of this chapter.

3.3 Write programs to implement the algorithms of this chapter (compare Section 2.15). Then apply the algorithms to some specific input matrices and/or polynomials. For example, compute $r_i(x) = v(x) \bmod m_i(x)$, $i = 1, \ldots, k$, for some specific polynomials $v(x), m_1(x), \ldots, m_k(x)$, $\deg v(x) < \sum_{i=1}^{k} \deg m_i(x)$, then recover the polynomial $v(x)$ from the set of polynomials $\{r_i(x), m_i(x)\}_{i=1}^{k}$. Next, rewrite the same polynomials via structured matrices and apply the matrix methods to the solution of the same problems.

3.4 Estimate the work space required for the algorithms of this chapter.

3.5 Extend Figures 3.3–3.6 as much as you can. Examine the extended figures and the algorithms that support the problem reductions shown by the arrows. Estimate how many ops are required for each reduction.

3.6 (a) Let $n + 1 = 2^h$ for an integer h. Recall that $\omega^k = \sin \frac{2\pi k}{n} + \sqrt{-1} \cos \frac{2\pi k}{n}$ for $\omega = \omega_n$ of Definition 2.1.2. Extend FFT to the computation of the following sine and cosine transforms,

$$v_i = \frac{\sqrt{2}}{n} \sum_{j=1}^{n} p_j \sin(\pi i j/(n+1)), \qquad i = 1, \ldots, n,$$

$$w_j = \frac{\sqrt{2}}{n} \sum_{j=1}^{n} p_j \cos(\pi i j/(n+1)), \qquad i = 0, \ldots, n.$$

Perform these transforms in $O(n \log n)$ ops.

(b) Evaluate a real polynomial $p(x)$ of degree of at most n at the Chebyshev nodes $x_h = \cos(\frac{2h+1}{n+1}\pi)$, $h = 0, 1, \ldots, n$, and interpolate to $p(x)$ from its values at these nodes; in both cases produce the output by using $O(n \log n)$ real ops.

3.7 Prove that the inequality in (3.2.2) turns into an equation only where $d_i = n/k$, $i = 1, \ldots, n$.

3.8 (a) Given $2n + 1$ scalars $c_0, \ldots, c_{n-1}$; $p_0, \ldots, p_n$, consider a polynomial $p(x) = \sum_{i=0}^n p_i x^i$ and its Newton representation: $p(x) = d_0 + d_1(x - c_0) + d_2(x - c_0)(x - c_1) + \cdots + d_n(x - c_0)(x - c_1) \cdots (x - c_{n-1})$. Compute the *divided differences* $d_0, \ldots, d_n$.

   (b) Given the values $c_0, \ldots, c_{n-1}$; $d_0, \ldots, d_n$, compute the coefficients $p_0, \ldots, p_n$ (precompute appropriate "supermoduli" and then apply the fan-out techniques).

   (c) Given the values $c_0, \ldots, c_{n-1}$; $d_0, \ldots, d_n$; $x_0, \ldots, x_n$, compute the values $p(x_i)$, $i = 0, 1, \ldots, n$.

   In all cases use $O(M(n) \log n)$ ops.

3.9 Let $m_i(x)$, for $i = 1, \ldots, k$, be cubic polynomials. Let $u(x) = \prod_{i=1}^k m_i(x)$. To compute polynomials $u_i(x) = (u(x)/m_i(x)) \bmod m_i(x)$, $i = 1, \ldots, k$, we may first perform division of the polynomial $u(x)$ by $m_i(x)$ (with the remainders equal to 0) and then reduce the results modulo $m_i(x)$ for $i = 1, \ldots, k$. Alternatively, we may proceed as in the proof of Theorem 3.5.4. In both cases estimate the computational cost as a function in k (as $k \to \infty$).

3.10 (a) [FHR93]. Prove the following modification of (3.4.1):

$$V^{-1}(\mathbf{t}) = JZ(J\mathbf{u_t})V^T(\mathbf{t})D'(\mathbf{t})$$

   for $D'(\mathbf{t})$ in (3.6.1).

   (b) [FHR93]. Combine this equation with (3.6.5) to obtain that

$$C(\mathbf{s}, \mathbf{t}) = D^{-1}(\mathbf{s}, \mathbf{t})V(\mathbf{s})JZ(J\mathbf{u_t})V^T(\mathbf{t}).$$

   (c) Compare the matrix and polynomial versions of the algorithms of Section 3.6 that reduce polynomial interpolation and multipoint evaluation Problems 3.1.1 and 3.3.1 to Problems 3.6.1 and 3.6.6 of Cauchy matrix computations.

   (d) Verify that $V^T D V$ is a Hankel matrix for any pair of $n \times n$ diagonal matrix D and Vandermonde matrix V.

   (e) [HR84], [T94]. Prove that every non-singular Hankel matrix H can be represented as a product $V^T D V$ (as in part (d)); furthermore, D can be assumed to be a positive definite diagonal matrix if H is a real positive definite Hankel matrix. (On positive definite matrices, see Definition 4.6.3.)

3.11 Devise a superfast loss(erasure)-resilient encoding/decoding algorithm for the generator matrix $\begin{pmatrix} I \\ C(\mathbf{s}, \mathbf{t}) \end{pmatrix}$ (see Section 3.7 or [MS77], [WB94], [P00c], [Pb]). Exploit Theorems 3.4.1 and 3.7.1, matrix equations (3.4.1), (3.6.5), (3.6.7)–(3.6.9), and the matrix equations of Exercise 3.10. Compare the ops count for this customary algorithm and for that of Section 3.7 for specific choices of the input parameters m and n. Assume that the dimension l of the vectors $\mathbf{p}$ and $\mathbf{r}$ of the customary approach is either

(a) randomly sampled from the interval $(0, \min\{n - m, m\})$ under the uniform probability distribution on this interval or

(b) $\min\{n - m, m\}$.

3.12 [OS00a, OS00b]. Specify the tangential confluent boundary Nevanlinna–Pick problem. Specify its solvability conditions, the global State-Space representation of its transfer matrix function, and the associated matrix equation of Sylvester type.

3.13 Estimate the arithmetic complexity R of root-finding for a polynomial $\prod_{i=1}^{n}(x - r_i)$ with only integer roots $r_1, \ldots, r_n$. Express R as a function in $\max_i |r_i|$ and n.

3.14 [GO94]. Specify a matrix equation of Sylvester type for the matrix function R in (3.9.2).

3.15 Deduce (3.11.2) from (3.11.1).

3.16 Show uniqueness up to scaling of the eigenvectors in (3.11.1) where all the n eigenvalues $z_0, \ldots, z_{n-1}$ are distinct.

3.17 Show how to represent every unreduced upper Hessenberg matrix H as a confederate matrix for a fixed polynomial basis $\mathbf{Q}$.

3.18 Prove that $HK = KF_u$ where $u(x) = c_H(x) = \det(xI - H)$, $K = K(H, \mathbf{e}_0)$ is a non-singular triangular matrix, and H is an unreduced Hessenberg matrix.

3.19 A straightforward extension of Theorem 3.11.1 is to the matrix algebras generated by a matrix of the form $M = W^{-1}F_u W$, for an $n \times n$ non-singular matrix W and a polynomial $u(x)$ in (3.11.3) with n distinct zeros. (3.11.4) and the latter expression for M together imply that $M = W^{-1}V^{-1}(\mathbf{z})D(\mathbf{z})V(\mathbf{z})W$ and consequently $v(M) = W^{-1}V^{-1}(\mathbf{z})v(D(\mathbf{z}))V(\mathbf{z})W$ for any polynomial $v(x) = \sum_{i=0}^{n-1} v_i x^i$.

(a) Show some matrix algebras $A(M)$ whose generator matrices M are diagonalized by the matrices $(V(\mathbf{z})W)^{-1}$ and $V(\mathbf{z})W$ that are rapidly multiplied by vectors in a numerically stable way.

(b) Choose any $\nabla_{Z_f, D(\mathbf{y})}$-like matrix W (e.g., $W = V^T(\mathbf{y})$), apply Theorem 1.5.4, and observe that in this case $V(\mathbf{z})W$ is a $\nabla_{D(\mathbf{z}), D(\mathbf{y})}$-like matrix with Cauchy type structure. Can this choice produce interesting matrix algebras $A(W^{-1}F_u W)$ for appropriate vectors $\mathbf{y}$?

(c) Specify the matrices $V(\mathbf{z})$ and W in the above diagonalization and the matrices U, K, and $V(\mathbf{z})$ in (3.11.7) for the matrices $M = H_{\mathbf{Q}}$ of Table 3.5.

(d) Try to define effective novel preconditioners for the conjugate gradient algorithm applied to Toeplitz linear systems [CN96], [CN99], [KO96] based on the above diagonalizations of M and on (3.11.7).

3.20 (R. C. Melville).

    (a) Block diagonalize a block circulant matrix $Z(\mathbf{v})$ (where $\mathbf{v}$ is a block vector) by a similarity transform $Z(\mathbf{v}) = W^{-1} D(\mathbf{z}) W$ where a matrix W does not depends on the block vector $\mathbf{v}$.

    (b) Try to yield a similar block diagonalization of a block matrix $Z(\mathbf{v}) + D(\mathbf{u})$ where the coordinates of the vectors $\mathbf{u}$ and $\mathbf{v}$ are $k \times k$ matrices and $k > 1$. Can you achieve this by using a matrix W of similarity transformation that does not depend on the block vectors $\mathbf{u}$ and $\mathbf{v}$?

3.21 [BF93]. For the Hartley matrix $H_1 = \left(\cos \frac{2kj\pi}{n} + \sin \frac{2kj\pi}{n} \right)_{k,j=0}^{n-1}$, prove that

    (a) $H_1^T H_1 = I$

and

    (b) for every real vector $\mathbf{v}$, the matrix $H_1 D(\mathbf{v}) H_1$ is the sum of a symmetric circulant matrix and a Hankel matrix.

3.22   (a) What is the minimum number of real ops required for multiplication of an $n \times n$ real Toeplitz matrix by a vector?

Answer the same question for

    (b) a complex Toeplitz matrix,

    (c) a real Toeplitz+Hankel matrix,

    (d) a complex Toeplitz+Hankel matrix.

Chapter 4

Structured Matrices and Displacement Operators

Summary In Chapter 1, we outlined the displacement rank approach (COMPRESS, OPERATE, DECOMPRESS) to computations with structured matrices and in more detail covered its OPERATE stage. In this chapter, we systematically cover the basic techniques required at the COMPRESS and DECOMPRESS stages of this approach. We present these techniques in a unified way but also detail the decompression techniques separately for each of the most popular classes of structured matrices. As an immediate result, we obtain superfast algorithms for multiplying these matrices by vectors and by each other. We accentuate the power of the approach based on the displacement transformations of two kinds that extend successful algorithms from one class of structured matrices to various other classes. We also briefly comment on parallel implementation of computations with structured matrices.

Graph for selective reading

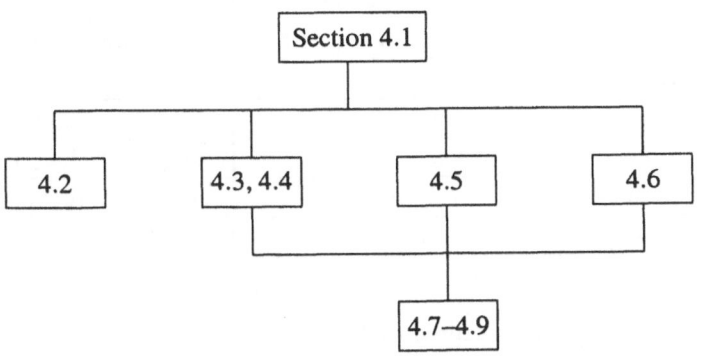

4.1 Some definitions and basic results

Our next goal is to revisit the displacement rank approach and to elaborate its outline presented in Chapter 1. Because of the two-chapter break, we start with reproducing some of the most used items, i.e., Tables 1.1 and 1.2, which cover the four basic classes of structured matrices and are now reproduced as Tables 4.1 and 4.2, respectively, and some definitions from Sections 1.3 and 2.1. The estimates in the third column of Table 1.2/4.2 have been proved in Sections 2.4, 2.6, 3.1, and 3.6.

Table 4.1: Four classes of structured matrices

Toeplitz matrices $\left(t_{i-j}\right)_{i,j=0}^{n-1}$

$$\begin{pmatrix} t_0 & t_{-1} & \cdots & t_{1-n} \\ t_1 & t_0 & \ddots & \vdots \\ \vdots & \ddots & \ddots & t_{-1} \\ t_{n-1} & \cdots & t_1 & t_0 \end{pmatrix}$$

Hankel matrices $\left(h_{i+j}\right)_{i,j=0}^{n-1}$

$$\begin{pmatrix} h_0 & h_1 & \cdots & h_{n-1} \\ h_1 & h_2 & \cdot^{\cdot^{\cdot}} & h_n \\ \vdots & \cdot^{\cdot^{\cdot}} & \cdot^{\cdot^{\cdot}} & \vdots \\ h_{n-1} & h_n & \cdots & h_{2n-2} \end{pmatrix}$$

Vandermonde matrics $\left(t_i^j\right)_{i,j=0}^{n-1}$

$$\begin{pmatrix} 1 & t_0 & \cdots & t_0^{n-1} \\ 1 & t_1 & \cdots & t_1^{n-1} \\ \vdots & \vdots & & \vdots \\ 1 & t_{n-1} & \cdots & t_{n-1}^{n-1} \end{pmatrix}$$

Cauchy matrices $\left(\frac{1}{s_i-t_j}\right)_{i,j=0}^{n-1}$

$$\begin{pmatrix} \frac{1}{s_0-t_0} & \cdots & \frac{1}{s_0-t_{n-1}} \\ \frac{1}{s_1-t_0} & \cdots & \frac{1}{s_1-t_{n-1}} \\ \vdots & & \vdots \\ \frac{1}{s_{n-1}-t_0} & \cdots & \frac{1}{s_{n-1}-t_{n-1}} \end{pmatrix}$$

Table 4.2: Parameter and flop count for representation of a structured matrix and for its multiplication by a vector

Matrices M	Number of parameters per an $m \times n$ matrix M	Number of flops for computation of $M\mathbf{v}$
general	mn	$2mn - n$
Toeplitz	$m + n - 1$	$O((m + n)\log(m + n))$
Hankel	$m + n - 1$	$O((m + n)\log(m + n))$
Vandermonde	m	$O((m + n)\log^2(m + n))$
Cauchy	$m + n$	$O((m + n)\log^2(m + n))$

The next two definitions reproduce Definitions 2.1.1 and 2.1.2, respectively.

Definition 4.1.1. W^T, W^*, $\mathbf{v}^T$, and $\mathbf{v}^*$ are the *transposes* and the *Hermitian (conjugate) transposes* of a matrix W and a vector $\mathbf{v}$, respectively. $(W_1, \ldots, W_n)$ is the $1 \times n$

block matrix with the blocks $W_1, \ldots, W_n$. $D(\mathbf{v}) = \mathrm{diag}(\mathbf{v}) = \mathrm{diag}(v_0, \ldots, v_{n-1})$ being an $n \times n$ *diagonal* matrix in (1.3.2). $\mathbf{e}_i$ is the i-th coordinate vector, having its i-th coordinate 1 and all other coordinates 0, so $\mathbf{e}_0 = (1, 0, \ldots, 0)^T$. $\mathbf{1} = (1, 1, \ldots, 1)^T = \sum_i \mathbf{e}_i$. The null matrices are denoted by 0 and O, and the $n \times n$ null matrix is also denoted by 0_n and O_n. I and I_n denote the $n \times n$ *identity* matrix, $I = \begin{pmatrix} 1 & & \\ & \ddots & \\ & & 1 \end{pmatrix} = (\mathbf{e}_0, \ldots, \mathbf{e}_{n-1})$. $J = \begin{pmatrix} & & 1 \\ & \cdots & \\ 1 & & \end{pmatrix} = (\mathbf{e}_{n-1}, \ldots, \mathbf{e}_0)$ is the $n \times n$ *reflection* matrix. (Here and hereafter, a blank space in matrix representation is assumed to be filled with zeros.) $Z_f = (\mathbf{e}_1, \ldots, \mathbf{e}_{n-1}, f\mathbf{e}_0)$ is the *unit f-circulant* matrix of equation (1.3.1), $Z_0 = Z = (\mathbf{e}_1, \ldots, \mathbf{e}_{n-1}, \mathbf{0})$ is the *unit lower triangular Toeplitz* matrix. $Z_f(\mathbf{u}) = \sum_i u_i Z_f^i$ is an *f-circulant matrix* for a vector $\mathbf{u} = (u_i)_{i=0}^{n-1}$ (see (2.6.1)). $Z_0(\mathbf{u}) = Z(\mathbf{u}) = \sum_i u_i Z^i$ is a lower triangular Toeplitz matrix. $K(M, \mathbf{v}, k) = (M^i \mathbf{v})_{i=0}^{k-1}$ is the $n \times k$ *Krylov matrix* defined by the triple of a natural k, an $n \times n$ matrix M, and a vector $\mathbf{v}$ of dimension n. The same triple $(M, \mathbf{v}, k)$ defines the *Krylov space* of vectors with the basis formed by the columns of the matrix $K(M, \mathbf{v}, k)$. $K(M, \mathbf{v})$ denotes the matrix $K(M, \mathbf{v}, n)$.

Definition 4.1.2. ω_n is a primitive *n-th root* of 1, i.e., $\omega_n^n = 1$, $\omega_n^s \neq 1$, $s = 1, 2, \ldots, n - 1$; e.g., $\omega_n = \exp(2\pi\sqrt{-1}/n)$ in the complex number field $\mathbb{C}$. $\mathbf{w}_n = (\omega_n^i)_{i=0}^{n-1}$ is the vector of all the n-th roots of 1. $\Omega_n = (\omega_n^{ij})_{i,j=0}^{n-1}$ is the $n \times n$ matrix of the **discrete Fourier transform (DFT)**, which is a special case of Vandermonde matrices (see Table 1.1). *The discrete Fourier transform of a vector $\mathbf{p}$ of dimension n is the vector $DFT(\mathbf{p}) = \Omega_n \mathbf{p}$.* DFT_n denotes the computation of this vector provided that the vectors $\mathbf{p}$ and $\mathbf{w}_n$ are given (the vector $\mathbf{w}_n$ can be immediately obtained in $n - 2$ ops if just ω_n, a primitive n-th root of 1, is given). We write $DFT = DFT_n$, $\omega = \omega_n$, $\mathbf{w} = \mathbf{w}_n$, $\Omega = \Omega_n$ where n is known from the context. The **inverse DFT problem** is the problem of computing the vector $\mathbf{p} = \Omega^{-1}\mathbf{v}$ satisfying the equation $\Omega\mathbf{p} = \mathbf{v}$ for two given vectors $\mathbf{v} = (v_i)_{i=0}^{n-1}$ and $\mathbf{w}_n$.

Let us now recall the definition of two linear operators, of Sylvester and Stein type (see Section 1.3).

Definition 4.1.3. For any fixed field $\mathbb{F}$ and a fixed pair $\{A, B\}$ of operator matrices, we define the linear *displacement operators* $L : \mathbb{F}^{m \times n} \to \mathbb{F}^{m \times n}$ of *Sylvester type*, $L = \nabla_{A,B}$:

$$L(M) = \nabla_{A,B}(M) = AM - MB,$$

and *Stein type*, $L = \Delta_{A,B}$:

$$L(M) = \Delta_{A,B}(M) = M - AMB.$$

The image $L(M)$ of the operator L is called the *displacement* of the matrix M.

We mostly stay with square matrices M and refer the reader to [PWa] on the rectangular matrix case.

Table 4.3: Some pairs of operators $\nabla_{A,B}$ and associated structured matrices

operator matrices		class of structured	rank of
A	B	matrices M	$\nabla_{A,B}(M)$
Z_1	Z	Toeplitz	≤ 2
Z_1	Z	Sylvester	≤ 2
Z_1	Z	Frobenius (companion)	≤ 2
Z_1	Z	$k \times l$ block matrices with Toeplitz blocks	$\leq k+l$
Z_1	Z^T	Hankel, Bezout	≤ 2
$Z + Z^T$	$Z + Z^T$	Toeplitz+Hankel	≤ 4
$D(t)$	Z	Vandermonde	≤ 1
Z^T	$D(t)$	transposed Vandermonde	≤ 2
$D(s)$	$D(t)$	Cauchy	≤ 1
$D(s)$	$D(t)$	Loewner	≤ 2

4.2   Displacements of basic structured matrices

Let us specify the COMPRESS stage of the displacement rank approach for the basic structured matrices in Examples 1.3.2–1.3.5 showing the compression in the form convenient for its further extension. We have

$$\nabla_{Z_f, Z_e}(T) = Z_f T - T Z_e = (Z J \mathbf{t}_- - e\mathbf{t})\mathbf{e}_{n-1}^T + \mathbf{e}_0 (f J \mathbf{t} - Z^T \mathbf{t}_-)^T \qquad (4.2.1)$$

for $T = (t_{i-j})_{i,j=0}^{n-1}$, $\mathbf{t} = (t_i)_{i=0}^{n-1}$, $\mathbf{t}_- = (t_{-i})_{i=0}^{n-1}$, and any pair of scalars e and f;

$$\nabla_{Z_f, Z_e^T}(H) = Z_f H - H Z_e^T = (Z\mathbf{h} - e\tilde{\mathbf{h}})\mathbf{e}_0^T + \mathbf{e}_0(f\tilde{\mathbf{h}} - Z\mathbf{h})^T \qquad (4.2.2)$$

for $H = (h_{i+j})_{i,j=0}^{n-1}$, $\mathbf{h} = (h_i)_{i=0}^{n-1}$, $\tilde{\mathbf{h}} = (h_{n-1+i})_{i=0}^{n-1}$, and any pair of scalars e and f;

$$\nabla_{D(\mathbf{t}), Z_f}(V) = D(\mathbf{t})V - V Z_f = (\mathbf{t}^n - f\mathbf{1})\mathbf{e}_{n-1}^T \qquad (4.2.3)$$

for $V = V(\mathbf{t}) = (t_i^j)_{i,j=0}^{n-1}$, $\mathbf{t} = (t_i)_{i=0}^{n-1}$, $\mathbf{t}^n = (t_i^n)_{i=0}^{n-1}$, and any scalar f, and

$$\nabla_{D(\mathbf{s}), D(\mathbf{t})}(C) = D(\mathbf{s})C - C D(\mathbf{t}) = \mathbf{1}\,\mathbf{1}^T \qquad (4.2.4)$$

for

$$C = C(\mathbf{s}, \mathbf{t}) = \left(\frac{1}{s_i - t_j}\right)_{i,j=0}^{n-1}, \quad \mathbf{s} = (s_i)_{i=0}^{n-1}, \ \mathbf{t} = (t_i)_{i=0}^{n-1}, \ \mathbf{1} = (1)_{i=0}^{n-1}. \qquad (4.2.5)$$

We observe that the above matrices T and H can be immediately recovered from the respective displacement matrices, whereas the matrices V and C can be immediately recovered from the operator matrices. (4.2.1)–(4.2.5) demonstrate that the displacements of the matrices of Table 1.1/4.1 have small rank for appropriate displacement operators. We have already cited this basic property in Section 1.1. We can see it also in Table 4.3, extending Table 1.3. Here are our pointers to some extensions of Table 4.3 and (4.2.1)–(4.2.5).

1. If $t_i \neq 0, i = 0, 1, \ldots, n - 1$, then we may fix any scalar f and observe that

$$D^{-1}(\mathbf{t})V - V Z_f^T = (t_i^{-1} - f t_i^{n-1})_{i=0}^{n-1} \mathbf{e}_0^T. \qquad (4.2.6)$$

2. Transposition of the matrices in (4.2.1)–(4.2.2) produces distinct equations where T^T and $H = H^T$, replacing T and H, are still Toeplitz and Hankel matrices, respectively, whereas the transposition of the matrices on both sides of (4.2.3) and (4.2.6) defines displacements of small rank for the transpose V^T of a Vandermonde matrix V.

3. Table 4.3 can be extended by the column "rank$(\Delta_{A,B}(M))$" and similarly (4.2.1)–(4.2.6) can be extended (see Theorem 1.3.1 and Exercise 4.4).

4. Since the appearance of the seminal paper [KKM79], particularly celebrated have become the linear operators Δ_{Z,Z^T} and $\Delta_{Z^T,Z}$, associated with Toeplitz and Toeplitz-like matrices. Note that

$$\Delta_{Z,Z^T}(T) = \mathbf{t}\mathbf{e}_0^T + \mathbf{e}_0\mathbf{t}_-^T - t_0\mathbf{e}_0\mathbf{e}_0^T, \qquad (4.2.7)$$

$$\Delta_{Z^T,Z}(T) = \mathbf{e}_{n-1}J\mathbf{t}^T + J\mathbf{t}_-\mathbf{e}_{n-1}^T - t_0\mathbf{e}_{n-1}\mathbf{e}_{n-1}^T, \qquad (4.2.8)$$

for a matrix T and two vectors $\mathbf{t}$ and $\mathbf{t}_-$ in (4.2.1).

5. Consider the folowing example.

Example 4.2.1. Displacement of a Toeplitz+Hankel matrix $T + H$.

$$\nabla_{Y_{00},Y_{11}}(T + H) = ((Z^T - I)\mathbf{t} + (Z - I)\mathbf{h})\mathbf{e}_0^T + ((Z - I)J\mathbf{t}_- + (Z^T - I)\tilde{\mathbf{h}})\mathbf{e}_{n-1}^T$$
$$- \mathbf{e}_0(\mathbf{t}_-^T Z + (Z\mathbf{h})^T) - \mathbf{e}_{n-1}((ZJ\mathbf{t})^T + \tilde{\mathbf{h}}^T Z) \qquad (4.2.9)$$

for the matrices $T = (t_{i-j})_{i,j=0}^{n-1}$, $H = (h_{i+j})_{i,j=0}^{n-1}$, $Y_{00} = Z + Z^T$, $Y_{11} = Y_{00} + \mathbf{e}_0\mathbf{e}_0^T + \mathbf{e}_{n-1}\mathbf{e}_{n-1}^T$, and the vectors $\mathbf{t} = (t_i)_{i=0}^{n-1}$ and $\mathbf{t}_- = (t_{-i})_{i=0}^{n-1}$ in (4.2.1) and $\mathbf{h}$ and $\tilde{\mathbf{h}}$ in (4.2.2). We have expressed the displacement of Toeplitz+Hankel matrix as a sum of at most four outer products of vectors because all entries of the displacement are zeros except for the entries of two rows and two columns (the first and the last ones).

To conclude this section, we recall the two-term bilinear expressions in (2.11.4) and Exercise 2.24 (b), (c) for the inverses of Toeplitz matrices and extend them to the inverses of the matrices of this section. The expressions follow immediately because Theorem 1.5.3 expresses the displacements $\nabla_{B,A}(M^{-1})$ of the inverse matrices M^{-1} via the displacements $\nabla_{A,B}(M)$. By combining these expressions with (4.2.1)–(4.2.6) and Example 4.2.1, we obtain that

$$\nabla_{Z_e,Z_f}(T^{-1}) = T^{-1}(\mathbf{e}\mathbf{t} - ZJ\mathbf{t}_-)\mathbf{e}_{n-1}^T T^{-1} + T^{-1}\mathbf{e}_0(\mathbf{t}_-^T Z - f\mathbf{t}^T J)T^{-1} \qquad (4.2.10)$$

for T, $\mathbf{t}$, and $\mathbf{t}_-$ in (4.2.1);

$$\nabla_{Z_e^T, Z_f}(H^{-1}) = H^{-1}(e\tilde{\mathbf{h}} - Z\mathbf{h})\mathbf{e}_0^T H^{-1} + H^{-1}\mathbf{e}_0(\mathbf{h}^T Z^T - f\tilde{\mathbf{h}}^T)H^{-1} \qquad (4.2.11)$$

for H, $\mathbf{h}$, and $\tilde{\mathbf{h}}$ in (4.2.2);

$$\nabla_{Z_f, D(\mathbf{t})}(V^{-1}(\mathbf{t})) = V^{-1}(\mathbf{t})(f\mathbf{1} - \mathbf{t}^n)\mathbf{e}_{n-1}^T V^{-1}(\mathbf{t})$$

for $V = V(\mathbf{t})$, $\mathbf{t}$, and $\mathbf{t}^n$ in (4.2.3), and

$$\nabla_{D(\mathbf{t}), D(\mathbf{s})}(C^{-1}) = -C^{-1}\mathbf{1}\,\mathbf{1}^T C^{-1}$$

for C, $\mathbf{s}$, $\mathbf{t}$, and $\mathbf{1}$ in (4.2.4) and (4.2.5);

$$\nabla_{Z_f^T, D^{-1}(\mathbf{t})}(V^{-1}) = V^{-1}(f\mathbf{t}^{n-1} - \mathbf{t}^{-1})\mathbf{e}_0^T V^{-1}$$

for V and $\mathbf{t}^a = (t_i^a)_{i=0}^{n-1}$, $a = -1$ and $a = n - 1$ in (4.2.6), and

$$\begin{aligned}
\nabla_{Y_{11}, Y_{00}}((T + H)^{-1}) = {}&(T + H)^{-1}((I - Z^T)\mathbf{t} + (I - Z)\mathbf{h})\mathbf{e}_0^T(T + H)^{-1} \\
&+ (T + H)^{-1}((I - Z)J\mathbf{t}_- + (I - Z^T)\tilde{\mathbf{h}})\mathbf{e}_{n-1}^T(T + H)^{-1} \\
&+ (T + H)^{-1}\mathbf{e}_0(\mathbf{t}_-^T Z + (Z\mathbf{h})^T)(T + H)^{-1} \qquad (4.2.12) \\
&+ (T + H)^{-1}\mathbf{e}_{n-1}((ZJ\mathbf{t})^T + \tilde{\mathbf{h}}^T Z)(T + H)^{-1}
\end{aligned}$$

for T, H, Y_{00}, Y_{11}, $\mathbf{t}$, $\mathbf{t}_-$, $\mathbf{h}$, and $\tilde{\mathbf{h}}$ in Example 4.2.1.

In the latter examples, each displacement of the inverse matrix is bilinearly expressed via its two, four or eight products by vectors.

4.3 Inversion of the displacement operators

Our next subject is the DECOMPRESS stage of the displacement rank approach, i.e., the inversion of linear operators $\nabla_{A,B}$ and $\Delta_{A,B}$. In Section 1.4 we elaborated on this stage for Cauchy-like matrices when we deduced (1.4.4),

$$M = \sum_{k=1}^{l} D(\mathbf{g}_k)C(\mathbf{s}, \mathbf{t})D(\mathbf{h}_k)$$

from (1.4.1) and (1.4.2), that is,

$$L(M) = D(\mathbf{s})M - MD(\mathbf{t}) = GH^T = \sum_{k=1}^{l} \mathbf{g}_k\mathbf{h}_k^T.$$

Now we are going to obtain such expressions for various other classes of structured matrices. We start with the general results and then specify the inversion formulae for the operators already studied.

Definition 4.3.1. A linear operator L is *non-singular* if the matrix equation $L(M) = 0$ implies that $M = 0$.

For instance, the operator $L(M) = D(s)M - MD(t)$ is non-singular provided that all coordinates of the vectors **s** and **t** are distinct.

Theorem 4.3.2. *Let* $\{\alpha_1, \ldots, \alpha_m\}$ = spectrum(A), $\{\beta_1, \ldots, \beta_n\}$ = spectrum(B). *Then the operator* $\nabla_{A,B}$ *of Sylvester type is non-singular if and only if* $\alpha_i \neq \beta_j$ *for all pairs* (i, j), *and the operator* $\Delta_{A,B}$ *of Stein type is non-singular if and only if* $\alpha_i \beta_j \neq 1$ *for all pairs* (i, j).

Proof. For any matrix $M = (\mathbf{m}_1, \mathbf{m}_2, \ldots)$, let $\overrightarrow{M}$ denote the column vector $\begin{pmatrix} \mathbf{m}_1 \\ \mathbf{m}_2 \\ \vdots \end{pmatrix}$,

and let $\otimes$ denote the Kronecker product. Then we have

$$\overrightarrow{\nabla_{A,B}(M)} = (I \otimes A - B^T \otimes I)\overrightarrow{M},$$
$$\overrightarrow{\Delta_{A,B}(M)} = (I - B^T \otimes A)\overrightarrow{M}.$$

The matrix $\nabla_{A,B}$ is non-singular if and only if the matrix $I \otimes A - B^T \otimes I$ is also. The matrix $\Delta_{A,B}$ is non-singular if and only if the matrix $I - B^T \otimes A$ is also. The theorem follows because spectrum($I \otimes A - B^T \otimes I$) = $\{\alpha_i - \beta_j,$ for all $i, j\}$, spectrum($I - B^T \otimes A$) = $\{1 - \alpha_i \beta_j,$ for all $i, j\}$. $\qquad\square$

Corollary 4.3.3. *If the operator* $\nabla_{A,B}$ *is non-singular, then at least one of the operator matrices A and B is non-singular.*

Theorem 4.3.2 only specifies a sufficient condition for the non-singularity of an operator L, but we seek sufficient conditions for this operator to have a *linear inverse*. Let us start with a simple result of independent interest.

Theorem 4.3.4. *Let* $\widehat{A} = VAV^{-1}, \widehat{B} = W^{-1}BW$ *for some non-singular matrices V and W. Then we have*

$$\nabla_{\widehat{A},\widehat{B}}(VMW) = V\nabla_{A,B}(M)W,$$
$$\Delta_{\widehat{A},\widehat{B}}(VMW) = V\Delta_{A,B}(M)W.$$

The next theorem applies to the case where the matrices A and B are diagonalized by some similarity transformations (see Section 3.11), that is, where

$$A = PD(s)P^{-1}, \qquad B = QD(t)Q^{-1} \tag{4.3.1}$$

for some non-singular matrices P and Q and for $D(s) = \text{diag}(s_1, \ldots, s_m)$, $D(t) = \text{diag}(t_1, \ldots, t_n)$. This is the case where $P = Q = I$ and the diagonal operator matrices A and B are associated with Cauchy-like matrices (see Example 1.4.1) as well as for various operator matrices A and B whose diagonalization is well known.

Theorem 4.3.5. *Let (4.3.1) hold. Then*

$$M = P \left(\frac{(P^{-1} \nabla_{A,B}(M)Q)_{i,j}}{s_i - t_j} \right)_{i,j=1}^{m,n} Q^{-1}, \quad \text{where } s_i \neq t_j \text{ for all } i, j, \quad (4.3.2)$$

$$M = P \left(\frac{(P^{-1} \Delta_{A,B}(M)Q)_{i,j}}{1 - s_i t_j} \right)_{i,j=1}^{m,n} Q^{-1}, \quad \text{where } s_i t_j \neq 1 \text{ for all } i, j. \quad (4.3.3)$$

Proof. We have already had this result for the diagonal operator matrices $A = D(\mathbf{s})$ and $B = D(\mathbf{t})$ (see Example 1.4.1), that is, for the matrices A and B in (4.3.1) where $P = Q = I$. The extension to the general case follows by application of Theorem 4.3.4 and the observation that $P^{-1} \nabla_{A,B}(M)Q = \nabla_{D(\mathbf{s}),D(\mathbf{t})}(P^{-1}MQ)$, $P^{-1} \Delta_{A,B}(M)Q = \Delta_{D(\mathbf{s}),D(\mathbf{t})}(P^{-1}MQ)$. $\qquad \square$

The next simple theorem is our basic tool. It applies where $A^p = aI$ and/or $B^q = bI$ for two operator matrices A and B, two scalars a and b, and two positive integers p and q.

Theorem 4.3.6. *For all $k \geq 1$, we have*

$$M = A^k M B^k + \sum_{i=0}^{k-1} A^i \Delta_{A,B}(M) B^i.$$

Corollary 4.3.7. *If $A^p = aI$ and/or $B^q = bI$ (i.e., if A is an a-potent matrix of order p and/or B is a b-potent matrix of order q (see (2.6.3))), then*

$$M = \left(\sum_{i=0}^{p-1} A^i \Delta_{A,B}(M) B^i \right) (I - aB^p)^{-1}$$

and/or

$$M = (I - bA^q)^{-1} \left(\sum_{i=0}^{q-1} A^i \Delta_{A,B}(M) B^i \right),$$

respectively.

The approach based on the latter equations complements our first approach, based on Theorems 4.3.4 and 4.3.5, because now we cover the f-circulant operator matrices Z_f and their transposes, in particular the *nilpotent* matrices Z_0 and Z_0^T, which are not diagonalizable. One may, however, define desired expressions by replacing these matrices by Z_ϵ and Z_ϵ^T for $\epsilon \neq 0$, then apply them for $\epsilon \to 0$, and arrive at the limiting expressions with Z_0 and Z_0^T. Also the displacement changes by at most a rank-one matrix where the operator matrix is changed from Z_e to Z_f for $e \neq f$.

A common feature of the matrix representations of Theorems 4.3.5 and 4.3.6 and Corollary 4.3.7 is that they represent the matrix M linearly via $L(M)$, that is, the *inverse operator L^{-1} is also linear.*

Definition 4.3.8. A non-singular linear operator L defined on the space of $m \times n$ matrices is *regular* if it has a linear inverse.

For regular operators L, every $m \times n$ matrix M can be expressed linearly in the displacement $L(M)$ and bilinearly in its every L-*generator* G, H *(of length l)*. That is, every $m \times n$ matrix M can be expressed as a bilinear form in l terms where the j-th term is the product of two matrices with the entries expressed linearly via the entries of the j-th column vectors, $\mathbf{g}_j$ of the matrix G and $\mathbf{h}_j$ of the matrix H, respectively, $j = 1, \ldots, l$. Example 1.4.1 and several further examples in the next two sections demonstrate bilinear representation of various structured matrices via their displacement generators. Compression of the displacement generators to their minimum length (equal to $\operatorname{rank}(L(M))$) is studied in Section 4.6.

4.4   Compressed bilinear expressions for structured matrices

Let us next apply the techniques of the previous section to specify expressions of various structured matrices via their displacement generators provided that $L(M) = GH^T = \sum_{k=1}^{l} \mathbf{g}_k \mathbf{h}_k^T$ (see (1.4.2), (1.4.3)). The derivation of these expressions is rather straightforward, except for Example 4.4.9. For simplicity we assume that $m = n$ and in some cases skip the derivation and only show the final expressions for the matrix M via the displacement $L(M)$.

Example 4.4.1. Toeplitz-like matrix M, Stein type operators (see Corollary 4.3.7).

a) $L = \Delta_{Z_e, Z_f^T}, \, ef \neq 1$,

$$(1 - ef)M = \sum_{i=0}^{n-1}(Z_e^i G)(Z_f^i H)^T = \sum_{j=1}^{l}\left(\sum_{i=0}^{n-1}(Z_e^i \mathbf{g}_j)(Z_f^i \mathbf{h}_j)^T\right)$$

$$= \sum_{j=1}^{l} \begin{pmatrix} g_{1,j} & eg_{n,j} & \ddots & eg_{2,j} \\ g_{2,j} & g_{1,j} & \ddots & \ddots \\ \ddots & \ddots & \ddots & eg_{n,j} \\ g_{n,j} & \ddots & g_{2,j} & g_{1,j} \end{pmatrix} \begin{pmatrix} h_{1,j} & h_{2,j} & \ddots & h_{n,j} \\ fh_{n,j} & h_{1,j} & \ddots & \ddots \\ \ddots & \ddots & \ddots & h_{2,j} \\ fh_{2,j} & \ddots & fh_{n,j} & h_{1,j} \end{pmatrix}.$$

b) $L = \Delta_{Z_e^T, Z_f}, \, ef \neq 1$,

$$(1 - ef)M = \sum_{j=1}^{l}\sum_{i=0}^{n-1}((Z_e^i)^T \mathbf{g}_j)(\mathbf{h}_j^T Z_f^i)$$

$$= \sum_{j=1}^{l} \begin{pmatrix} g_{1,j} & \iddots & g_{n-1,j} & g_{n,j} \\ \iddots & \iddots & g_{n,j} & eg_{1,j} \\ g_{n-1,j} & \iddots & \iddots & \iddots \\ g_{n,j} & eg_{1,j} & \iddots & eg_{n-1,j} \end{pmatrix} \begin{pmatrix} h_{1,j} & \iddots & h_{n-1,j} & h_{n,j} \\ \iddots & \iddots & h_{n,j} & fh_{1,j} \\ h_{n-1,j} & \iddots & \iddots & \iddots \\ h_{n,j} & fh_{1,j} & \iddots & fh_{n-1,j} \end{pmatrix}.$$

Example 4.4.2. Toeplitz-like matrix M, Sylvester type operators.

a) $L = \nabla_{Z_e, Z_f}$, $e \neq f$, $e \neq 0$. By combining Theorems 1.3.1 and 2.1.3, we obtain that $\Delta_{Z_e^{-1}, Z_f} = Z_e^{-1} \nabla_{Z_e, Z_f}$, $Z_e^{-1} = Z_{1/e}^T$, and, therefore, $\Delta_{Z_{1/e}^T, Z_f}(M) = Z_{1/e}^T \nabla_{Z_e, Z_f}(M) = Z_{1/e}^T G H^T$. Combine the latter expression with Example 4.4.1 b) and deduce that

$$(1 - f/e)M = \sum_{j=1}^{l} \sum_{i=0}^{n-1} ((Z_{1/e}^{i+1})^T \mathbf{g}_j)(\mathbf{h}_j^T Z_f^i)$$

$$= \sum_{j=1}^{l} \begin{pmatrix} g_{2,j} & \iddots & g_{n,j} & e^{-1}g_{1,j} \\ \iddots & \iddots & e^{-1}g_{1,j}e^{-1}g_{2,j} \\ g_{n,j} & \iddots & \iddots & \iddots \\ e^{-1}g_{1,j}e^{-1}g_{2,j} & \iddots & e^{-1}g_{n,j} \end{pmatrix} \begin{pmatrix} h_{1,j} & \iddots & h_{n-1,j} & h_{n,j} \\ \iddots & \iddots & h_{n,j} & fh_{1,j} \\ h_{n-1,j} & \iddots & \iddots & \iddots \\ h_{n,j} & fh_{1,j} & \iddots & fh_{n-1,j} \end{pmatrix}.$$

Multiply these equations by e, insert $J^2 = I$ between the factor-matrices in each of l terms, and deduce that $(e - f)M = \sum_{j=1}^{l} Z_e(\mathbf{g}_j) Z_f(J\mathbf{h}_j)$ (even for $e = 0$, $f \neq 0$).

b) $L = \nabla_{Z_e^T, Z_f^T}$, $e \neq f$. In this case, the desired bilinear expressions for the matrix M are deduced immediately from part a) combined with the next equations, implied by Theorem 1.5.2,

$$(\nabla_{A,B}(M))^T = -\nabla_{B^T, A^T}(M^T).$$

Example 4.4.3. Hankel-like matrix M, Stein type operators.

a) $L = \Delta_{Z_e, Z_f}$, $ef \neq 1$. By applying Corollary 4.3.7, we obtain that

$$(1 - ef)M = \sum_{i=0}^{n-1} Z_e^i G H^T Z_f^i = \sum_{j=1}^{l} \sum_{i=0}^{n-1} (Z_e^i \mathbf{g}_j)(\mathbf{h}_j^T Z_f^i)$$

$$= \sum_{j=1}^{l} \begin{pmatrix} g_{1,j} & eg_{n,j} & \ddots & eg_{2,j} \\ g_{2,j} & g_{1,j} & \ddots & \ddots \\ \ddots & \ddots & \ddots & eg_{n,j} \\ g_{n,j} & \ddots & g_{2,j} & g_{1,j} \end{pmatrix} \begin{pmatrix} h_{1,j} & \iddots & h_{n-1,j} & h_{n,j} \\ \iddots & \iddots & h_{n,j} & fh_{1,j} \\ h_{n-1,j} & \iddots & \iddots & \iddots \\ h_{n,j} & fh_{1,j} & \iddots & fh_{n-1,j} \end{pmatrix}.$$

Apply the equation $(\Delta_{A,B}(M))^T = \Delta_{B^T, A^T}(M^T)$ of Theorem 1.5.2 to extend the latter expression to the case where

b) $L = \Delta_{Z_e^T, Z_f^T}$, $ef \neq 1$.

Example 4.4.4. Hankel-like matrix M, Sylvester type operators.

a) $L = \nabla_{Z_e, Z_f^T}, e \neq f$;   b) $L = \nabla_{Z_e^T, Z_f}, e \neq f$.

Along the line of Example 4.4.2 a), we could have reduced this case to that of the Stein type operators but we use another reduction — to the Toeplitz-like case of Example 4.4.2. By combining the simple equations

$$J^2 = I, \quad JZ_eJ = Z_e^T$$

of Theorems 2.1.3 and 2.1.4, we obtain the following implications:

Theorem 4.4.5.

a) $\nabla_{Z_e, Z_f^T}(M) = \nabla_{Z_e, Z_f}(MJ)J$,   b) $\nabla_{Z_e^T, Z_f}(M) = J\nabla_{Z_e, Z_f}(JM)$.

Now consider case a) of Example 4.4.4, where $L = \nabla_{Z_e, Z_f^T}, e \neq f$. Apply Theorem 4.4.5 a) and obtain that $L(M) = \nabla_{Z_e, Z_f}(MJ)J$. Recall Example 4.4.2 a) and obtain that

$$(e - f)M = \sum_{j=1}^{l} Z_e(\mathbf{g}_j)Z_f(\mathbf{h}_j)J$$

$$= \sum_{j=1}^{l} \begin{pmatrix} g_{1,j} & eg_{n,j} & \cdots & eg_{2,j} \\ g_{2,j} & g_{1,j} & \ddots & \ddots \\ \vdots & \ddots & \ddots & eg_{n,j} \\ g_{n,j} & \cdots & g_{2,j} & g_{1,j} \end{pmatrix} \begin{pmatrix} fh_{2,j} & \cdots & fh_{n,j} & h_{1,j} \\ \cdots & \cdots & h_{1,j} & h_{2,j} \\ fh_{n,j} & \cdots & \cdots & \vdots \\ h_{1,j} & h_{2,j} & \cdots & h_{n,j} \end{pmatrix}.$$

Alternatively, we could have applied Theorem 4.4.5 b) and reduced the problem to Example 4.4.2 b).

Now consider case b) $L = \nabla_{Z_e^T, Z_f}, e \neq f$. By combining Theorem 4.4.5 b) and Example 4.4.2 a), we obtain that $(e - f)M = J\sum_{j=1}^{l} Z_e(J\mathbf{g}_j)Z_f(J\mathbf{h}_j)$. $\quad\square$

Example 4.4.6. Vandermonde-like matrix M.

a) $L = \Delta_{D(\mathbf{t}), Z_f}, t_i^n f \neq 1$ for all i. By applying Corollary 4.3.7, we obtain that

$$M = (I - fD^n(\mathbf{t}))^{-1} \sum_{j=1}^{l} \sum_{i=0}^{n-1} (D^i(\mathbf{t})\mathbf{g}_j)(\mathbf{h}_j^T Z_f^i)$$

$$= (I - fD^n(\mathbf{t}))^{-1} \sum_{j=1}^{l} D(\mathbf{g}_j)V(\mathbf{t})JZ_f(J\mathbf{h}_j)$$

$$= \sum_{j=1}^{l} \begin{pmatrix} \frac{g_{1,j}}{1-t_1^n f} & & & \\ & \frac{g_{2,j}}{1-t_2^n f} & & \\ & & \ddots & \\ & & & \frac{g_{m,j}}{1-t_m^n f} \end{pmatrix} \begin{pmatrix} 1 & t_1 & \cdots & t_1^{n-1} \\ 1 & t_2 & \cdots & t_2^{n-1} \\ \vdots & \vdots & & \vdots \\ 1 & t_m & \cdots & t_m^{n-1} \end{pmatrix} \begin{pmatrix} h_{1,j} & \cdot^{\cdot^{\cdot}} & h_{n-1,j} & h_{n,j} \\ \cdot^{\cdot^{\cdot}} & \cdot^{\cdot^{\cdot}} & h_{n,j} & fh_{1,j} \\ h_{n-1,j} & \cdot^{\cdot^{\cdot}} & \cdot^{\cdot^{\cdot}} & \cdot^{\cdot^{\cdot}} \\ h_{n,j} & fh_{1,j} & \cdot^{\cdot^{\cdot}} & fh_{n-1,j} \end{pmatrix}.$$

b) $L = \Delta_{D(\mathbf{t}),Z_f^T}$, $t_i^n f \neq 1$ for all i. By applying Corollary 4.3.7, obtain that

$$M = (I - fD^n(\mathbf{t}))^{-1} \sum_{j=1}^{l} \sum_{i=0}^{n-1} (D^i(\mathbf{t})\mathbf{g}_j)(Z_f^i \mathbf{h}_j)^T$$

$$= (I - fD^n(\mathbf{t}))^{-1} \sum_{j=1}^{l} D(\mathbf{g}_j)V(\mathbf{t})(Z_f(\mathbf{h}_j))^T$$

$$= \sum_{j=1}^{l} \begin{pmatrix} \frac{g_{1,j}}{1-t_1^n f} & & & \\ & \frac{g_{2,j}}{1-t_2^n f} & & \\ & & \ddots & \\ & & & \frac{g_{m,j}}{1-t_m^n f} \end{pmatrix} \begin{pmatrix} 1 & t_1 & \cdots & t_1^{n-1} \\ 1 & t_2 & \cdots & t_2^{n-1} \\ \vdots & \vdots & & \vdots \\ 1 & t_m & \cdots & t_m^{n-1} \end{pmatrix} \begin{pmatrix} h_{1,j} & h_{2,j} & \cdot^{\cdot^{\cdot}} & h_{n,j} \\ fh_{n,j} & h_{1,j} & \cdot^{\cdot^{\cdot}} & \cdot^{\cdot^{\cdot}} \\ \cdot^{\cdot^{\cdot}} & \cdot^{\cdot^{\cdot}} & \cdot^{\cdot^{\cdot}} & h_{2,j} \\ fh_{2,j} & \cdot^{\cdot^{\cdot}} & fh_{n,j} & h_{1,j} \end{pmatrix}.$$

c) $L = \nabla_{D^{-1}(\mathbf{t}),Z_f}$, $t_i \neq 0$, $t_i^n f \neq 1$ for all i. By Theorem 1.3.1, we have $\Delta_{D(\mathbf{t}),Z_f}(M) = D(\mathbf{t})GH^T$. We recall Example 4.4.6 a) and obtain that

$$M = (I - fD^n(\mathbf{t}))^{-1}D(\mathbf{t}) \sum_{j=1}^{l} \sum_{i=0}^{n-1} (D^i(\mathbf{t})\mathbf{g}_j)(\mathbf{h}_j^T Z_f^i)$$

$$= \sum_{j=1}^{l} \begin{pmatrix} \frac{g_{1,j}t_1}{1-t_1^n f} & & & \\ & \frac{g_{2,j}t_2}{1-t_2^n f} & & \\ & & \ddots & \\ & & & \frac{g_{m,j}t_m}{1-t_m^n f} \end{pmatrix} \begin{pmatrix} 1 & t_1 & \cdots & t_1^{n-1} \\ 1 & t_2 & \cdots & t_2^{n-1} \\ \vdots & \vdots & & \vdots \\ 1 & t_m & \cdots & t_m^{n-1} \end{pmatrix} \begin{pmatrix} h_{1,j} & \cdot^{\cdot^{\cdot}} & h_{n-1,j} & h_{n,j} \\ \cdot^{\cdot^{\cdot}} & \cdot^{\cdot^{\cdot}} & h_{n,j} & fh_{1,j} \\ h_{n-1,j} & \cdot^{\cdot^{\cdot}} & \cdot^{\cdot^{\cdot}} & \cdot^{\cdot^{\cdot}} \\ h_{n,j} & fh_{1,j} & \cdot^{\cdot^{\cdot}} & fh_{n-1,j} \end{pmatrix}.$$

d) $L = \nabla_{D(\mathbf{t}),Z_f^T}$, $t_i^n \neq f$, $t_i \neq 0$ for all i. By Theorem 1.3.1, we have

$\triangle_{D^{-1}(\mathbf{t}),Z_f^T}(M) = D^{-1}(\mathbf{t})GH^T$. We recall Example 4.4.6 b) and obtain that

$$M = (I - fD^{-n}(\mathbf{t}))^{-1} \sum_{j=1}^{l} \sum_{i=0}^{n-1} (D^{-i-1}(\mathbf{t})\mathbf{g}_j)(Z_f^i \mathbf{h}_j)^T$$

$$= \sum_{j=1}^{l} \begin{pmatrix} \frac{g_{1,j}t_1^{n-1}}{t_1^n-f} & & & \\ & \frac{g_{2,j}t_2^{n-1}}{t_2^n-f} & & \\ & & \ddots & \\ & & & \frac{g_{m,j}t_m^{n-1}}{t_m^n-f} \end{pmatrix} \begin{pmatrix} 1 & t_1^{-1} & \cdots & t_1^{1-n} \\ 1 & t_2^{-1} & \cdots & t_2^{1-n} \\ \vdots & \vdots & & \vdots \\ 1 & t_m^{-1} & \cdots & t_m^{1-n} \end{pmatrix} \begin{pmatrix} h_{1,j} & h_{2,j} & \cdots & h_{n,j} \\ fh_{n,j} & h_{1,j} & \ddots & \ddots \\ \ddots & \ddots & \ddots & h_{2,j} \\ fh_{2,j} & \ddots & fh_{n,j} & h_{1,j} \end{pmatrix}.$$

e) $L = \nabla_{D(\mathbf{t}),Z_{1/f}^T}$, $f \neq 0$, $t_i^n f \neq 1$ for all i. Recall that $(Z_{1/f}^T)^{-1} = Z_f$, apply Theorem 1.3.1, and obtain that $\triangle_{D(\mathbf{t}),Z_f}(M) = -\nabla_{D(\mathbf{t}),Z_{1/f}^T}(M)Z_f = -GH^T Z_f$. Recall Example 4.4.6 a) and obtain that

$$M = (fD^n(\mathbf{t}) - I)^{-1} \sum_{j=1}^{l} \sum_{i=0}^{n-1} D^i(\mathbf{g}_j)V(\mathbf{t})(\mathbf{h}_j^T Z_f^{i+1})$$

$$= \sum_{j=1}^{l} \begin{pmatrix} \frac{g_{1,j}}{1-t_1^n f} & & & \\ & \frac{g_{2,j}}{1-t_2^n f} & & \\ & & \ddots & \\ & & & \frac{g_{m,j}}{1-t_m^n f} \end{pmatrix} \begin{pmatrix} 1 & t_1 & \cdots & t_1^{n-1} \\ 1 & t_2 & \cdots & t_2^{n-1} \\ \vdots & \vdots & & \vdots \\ 1 & t_m & \cdots & t_m^{n-1} \end{pmatrix} \begin{pmatrix} h_{2,j} & \cdots & h_{n,j} & fh_{1,j} \\ \cdots & \cdots & \cdots & fh_{1,j} & fh_{2,j} \\ h_{n,j} & \cdots & \cdots & \cdots \\ fh_{1,j} & fh_{2,j} & \cdots & fh_{n,j} \end{pmatrix}.$$

For $t_i f \neq 0$, $i = 1, \ldots, n$, parts d) and e) give us two distinct expressions for the same matrix M.

By combining parts a)–e) with Theorem 1.5.2, we immediately obtain bilinear expressions for a matrix M via the generators of its displacements $\triangle_{Z_f^T,D(\mathbf{t})}$, $\triangle_{Z_f,D(\mathbf{t})}$, $\nabla_{Z_f^T,D^{-1}(\mathbf{t})}$, and $\nabla_{Z_f,D(\mathbf{t})}$. $\square$

Our next example complements Example 1.4.1.

Example 4.4.7. (See Theorem 4.3.5.) $m \times n$ **Cauchy-like matrix** M. $L = \triangle_{D(\mathbf{s}),D(\mathbf{t})}$, $s_i t_j \neq 1$ for all pairs i, j. By inspection, we easily verify that

$$M = \sum_{j=1}^{l} \begin{pmatrix} g_{1,j} & & & \\ & g_{2,j} & & \\ & & \ddots & \\ & & & g_{m,j} \end{pmatrix} \begin{pmatrix} \frac{1}{1-s_1 t_1} & \frac{1}{1-s_1 t_2} & \cdots & \frac{1}{1-s_1 t_n} \\ \frac{1}{1-s_2 t_1} & \frac{1}{1-s_2 t_2} & \cdots & \frac{1}{1-s_2 t_n} \\ \vdots & \vdots & & \vdots \\ \frac{1}{1-s_m t_1} & \frac{1}{1-s_m t_2} & \cdots & \frac{1}{1-s_m t_n} \end{pmatrix} \begin{pmatrix} h_{1,j} & & & \\ & h_{2,j} & & \\ & & \ddots & \\ & & & h_{n,j} \end{pmatrix}.$$

If $s_i \neq 0$ for all i and/or $t_j \neq 0$ for all j, the latter expression can be also deduced from Theorem 1.3.1 and Example 1.4.1.

Example 4.4.8. Toeplitz+Hankel-like matrix M.

a) $L = \nabla_{Y_{00}, Y_{11}}$. Assume computations over the field of real numbers. Combine Theorems 3.11.6 and 4.3.5 for $A = Y_{00}$, $P = S$, $D(\mathbf{s}) = D_S$, $B = Y_{11}$, $Q = C$, $D(\mathbf{t}) = D_C$, and obtain that

$$M = S \left(\frac{SGH^T C}{s_i - t_j} \right)^{n-1}_{i,j=0} C^T,$$

where $\mathbf{s} = (s_i)_{i=0}^{n-1}$, $\mathbf{t} = (t_j)_{j=0}^{n-1}$. Note that $SMC = \left(\dfrac{SGH^T C}{s_i - t_j} \right)^{n-1}_{i,j=0}$ is a Cauchy-like matrix of displacement rank l. (In particular, $l \le 4$ if $M = T + H$ where $T = (t_{i-j})_{i,j}$ is a Toeplitz matrix and $H = (h_{i+j})_{i,j}$ is a Hankel matrix (see Example 4.2.1).) By combining the latter equation with the expression in Example 1.4.1 for Cauchy-like matrices, we obtain a bilinear expression of the matrix M via the columns of the matrices G and H (see Exercise 4.5).

b) $L = \nabla_{Y_{11}, Y_{00}}$. By applying Theorem 1.5.2, we immediately reduce this case to part a) because the matrices Y_{00} and Y_{11} are real symmetric.

Example 4.4.9. Confluent matrix M. Sylvester type operators. [PWa]. $L = \nabla_{A,B}$, $A = \lambda I_m + Z_e$, $B = \mu I_n + Z_f$. The latter operators cover the confluent matrices involved in the tangential confluent Nevanlinna–Pick Problem 3.8.3 of rational interpolation. Observe immediately that $\nabla_{A,B} = \nabla_{(\lambda-\mu)I_m + Z_e, Z_f} = \nabla_{Z_e, (\mu-\lambda)I_n + Z_f}$.

We consider separately several cases depending on the values e, f, and $\lambda - \mu$.

1. $e = f = 0$.

   (a) For $\lambda \ne \mu$, we deduce from the equation

   $$\nabla_{A,B}(M) = ((\lambda - \mu)I + Z)M - MZ = GH^T$$

   that

   $$M = \sum_{k=1}^{l} Z(\mathbf{g}_k) \Theta_0 (\lambda - \mu) Z^T (J \mathbf{h}_k) J,$$

   where $\Theta_0(s) = \left(\dfrac{(-1)^{i-1}(i+j-2)!}{(i-1)! s^{i+j-1}(j-1)!} \right)_{\substack{1 \le i \le n \\ 1 \le j \le n}}$ is an $n \times n$ matrix.

   $\Theta_0(s) = \text{diag} \left(\dfrac{(-1)^{i-1}}{(i-1)!} \right)_{1 \le i \le n} H \, \text{diag} \left(\dfrac{1}{(j-1)!} \right)_{1 \le j \le n}$,

   $H = \left(\dfrac{(i+j-2)!}{s^{i+j-1}} \right)_{\substack{1 \le i \le n \\ 1 \le j \le n}}$ is an $n \times n$ Hankel matrix.

   (b) If $\lambda = \mu$, we arrive at the operator ∇_{Z_e, Z_f} of part (a) of Example 4.4.2.

2. $e \ne 0$, $f = 0$.

   (a) For $(\mu - \lambda)^n \ne e$, we apply the equation

   $$\nabla_{A,B}(M) = ((\lambda - \mu)I + Z_e)M - MZ = GH^T$$

to deduce that

$$M = \sum_{k=1}^{l} Z_e(\mathbf{g}_k)\Theta_1(\lambda - \mu)Z^T(J\mathbf{h}_k)J,$$

where $\Theta_1(s) = V^{-1}(\mathbf{t})\left(\left(\frac{1}{s+t_i}\right)^j\right)_{\substack{1 \le i \le n \\ 1 \le j \le n}}$, $\mathbf{t} = (t_i)_{1 \le i \le n}$, and $t_1, \ldots, t_n$ are all the

n-th roots of e. We have $\Theta_1(s) = \frac{1}{n}V^T(\mathbf{t}^{-1})V((\mathbf{t}+s\mathbf{1})^{-1})D(\mathbf{t}+s\mathbf{1})^{-1}$.

(b) If $(\mu - \lambda)^n = e$, then the operator L is singular. In this case, we rely on the equation

$$((\lambda - \mu)I + Z)M - MZ = \nabla_{A,B}(M) + (Z - Z_e)M = GH^T - ee_0e_{n-1}^T M$$

and, similarly to part 1, deduce that

$$M = \sum_{k=1}^{l} Z(\mathbf{g}_k)\Theta_0(\lambda - \mu)Z^T(J\mathbf{h}_k)J - e\Theta_0(\lambda - \mu)Z^T(JM^T\mathbf{e}_{n-1})J.$$

The matrix M is expressed via its L-generator and last row $\mathbf{e}_{n-1}^T M$.

3. $e = 0$, $f \ne 0$.

  (a) If $(\lambda - \mu)^n \ne f$, then we deduce from the equation

$$\nabla_{A,B}(M) = ZM - M((\mu - \lambda)I_n + Z_f) = GH^T$$

that

$$M = \sum_{k=1}^{l} Z(\mathbf{g}_k)\Theta_2(\mu - \lambda)Z_{1/f}^T(\mathbf{h}_k),$$

where $\Theta_2(s) = -\frac{1}{n}\left(\left(\frac{1}{s+t_j}\right)^i\right)_{\substack{1 \le i \le n \\ 1 \le j \le n}} V(\mathbf{t})$, $\Theta_2(s) = -\frac{1}{n}D^{-1}(\mathbf{t}+s\mathbf{1})V^T((\mathbf{t}+$

$s\mathbf{1})^{-1})V(\mathbf{t})$, $\mathbf{t} = (t_i)_{1 \le i \le n}$, and $t_1, \ldots, t_n$ are all the n-th roots of f.

(b) If $(\lambda - \mu)^n = f$, then the operator L is singular. In this case, by relying on the equation

$$((\lambda - \mu)I + Z)M - MZ = \nabla_{A,B}(M) + M(Z_f - Z) = GH^T + eM\mathbf{e}_0\mathbf{e}_{n-1}^T,$$

we deduce similar to part 1 that

$$M = \sum_{k=1}^{l} Z(\mathbf{g}_k)\Theta_0(\lambda - \mu)Z^T(J\mathbf{h}_k)J + fZ(M\mathbf{e}_0)\Theta_0(\lambda - \mu)J.$$

The matrix M is expressed via its L-generator and first column, $M\mathbf{e}_0$.

4. $ef \ne 0$.

(a) If the operator L is non-singular, then both matrices $I - f((\lambda - \mu)I + Z_e)^n$ and $I - e((\mu - \lambda)I + Z_f)^n$ are non-singular, and we obtain that

$$M = \left(\sum_{k=1}^{l} J Z_{1/e}(J Z_e^{-1} \mathbf{g}_k) \Theta_3(\mu - \lambda) Z_f^T(J\mathbf{h}_k) J \right) (I - e((\mu - \lambda)I + Z_f)^n)^{-1}$$

where $\Theta_3(s) = \left(\frac{(i-1)! s^{i-j}}{(j-1)!(i-j)!} \right)_{\substack{1 \le i \le n \\ 1 \le j \le i}}$ is an $n \times n$ lower triangular matrix.

$\Theta_3(s) = \operatorname{diag}((i-1)!)_{1 \le i \le n} \left(\frac{s^{i-j}}{(i-j)!} \right)_{\substack{1 \le i \le n \\ 1 \le j \le i}} \operatorname{diag}\left(\frac{1}{(j-1)!} \right)_{1 \le j \le n}$. We also obtain that

$$M = (I - f((\lambda - \mu)I + Z_e)^n)^{-1} \left(\sum_{k=1}^{l} Z_e(\mathbf{g}_k) \Theta_4(\lambda - \mu) Z_{1/f}(Z_{1/f}\mathbf{h}_k) \right),$$

where $\Theta_4(s) = -\left(\frac{(j-1)! s^{j-i}}{(i-1)!(j-i)!} \right)_{\substack{1 \le i \le j \\ 1 \le j \le n}}$ is an $n \times n$ upper triangular matrix,

$\Theta_4(s) = -\operatorname{diag}\left(\frac{1}{(i-1)!} \right)_{1 \le i \le n} \left(\frac{s^{j-i}}{(j-i)!} \right)_{\substack{1 \le i \le j \\ 1 \le j \le n}} \operatorname{diag}((j-1)!)_{1 \le j \le n}$.

(b) If the operator L is singular, then for any 4-tuple (λ, μ, e, f), we apply the equations

$$(\lambda I + Z_e)M - M(\mu I + Z) = \nabla_{A,B}(M) + M(Z_f - Z) = GH^T + f M e_0 e_{n-1}^T$$

and

$$(\lambda I + Z)M - M(\mu I + Z_f) = \nabla_{A,B}(M) + (Z - Z_e)M = GH^T - e e_0 e_{n-1}^T M$$

and, as in parts 2 and 3, deduce that

$$M = \sum_{k=1}^{l} Z_e(\mathbf{g}_k) \Theta_1(\lambda - \mu) Z^T(J\mathbf{h}_k) J + f Z_e(M e_1) \Theta_1(\lambda - \mu) J$$

and

$$M = \sum_{k=1}^{l} Z(\mathbf{g}_k) \Theta_2(\mu - \lambda) Z_{1/f}^T(\mathbf{h}_k) - e \Theta_2(\mu - \lambda) Z_{1/f}^T(J M^T e_{n-1}) J.$$

Then again, the matrix M is expressed via its L-generator and either the first column, $M e_0$ or the last row, $e_{n-1}^T M$. $\square$

Remark 4.4.10.

a) Assume that $ef \ne 0$ and write $V_e = \left(s_i^{j-1} \right)_{\substack{1 \le i \le m \\ 1 \le j \le m}}$, $V_f = \left(t_i^{j-1} \right)_{\substack{1 \le i \le n \\ 1 \le j \le n}}$, where $s_1, \ldots, s_m$ are all the m-th roots of e, and $t_1, \ldots, t_n$ are all the n-th roots of f, so that the Vandermonde matrices V_e and V_f are non-singular. Now, based on Theorem 2.6.4, we obtain the following equations:

$$(I_m - f((\lambda - \mu)I_m + Z_e)^{-n})^{-1} = V_e^{-1} \operatorname{diag}\left(\frac{(\lambda - \mu + s_i)^n}{(\lambda - \mu + s_i)^n - f}\right)_{1 \le i \le m} V_e,$$

$$(I_n - e((\mu - \lambda)I_n + Z_f)^{-m})^{-1} = V_f^{-1} \operatorname{diag}\left(\frac{(\mu - \lambda + t_i)^m}{(\mu - \lambda + t_i)^m - e}\right)_{1 \le i \le n} V_f.$$

We may substitute these expressions (for $m = n$) in part 4 a) of Example 4.4.9.

b) By virtue of Theorem 2.1.3 and 2.1.4, we have

$$\nabla_{\lambda I + Z_e, \mu I + Z_f}(MJ) = \nabla_{\lambda I + Z_e, \mu I + Z_f^T}(M)J = G(JH)^T,$$

$$\nabla_{\lambda I + Z_e, \mu I + Z_f}(JM) = J\nabla_{\lambda I + Z_e^T, \mu I + Z_f}(M) = (JG)H^T,$$

$$\nabla_{\lambda I + Z_e, \mu I + Z_f}(JMJ) = J\nabla_{\lambda I + Z_e^T, \mu I + Z_f^T}(M)J = (JG)(JH)^T.$$

Based on the latter equations, we extend the results of Example 4.4.9 to express structured matrices M via their displacements $L(M)$ for the Sylvester type operators $L = \nabla_{\lambda I + Z_e, \mu I + Z_f^T}$, $L = \nabla_{\lambda I + Z_e^T, \mu I + Z_f}$, and $L = \nabla_{\lambda I + Z_e^T, \mu I + Z_f^T}$.

c) For a structured matrix M associated with a Sylvester type operator $L = \nabla_{A,B}$, suppose we know the Jordan blocks of the operator matrices A and B,

$$PAP^{-1} = \operatorname{diag}(\lambda_i I_{m_i} + Z)_{1 \le i \le p}, \quad QBQ^{-1} = \operatorname{diag}(\mu_j I_{n_j} + Z)_{1 \le j \le q}.$$

For $P = I$, $Q = I$, this covers the matrices associated with the tangential confluent Nevanlinna–Pick Problem 3.8.3. In this case, we recover the matrix M from its displacement $L(M) = GH^T$ by applying the following steps.

(i) Write $A_i = \lambda_i I_{m_i} + Z, i = 1, \ldots, p; B_j = \mu_j I_{n_j} + Z, j = 1, \ldots, q.$

(ii) Represent the matrix PMQ^{-1} as a $p \times q$ block matrix with blocks $M_{i,j}$ of size $m_i \times n_j$; represent the matrix PG as a $p \times 1$ block matrix with blocks G_i of size $m_i \times l$; represent the matrix $H^T Q^{-1}$ as a $1 \times q$ block matrix with blocks H_j^T of size $l \times n_j$.

(iii) Replace the matrix equation $\nabla_{A,B}(M) = GH^T$ by the set of block equations

$$\nabla_{A_i, B_j}(M_{i,j}) = G_i H_j^T$$

for all pairs (i, j), $i = 1, \ldots, p; j = 1, \ldots, q.$

(iv) Recover the blocks $M_{i,j}$ from their displacement generators (G_i, H_j) as in Example 4.4.9.

(v) Finally, compute the matrix $M = P^{-1}(M_{i,j})_{\substack{1 \le i \le p \\ 1 \le j \le q}} Q.$

Remark 4.4.11. Theorem 1.5.3 expresses the displacements of the matrix M^{-1} via the displacements of a non-singular matrix M, that is, via $\nabla_{A,B}(M)$ and $\Delta_{A,B}(M)$.

By combining these expressions with Examples 4.4.1–4.4.9, we obtain l-term bilinear expressions of the matrix M^{-1} via the l pairs of the vectors $M^{-1}\mathbf{g}_j$ and $\mathbf{h}_j^T M^{-1}$, $j = 1, \ldots, l$, where the vectors $\mathbf{g}_j$ and $\mathbf{h}_j$ generate the displacement of M (see (1.4.2), (1.4.3)). We showed such expressions for the basic classes of structured matrices at the end of Section 4.2.

Remark 4.4.12. The above examples, together with the estimates of Table 1.2/4.2 and the theorems of Sections 1.5 and 4.3, support the displacement rank approach (COMPRESS, OPERATE, DECOMPRESS) to computations with structured matrices. In particular, these theorems reduce various operations with structured matrices to their multiplication by vectors. The bilinear expressions in Examples 1.4.1 and 4.4.1–4.4.9 reduce the latter operation to multiplication of $O(l)$ basic structured matrices of Table 1.2/4.2 by $O(l)$ vectors, which can be done in nearly linear time $O(nl \log^d n)$, $d \leq 2$ (see more details in Section 4.7). Moreover, we may now extend the displacement rank approach and the superfast algorithms to the computations with other classes of structured matrices, such as the L-structured matrices for $L = \nabla_{Z_e^T, -Z_f}$ (see a sample application in Example 5.5.1 in Section 5.5).

4.5   Partly regular displacement operators

Some studies of structured matrices involve operators L that are singular but still allow simple recovery of a structured matrix M from only a small number of its entries and its displacement $L(M)$. We encounter such operators in parts 2(b), 3(b), and 4(b) of Example 4.4.9.

Generally, let a singular operator L be defined on the space $\mathbb{F}^{m \times n}$ of $m \times n$ matrices. In this space define the *null space (kernel)* of the operator L,

$$N(L) = \{M : L(M) = 0\},$$

and its orthogonal complement $C(L)$. Then every matrix M can be represented uniquely as the sum

$$M = M_N + M_C, \quad M_N \in N(L), \quad M_C \in C(L),$$

where the operator L is non-singular on the space $C(L)$ and every matrix M_C can be uniquely decompressed from the displacement, $L(M) = L(M_C)$. We deal only with those singular operators L for which every $n \times n$ matrix M can be linearly expressed via its displacement $L(M)$ and a fixed *irregularity set* $M_{I.S.}(L)$ of its entries. The positions of these entries in the matrix (such as the diagonal or the first or last column or row) are assumed to be fixed for the operator L and invariant in matrices M; furthermore, the cardinality of the irregularity set is assumed to be small relative to n^2. In these cases we say that the operator L is *partly regular* and defines the *L-generator* (or the *displacement generator*) of length l for a matrix M as the matrix pair (G, H) in (1.4.2) combined with the entry set $M_{I.S.}(L)$. In particular, such are the operators

$$L = \Delta_{Z_f, Z_{1/f}^T} \text{ for } f \neq 0, \quad L = \nabla_{Z_f, Z_f}, \text{ and } L = \nabla_{Z_f^T, Z_f^T},$$

whose null spaces are given by the classes of f-circulant matrices or their transposes, i.e., $L(M) = 0$ if and only if $M = Z_f(\mathbf{v})$ or $M = Z_f^T(\mathbf{v})$, for some vector $\mathbf{v}$. Here are some (bi)linear recovery formulae:

$$M = Z_f(M\mathbf{e}_{n-1}) + \frac{1}{1 - ef} \sum_{j=1}^{l} Z_f(\mathbf{g}_j) Z_e^T(\mathbf{h}_j) \tag{4.5.1}$$

where $ef \neq 1$, $\Delta_{Z_f, Z_{1/f}^T}(M) = GH^T$ for G, H in (1.4.2) (in this case $M_{I.S.}(L) = M\mathbf{e}_{n-1}$), and

$$M = Z(M\mathbf{e}_0) - \sum_{j=1}^{l} Z(\mathbf{g}_j) Z^T(Z\mathbf{h}_j) \tag{4.5.2}$$

where $\nabla_{Z,Z}(M) = GH^T$ for G, H in (1.4.2), (in this case $M_{I.S.}(L) = M\mathbf{e}_0$).

The next theorem shows similar expressions for the recovery of a matrix M from its first (or last) column $M\mathbf{e}_0$ (or $M\mathbf{e}_{n-1}$) or row $M^T\mathbf{e}_0$ (or $M^T\mathbf{e}_{n-1}$) and the displacement $L(M)$ for the partly regular operator $L = \Delta_{Z+Z^T, Z+Z^T}$. (The operator L defines the class of Toeplitz+Hankel-like matrices M such that rank$(L(M))$ is small.) That is, either of the vectors $M\mathbf{e}_0$, $M\mathbf{e}_{n-1}$, $M\mathbf{e}_0^T$, and $M\mathbf{e}_{n-1}^T$ can be selected to define an irregularity set $M_{I.S.}(L)$.

Theorem 4.5.1. *Define polynomials $Q(x) = 1$, $Q_1(x) = x$, $Q_{j+1}(x) = xQ_j(x) - Q_{j-1}(x)$, $j = 1, 2, \ldots, n-1$. Let $\tau(\mathbf{v})$ be the matrix of the algebra $\tau = A(Y_{00})$ generated by the matrix Y_{00} and having the vector $\mathbf{v}$ as its first column. Let $\nabla_{Y_{00}, Y_{00}}(M) = GH^T$ for the matrix $Y_{00} = Z^T + Z$ and matrices G, H in (1.4.3). Then we have simultaneously*

$$M = \tau(M\mathbf{e}_0) - \sum_{j=1}^{l} \tau(\mathbf{g}_j) Z^T(Z\mathbf{h}_j),$$

$$M = \tau(JM\mathbf{e}_{n-1}) - \sum_{j=1}^{l} \tau(J\mathbf{g}_j) Z(ZJ\mathbf{h}_j),$$

$$M = \tau(M^T\mathbf{e}_0) + \sum_{j=1}^{l} Z(Z\mathbf{g}_j)\tau(\mathbf{h}_j),$$

$$M = \tau(JM^T\mathbf{e}_{n-1}) + \sum_{j=1}^{l} Z^T(ZJ\mathbf{g}_j)\tau(Z\mathbf{h}_j).$$

Another example is given by the generalized Pick matrices associated with the tangential boundary Nevanlinna–Pick Problem 3.8.5 of rational interpolation. These matrices fall into the class of Cauchy-like matrices in Example 1.4.1 except that their first ρ diagonal entries are defined separately. The associated operator $\nabla_{D(s), D(t)}$ is singular but partly regular; it becomes regular when it is restricted to the off-diagonal entries and the last $n - \rho$ diagonal entries. Except for its ρ northwestern diagonal entries, the matrix $M = (m_{i,j})_{i,j}$ is expressed via its displacement $\nabla_{D(s),D(t)}(M)$, as in Example 1.4.1. In this case $M_{I.S.}(L) = \{m_{i,i}, i = 0, 1, \ldots, \rho - 1\}$. The following (bi)linear recovery formula (compare (1.4.1)) covers the latter case:

$$M = D + \sum_{k=1}^{l} D(\mathbf{g}_k) C_R(\mathbf{s}, \mathbf{t}) D(\mathbf{h}_k) \tag{4.5.3}$$

where $C_R(\mathbf{s}, \mathbf{t}) = (c_{i,j})$ is a regularized Cauchy matrix; $c_{i,j} = 0$ if $s_i = t_j$, $c_{i,j} = \frac{1}{s_i - t_j}$ otherwise; D is a diagonal matrix, $D = \text{diag}(d_i)_{i=0}^{n-1}$; $d_i = m_{i,i}$, $i = 0, 1, \ldots, \rho - 1$, and $d_i = 0$ for $i \geq \rho$. The above formula can be generalized easily to any operator $\nabla_{D(s),D(t)}$ where $s_i = t_j$ for some pairs i and j.

The next definition summarizes the definitions of this section, generalizes the latter examples, and is followed by two further examples and two remarks.

Definition 4.5.2. A linear operator L defined on the space of $m \times n$ matrices is called *partly regular* if every $m \times n$ matrix M can be expressed linearly via its displacement $L(M)$ and an *irregularity set* $M_{I.S.}(L)$ of its entries whose positions in the matrix are fixed for the operator L; the cardinality of this set is supposed to be small relatively to mn. The triple $(G, H, M_{I.S.}(L))$ for two matrices G and H in (1.4.2) is called an *L-generator* (or a *displacement generator*) *of length l for M*. $\text{rank}(L(M))$ is called the *L-rank* or the *displacement rank* of M. A partly regular operator L is *internally regular* if the associated irregularity set $M_{I.S.}(L)$ is restricted to all or some entries of the first row and column and the last row and column; a partly regular operator L is *off-diagonally regular* if the set $M_{I.S.}(L)$ lies on the matrix diagonal.

Example 4.5.3. For the matrix $A = Z_e$ and a real e, the operator Δ_{A,A^T} is regular if and only if $e \neq \pm 1$ and is internally regular for $e = \pm 1$ (see Example 4.4.1 a) and (4.5.1)).

Example 4.5.4. The operator $\Delta_{D(s),D^*(s)}$ is regular if and only if all entries of the vector $\mathbf{s}$ are non-real. Otherwise, the operator is off-diagonally regular.

Remark 4.5.5. In the study of Toeplitz-like and Hankel-like matrices, we may usually stay with regular displacement operators L such as Δ_{Z_e,Z_d^T}, $\Delta_{Z_e^T,Z_d}$, and ∇_{Z_e,Z_f} where $e \neq f$, $ed \neq 1$. Furthermore, we may regularize a singular operator via multiplicative transformations in Section 4.8. In Remark 5.4.4, however, we comment on an important algorithmic simplification which can be obtained by using the internally regular but singular operator ∇_{Z,Z^T}. Furthermore, in addition to the case of Problem 3.8.5 already cited, other applications of the off-diagonally regular operators $\nabla_{D(s),D(s)}$ and $\Delta_{D^{-1}(s),D(s)}$ are shown in [KO96]. The common null space of the latter operators is made up of the diagonal matrices.

Remark 4.5.6. In Section 5.3, we apply Theorem 1.5.6 to structured matrices associated with regular, off-diagonally regular, and internally regular displacement operators L. This defines projection of the operators L induced by the projection of the associated matrices into their blocks. Inspection immediately shows that these projections of the operators L in Examples 1.4.1, 4.4.1–4.4.9, and this section remain regular, off-diagonally regular, and/or internally regular if so is the original operator L. In fact, the projection even regularizes some internally regular operators in Example 4.4.9.

4.6   Compression of a generator

The length of displacement generators generally grows when we sum or multiply structured matrices and when we project them into their blocks (see Theorems 1.5.1, 1.5.4, and 1.5.6). A chain of operations of summation and multiplication may easily produce a rather long generator for a matrix having small displacement rank. In Chapters 5–7, we can see this phenomenon in both direct and iterative algorithms for the inversion of structured matrices. To keep the algorithms efficient, we must *compress longer generators* of matrices having smaller ranks. This is our subject in the present section.

Clearly, the choice of the matrices G and H in (1.4.2), (1.4.3) is not unique for a fixed matrix $L(M)$. For instance, we have $GH^T = G_1H_1^T$ where $G_1 = (G + R)V$, $H_1^T = V^{-1}(S^T + H^T)$, $RH^T = 0$, $(G + R)S^T = 0$, and V is an $n \times n$ nonsingular matrix. Thus we seek a non-unique generator of the minimum length, and this can be achieved in various ways. In the next two subsections, we separately study some techniques for *numerical and algebraic compression* of the generators.

4.6.1   SVD-based compression of a generator in numerical computations with finite precision

Let us next follow the papers [P92b], [P93], and [P93a] and rely on the SVD of the displacement $L(M)$.

Definition 4.6.1. SVD of a matrix and its orthogonal generators. A matrix U is *orthogonal* (or *unitary*) if $U^*U = I$ (see Definition 2.1.1/4.1.1). Every $m \times n$ matrix W has its *SVD (singular value decomposition)*:

$$W = U\Sigma V^*, \quad U \in \mathbb{C}^{m \times r}, \quad V \in \mathbb{C}^{n \times r}, \tag{4.6.1}$$

$$U^*U = V^*V = I_r, \quad \Sigma = \text{diag}(\sigma_1, \ldots, \sigma_r), \tag{4.6.2}$$

$$\sigma_1 \geq \sigma_2 \geq \ldots \geq \sigma_r > \sigma_{r+1} = \ldots = \sigma_n = 0, \tag{4.6.3}$$

where $r = \text{rank}(W)$, $\sigma_1, \ldots, \sigma_r$ are *singular values* of the matrix W, and U^* and V^* denote the Hermitian (conjugate) transposes of the orthogonal matrices U and V, respectively . The rows of the matrix U^* and the columns of the matrix V are the left and right *singular vectors* of the matrix W, respectively. The $n \times m$ matrix $W^+ = V\Sigma^{-1}U^*$ is the *Moore–Penrose generalized (pseudo) inverse* of the matrix W. $W^+ = W^{-1}$ if $m = n = r$. Based on the SVD of a matrix W, its *orthogonal generator of the minimum length* is defined by the following equations:

$$G = U\Sigma^{1/2}, \quad H = V\Sigma^{1/2}. \tag{4.6.4}$$

(In an alternative policy of diagonal scaling, we may write, say $G = U$, $H = V\Sigma$, or we may define generators by the triples (U, Σ, V).) If $W = L(M)$ for a matrix M, then G, H is an *orthogonal L-generator of the minimum length* for the matrix M. In numerical computation of the SVD with a fixed positive tolerance ϵ to the output errors caused by rounding, one truncates (that is, sets to zero) all the smallest singular values σ_i up to ϵ. This produces a *numerical orthogonal generator* (G_ϵ, H_ϵ) *of ϵ-length* r_ϵ for a matrix W, which is also a numerical orthogonal L-generator for M. Here, r_ϵ is the *ϵ-rank* of the matrix $W = L(M)$, equal to the minimum number of the singular values of the displacement $L(M)$ exceeding ϵ, $r_\epsilon \le r = \text{rank}(W)$.

In Chapter 6, we use the fact that the pair (G_ϵ, H_ϵ) forms a shortest L-generator for the matrices $\tilde{M}$ lying in the ϵ-neighborhood of $L(M)$, $\{\tilde{M}, \|L(\tilde{M}) - L(M)\| \le \epsilon\}$ under the 2-norm as well as under the Frobenius norm of matrices (see Definition 6.1.1 and Theorem 6.1.3). *The latter minimization, based on using orthogonal L-generators ensures numerical stability of the representation of the displacement matrix $L(M)$ via its generator and decreases the generator length to its numerical minimum.* These properties motivated the first study of orthogonal L-generators in the papers [P92b], [P93], and [P93a] and their application in the papers [P92b] and [P93a] to Newton's iteration for the inversion of Toeplitz and Toeplitz-like matrices, whose displacement ranks are generally tripled in each iteration step.

Shortest orthogonal L-generators can be recovered easily from short L-generators.

Theorem 4.6.2. $O(l^2 n + (r \log r) \log\log(\sigma_1/\epsilon))$ *ops are sufficient to compute within the output error bound $\epsilon > 0$ the SVD's of the matrices $G, H,$ and GH^T in (1.4.2), (1.4.3). Here the output error bound ϵ means the upper bound ϵ on all the output errors of the singular values σ_i and the entries of the computed matrices U and V in (4.6.1) and (4.6.2) for $W = L(M)$.*

Realistically, in the cost bound of Theorem 4.6.2 the first term, $l^2 n$ dominates and the second term can be ignored.

We next show simplification of orthogonal generators for an important class of matrices.

Definition 4.6.3. Eigendecomposition of a Hermitian matrix and its orthogonal symmetric generators. An $n \times n$ complex matrix W is *Hermitian* if $W = W^*$. A real Hermitian matrix is usually called real *symmetric*. A Hermitian matrix W has *eigendecomposition*

$$W = U\Lambda U^*, \quad U \in \mathbb{C}^{n \times r}, \tag{4.6.5}$$

where

$$U^* U = I_r, \quad \Lambda = \text{diag}(\lambda_1, \ldots, \lambda_r) \in \mathbb{R}^{r \times r}, \tag{4.6.6}$$

$$|\lambda_1| \ge \ldots \ge |\lambda_r| > 0, \quad \lambda_j = 0 \text{ for } j > r = \text{rank}(W), \tag{4.6.7}$$

$\lambda_1, \ldots, \lambda_r$ are real numbers called the *eigenvalues* of the matrix W, and the columns of the matrix U are the associated *eigenvectors*. The Hermitian or real symmetric matrix W in (4.6.5)–(4.6.7) is *semi-definite* or *non-negative definite* if $\lambda_1 \ge \ldots \ge \lambda_r > 0$ or equivalently if the matrix W allows its factorization of the form $W = VV^*$ for a matrix $V \in \mathbb{C}^{n \times r}$. If $W = VV^*$ for a non-singular matrix V, then W is a *positive definite* matrix. A semi-definite matrix is positive definite if $r = n$. The matrices not known to be semi-definite are called *indefinite*. For a semi-definite matrix W, its eigendecomposition (4.6.5)–(4.6.7) turns into the SVD in (4.6.1)–(4.6.3). Even in the case where $W = L(M)$ is an indefinite Hermitian or real symmetric matrix, its eigendecomposition is a natural substitution for the SVD in (4.6.1)–(4.6.3). The symmetric matrix W can be defined by the $(n + 1)r$ entries of the matrices U and Λ or, equivalently, by the r eigenvalue/eigenvector pairs of the matrix W (versus the $(2n + 1)r$ entries of the SVD of an $n \times n$ unsymmetric matrix). For a real symmetric (or Hermitian) matrix $W = L(M)$, (4.6.5) defines its *orthogonal real symmetric* (or *Hermitian*) *generator* (U, Λ), also called an *orthogonal real symmetric* (or *Hermitian*) *L-generator of the matrix M*. (Likewise, the concepts of numerical orthogonal generators and L-generators, their length minimization property in the ϵ-neighborhood of the displacement, and Theorem 4.6.2 can be extended.)

(4.6.1) implies that $W^* W = V \Sigma^2 V^*$, $W W^* = U \Sigma^2 U^*$. If W is a Hermitian (indefinite) matrix W, then $W^* W = W W^* = W^2$ is a semi-definite matrix, and equations (4.6.5)–(4.6.7) imply that

$$W^2 = U \Lambda^2 U^*, \quad \lambda_j^2 = \sigma_j, \quad j = 1, \ldots, n. \qquad (4.6.8)$$

For a real symmetric or Hermitian matrix $M = M^*$, the matrix $W = L(M)$ is real symmetric or Hermitian if $L = \nabla_{A,-A^*}$ or $L = \Delta_{A,A^*}$. Indeed,

$$(\nabla_{A,-A^*}(M))^* = (AM + MA^*)^* = AM^* + M^*A^* = AM + MA^*,$$
$$(\Delta_{A,A^*}(M))^* = (M - AMA^*)^* = M^* - AM^*A^* = M - AMA^*.$$

4.6.2 Elimination based compression of a generator in computations in an abstract field

The SVD/eigendecomposition approach is appropriate where the computations are performed numerically, with rounding to a fixed finite precision. *For computations in an abstract field*, one may compute a shortest L-generator of a matrix M by applying Gaussian elimination to the matrix $L(M)$ and using complete pivoting to avoid singularities. This produces factorization (1.4.2), (1.4.3) where $G = P_0 L_0$, $H = P_1 L_1$, P_0 and P_1 are permutation matrices, L_0 and L_1 are $n \times r$ lower triangular or lower trapezoidal matrices, and $l = r = \text{rank}(L(M))$. If we already have a short but not the shortest L-generator G, H of length l for the matrix M satisfying (1.4.1), (1.4.2), then the latter process (if performed properly) *compresses* the matrix $L(M)$ at a low cost, that is, outputs a shortest L-generator for the matrix M by performing $O(l^2 n)$ ops. The latter cost bound is no worse than that of Theorem 4.6.2, but the resulting L-generator must not be orthogonal. Thus, *the SVD/eigendecomposition approach has the advantage of numerical stability in computations with rounding, whereas the Gaussian elimination approach has the advantage of being applicable over any field.* Next, let us supply proofs and some details for the latter approach.

Theorem 4.6.4. *Given an L-generator G, H of length l (see (1.4.2), (1.4.3)) for a matrix M having L-rank $r \le l$, it is sufficient to use $O(l^2 n)$ ops to compute an L-generator of length r for the matrix M over any field.*

To prove Theorem 4.6.4, we first verify the two following theorems.

Theorem 4.6.5. *Let* $G = \begin{pmatrix} S & E \\ G_{10} & G_{11} \end{pmatrix}$ *be an* $n \times l$ *matrix of rank* q *and let* S *be a* $q \times q$ *non-singular submatrix of* G. *Let* $R = \begin{pmatrix} I & -S^{-1}E \\ 0 & I \end{pmatrix}$, $\widehat{G} = \begin{pmatrix} S \\ G_{10} \end{pmatrix}$. *Then* $GR = (\widehat{G}, 0)$ *where* $\widehat{G}$ *is an* $n \times q$ *matrix of full rank* q.

Proof. Observe that

$$GR = \begin{pmatrix} S & 0 \\ G_{10} & X \end{pmatrix}.$$

Theorem 4.6.5 follows because S is a non-singular matrix, whereas rank$(GR) \leq$ rank$(G) = q$ (which implies that $X = 0$).   □

Theorem 4.6.6. *For any* $n \times l$ *matrix* H *and for the matrices* G, $\widehat{G}$, S, *and* E *in Theorem 4.6.5, we have*

$$GH^T = \widehat{G}S^{-1}\widehat{H}^T, \quad \widehat{H}^T = (S, E)H^T, \tag{4.6.9}$$

$\widehat{H}$ *is an* $n \times q$ *matrix. Furthermore,*

$$\text{rank}(\widehat{H}) = \text{rank}(GH^T). \tag{4.6.10}$$

Proof. Let R be the matrix in Theorem 4.6.5 and deduce that

$$GH^T = (GR)R^{-1}H^T = (\widehat{G}, 0)R^{-1}H^T = \widehat{G}(I, S^{-1}E)H^T = \widehat{G}S^{-1}(S, E)H^T.$$

This is (4.6.9). (4.6.10) follows because the matrix $\widehat{G}S^{-1}$ has full rank.   □

Proof of Theorem 4.6.4. Theorems 4.6.5 and 4.6.6 enable us to compress the generator matrix G. We extend these results immediately to the matrix $\widehat{H}$ to compress it similarly, thus yielding a generator of the minimum length for a matrix GH^T. To achieve the required non-singularity of the $q \times q$ and $r \times r$ northwestern submatrices of the matrices G and $\widehat{H}$ where $q = \text{rank}(G)$ and $r = \text{rank}(\widehat{H}) = \text{rank}(GH^T)$, we interchange the rows and columns of the matrices G and $\widehat{H}$ in an appropriate way. We identify the desired interchanges when we apply Gaussian elimination with complete pivoting to the matrices G and $\widehat{H}$, which enables us to compute and invert some non-singular submatrices of these two matrices having maximum size. The length of the resulting generator $(\widetilde{G}, \widetilde{H})$ decreases to its minimum because the matrices $\widetilde{G}$ and $\widetilde{H}$ have full rank by their construction and because (4.6.10) holds. Clearly, the entire computation requires $O(l^2 n)$ ops to output the desired generator $(\widetilde{G}, \widetilde{H})$ of the minimum length r for the matrix GH^T. This proves Theorem 4.6.4.   □

Let us also specify the resulting expressions for the generator matrices $\widetilde{G}$ and $\widetilde{H}$. (We use these expressions in Section 7.9.) To simplify the notation, assume that no row or column interchange for the matrices G and $\widehat{H}$ is required. Our compression of the generator (G, H) produces the matrices $\widetilde{G}$, S^{-1}, T^{-1}, $\widetilde{H}$, E, F such that

$$\widetilde{G}\widetilde{H}^T = GH^T, \qquad \widetilde{G}^T = (T, F)(\widehat{G}S^{-1})^T,$$
$$\widehat{H} = \begin{pmatrix} T & F \\ \widehat{H}_{10} & \widehat{H}_{11} \end{pmatrix}, \qquad \widetilde{H} = \begin{pmatrix} T \\ \widehat{H}_{10} \end{pmatrix} T^{-1}, \tag{4.6.11}$$

the matrices $\widehat{G}$, S, E and $\widehat{H}$ are defined in Theorems 4.6.5 and 4.6.6, and $\widetilde{H}$ is an $n \times r$ matrix of full rank, $r = \text{rank}(\widehat{H}) = \text{rank}(GH^T)$. Up to the row and column interchange, the $n \times r$ matrices $\widetilde{G}$ and $\widetilde{H}$ are completely specified by (4.6.9) and (4.6.11).

Remark 4.6.7. In Theorem 4.6.6, it is possible to decrease the cost bound of $O(l^2 n)$ ops to yield the bound $O(\tilde{m}(l)n/l)$ where $\tilde{m}(l)$ ops are required for $l \times l$ matrix multiplication, $\tilde{m}(l) = o(l^3)$.

Remark 4.6.8. The algorithm supporting Theorem 4.6.4 is reduced to computing the SVD and applying Gaussian elimination to the input matrices of size $l \times n$. For $l = O(1)$, this is in turn reduced to the search for $O(1)$ maximums of n values, the computation of $O(1)$ sums and inner products of vectors of dimension n, and their multiplication by $O(1)$ scalars. These basic operations allow parallel implementation by using arithmetic parallel time $O(\log n)$ on n processors under some commonly used realistic models of parallel computing (see Remark 2.4.6). The algorithm that supports Theorem 4.6.2 uses the same basic blocks plus the solution of the real symmetric tridiagonal eigenproblem (required in computing the SVD of the product GH^T where G and H are $n \times l$ matrices). This problem can be solved in polylogarithmic arithmetic time on n processors.

4.7   Structured matrix multiplication

Superfast algorithms for structured matrices are ultimately reduced to multiplication of structured matrices by vectors and by each other. In Sections 2.4, 2.6, 3.1, and 3.6, we have shown superfast algorithms for multiplication by vectors of the basic structured matrices in Table 1.1/4.1. The resulting ops count was shown in Table 1.2/4.2. Bilinear expressions for a matrix via the entries of its L-generator enable us to extend these algorithms to more general classes of structured matrices and to their pairwise multiplication. *In this section we summarize the resulting complexity estimates.*

Combining bilinear expressions in Examples 1.4.1, 4.4.1–4.4.9, and Section 4.5 with the estimates in Table 1.2/4.2 immediately implies the cost estimates of $O(lM(n))$ or $O(lM(n)\log n)$ for the multiplication by a vector of $n \times n$ matrices in all these examples and in Section 4.5. More precisely, the bound of $O(lM(n)\log n)$ ops applies to the Cauchy-like and Vandermonde-like matrices, whereas the bound of $O(lM(n))$ applies to the other matrices.

Let us next briefly comment on the less straightforward cases of the matrices in Examples 4.4.8 and 4.4.9. (More comments are given in the proof of Theorem 4.7.2.) By combining the representations of Examples 4.4.8 and 1.4.1 and by applying Theorem 3.11.7, we obtain that $O(M(n))$ ops are sufficient to reduce the Toeplitz+Hankel-like problem to multiplication of the Cauchy matrix $C(\mathbf{s}, \mathbf{t}) = \left(\frac{1}{s_i - t_j}\right)_{i,j=0}^{n-1}$ by l vectors where $D(\mathbf{s}) = D_S$, $D(\mathbf{t}) = D_C$. On the solution of the latter problem in $O(lM(n))$ ops, see Exercise 4.5(a), (b) or the paper [HR98] where a Hankel+Toeplitz matrix is multiplied by a vector at a record low arithmetic cost.

Likewise, multiplication of every confluent matrix in Example 4.4.9 by a vector is immediately reduced by using $O(M(n)l)$ ops to multiplication of a matrix $\Theta_j(\mu - \lambda)$

by l or $l+1$ vectors for $0 \le j \le 4$. In Example 4.4.9, we display a factorization of this matrix for every j, $j = 0, 1, 2, 3, 4$, into the product of three matrices, each allowing its multiplication by a vector in $O(M(n))$ ops. (For the matrices $\theta_1(\mu - \lambda)$ and $\theta_2(\mu - \lambda)$, the ops bound follows because of the special structure of the Vandermonde matrices $V(\mathbf{t})$, $V(\mathbf{t}^{-1})$, $V(\mathbf{t}+s\mathbf{1})$, and $V((\mathbf{t}+s\mathbf{1})^{-1})$ involved [PWa].) Theorem 1.5.4 immediately enables us to extend these results to pairwise multiplication of the structured matrices in Examples 1.4.1, 4.4.1–4.4.9, and Section 4.5 as long as these matrices are pairwise associated with compatible displacement operators (see Definition 1.5.5).

Let us formally state the definitions and the estimates.

Definition 4.7.1. $v(M)$, $i(M)$, and $m(M, M_1)$ denote the numbers of ops required for multiplying a matrix M by a vector, for its inversion assuming non-singularity, and for pairwise multiplication of the matrices M and M_1, respectively. We write $v_{r,n}(L) = \max_M v(M)$, $i_{r,n}(L_-) = \max_M i(M)$, and $m_{r,r_1,n}(L, L_1) = \max_{M,M_1} m(M, M_1)$ for regular operators L and L_1 provided that

a) the maximization is over all $n \times n$ matrices M represented with their L-generators of length of at most r and over all $n \times n$ matrices M_1 represented with their L_1-generators of length r_1 where the operators L and L_1 are compatible according to Definition 1.5.5 and

b) the matrices M^{-1} and MM_1 are represented with their L_--generators of length r and L_*-generators of length of at most $r + r_1$, respectively; here we have $L = \nabla_{A,B}$, $L_- = \nabla_{B,A}$, $L_1 = \nabla_{B,C}$, and $L_* = \nabla_{A,C}$ in the Sylvester case, whereas in the Stein case we have $L = \Delta_{A,B}$, $L_- = \Delta_{B,A}$, A and/or B are non-singular matrices, $L_* = \Delta_{A,C}$, and L_1 can be either $\nabla_{B,C}$, or $\Delta_{B^{-1},C}$ provided that B is a non-singular matrix, or $\Delta_{B,C^{-1}}$ provided that C is a non-singular matrix (see Theorems 1.5.3 and 1.5.4).

For partly regular operators L and L_1, all matrix representations in both parts a) and b) additionally include all matrix entries on the irregularity sets of their associated displacement operators.

Theorem 4.7.2. *For the regular and partly regular compatible operators L and L_1 in Examples 1.4.1, 4.4.1–4.4.9, and Section 4.5 applied to $n \times n$ matrices and for integers $r \ge 1$, $r_1 \ge 1$, we have*

$$v_{r,n}(L) = O(v_{1,n}(L)r), \quad m_{r,r_1,n}(L, L_1) = O(m_{1,1,n}(L, L_1)rr_1),$$
$$m_{r,r_1,n}(L, L_1) = O(v_{r_1,n}(L_1)r + v_{r,n}(L)r_1) = O((v_{1,n}(L) + v_{1,n}(L_1))rr_1).$$

The same cost bounds apply where the operators L and L_1 can be chosen among the internally and off-diagonally regular compatible operators of Section 4.5.

Proof. The theorem is easily verified by inspection based on Theorem 1.5.4 and the bilinear expressions for the structured matrices in Examples 1.4.1, 4.4.1–4.4.9, Theorem 4.5.1, and (4.5.1)–(4.5.3). For a partly regular operator L associated with a product MM_1, consider this product as the sum of its regular part, decompressed from the displacement of the product, and the remaining irregular part, defined by the irregularity set $(MM_1)_{I.S.}(L)$. For every operator L in Examples 1.4.1, 4.4.1–4.4.9, and Section 4.5, the latter part is either the null matrix or is defined by $O(n)$ entries of the matrix MM_1. Each of these entries is obtained as the sum of $O(1)$ products of compatible pairs of the entries of the matrices M and M_1 where at least one entry of each pair is from the irregularity set of the respective entries of M or M_1.   □

By combining the bounds of Theorem 4.7.2 and Sections 2.4, 2.6, 3.1, and 3.6, we obtain the following estimates in terms of $M(n)$ in (2.4.1), (2.4.2).

Theorem 4.7.3.

a) $v_{1,n}(\nabla_{A,B}) = O(M(n))$, for $A, B \in \{Z_e, Z_f, Z_e^T, Z_f^T, D(\mathbf{s}), D(\mathbf{t})\}$ as well as for $(A, B) = (Y_{00}, Y_{11})$, for any 10-tuple of scalars $a, b, c, d, e, f, g, h, s, t$, and for the vectors $\left(\mathbf{s} = (s_i)_{i=0}^{n-1}, \mathbf{t} = (t_i)_{i=0}^{n-1}\right)$ with $s_i = cs^{2i} + bs^i + a$, $t_i = dt^{2i} + gt^i + h$, $i = 0, 1, \ldots, n-1$ (see Theorem 2.4.4).

b) $v_{1,n}(\nabla_{A,B}) = O(M(n) \log n)$, for $A, B \in \{Z_e, Z_f, Z_e^T, Z_f^T, D(\mathbf{s}), D(\mathbf{t})\}$, any pair of scalars e, f, and any pair of vectors $\mathbf{s}, \mathbf{t}$.

By combining Theorems 3.11.7 and 4.7.3 with the estimates of Table 1.2/4.2, we estimate the values $v_{1,n}(L)$ for the operators in Examples 1.4.1, 4.4.1–4.4.9, and Section 4.5 and observe that all of them satisfy the next assumption.

Assumption 4.7.4. $v_{1,n}(L) = O(n \log^d n)$, $d \leq 2$, over the fields supporting DFT_{n_+} for $n_+ = 2^{\lceil \log_2 n \rceil}$. $v_{1,n}(L) = O(M(n) \log n)$ over any field of constants.

Finally, we recall that multiplication by a vector of an $n \times n$ structured matrix M of any of the four basic classes can be reduced to performing polynomial multiplication a small number of times. This enables effective parallel implementation of the multiplication of structured matrices (see Remarks 2.4.5 and 2.4.6). We omit the detailed study but state the following estimates, assuming computations over the field of complex numbers.

Theorem 4.7.5. *For a displacement operator L in Examples 1.4.1, 4.4.1–4.4.9, or Section 4.5 and an $n \times n$ matrix given with its L-generator of length r, including its entries of the (possibly empty) irregularity set of the operator L, it is sufficient to use arithmetic time $O(\log n)$ and $O(r M(n) \log^{d-1} n)$ processors to multiply this matrix by a vector where $d \leq 2$. Furthermore, $d = 1$ for the operators $L = \nabla_{A,B}$ of part a) of Theorem 4.7.3.*

Theorem 4.7.6. *For two compatible operators L and L_1 in Examples 1.4.1, 4.4.1–4.4.9, and Section 4.5, let a pair of $n \times n$ matrices M and M_1 be represented with their compatible displacement generators $(G, H, M_{I.S.}(L))$ and $(G_1, H_1, (M_1)_{I.S.}(L_1))$ of length r and r_1, respectively, where $GH^T = L(M)$, $G_1 H_1^T = L_1(M_1)$. (The generators generally include the entries of the matrices M and M_1 on the irregularity sets of the operators L and L_1, respectively.) Then a displacement generator $(G_*, H_*, (MM_1)_{I.S.}(L_*))$ of rank rr_1 for the matrix MM_1 (including its entries on the irregularity set of the associated operator L_*) can be computed by using arithmetic time $O(\log n)$ and $O(rr_1 M(n) \log^{d-1} n)$ processors where $d \leq 2$. Furthermore,*

$d = 1$ *where both compatible operators* L *and* L_1 *are from the class covered in part a)
of Theorem 4.7.3.*

4.8   Algorithm design based on multiplicative transformation of operator matrix pairs

In this section we cover the following remarkable consequence of the displacement
rank approach: *multiplicative transformations of operator/matrix pairs* $\{L, M\}$ *based
on Theorem 1.5.4 and Table 1.4 enable extension of various successful algorithms devised for one class of structured matrices to other classes.* For a demonstration, assume
that we are given a successful algorithm for the inversion of a $\nabla_{A,C}$-like matrix and extend it to the inversion of a $\nabla_{A,B}$-like matrix. For simplicity, assume dealing with
regular displacement operators throughout this section.

Problem 4.8.1. INVERT(A, B) (see Theorem 1.5.3).

INPUT: Natural r and n, a pair of $n \times n$ matrices A and B, a $\nabla_{A,B}$-generator of length
r for an $n \times n$ non-singular matrix M, and three black box subroutines: INVERT(A, C),
MULTIPLY(A, B, C), and COMPRESS(B, A) (see below).

OUTPUT: A $\nabla_{B,A}$-generator of length of at most r for the matrix M^{-1}.

Subroutine INVERT(A, C)   (see Theorem 1.5.3).
   INPUT: Natural l and n, a pair of $n \times n$ matrices A and C, and a $\nabla_{A,C}$-generator of
length l for an $n \times n$ non-singular matrix R.
   OUTPUT: a $\nabla_{C,A}$-generator of length of at most l for the matrix R^{-1}.

Subroutine MULTIPLY(A, B, C)   (see Theorem 1.5.4 and Table 1.4).
   INPUT: Natural k, l, and n, a triple of $n \times n$ matrices A, B, and C, a $\nabla_{A,B}$-generator
of length k for a matrix M, and a $\nabla_{B,C}$-generator of length l for a matrix N.
   OUTPUT: a $\nabla_{A,C}$-generator of length of at most $k + l$ for the matrix MN.

Subroutine COMPRESS(B, A)   (see Theorems 4.6.2 and 4.6.4).
   INPUT: Natural l, n, and r, $r < l$, a pair of $n \times n$ matrices A and B, and a $\nabla_{B,A}$-
generator of length l for a matrix M where rank$(\nabla_{B,A}(M)) \le r$.
   OUTPUT: a $\nabla_{B,A}$-generator of length of at most r for the matrix M.

Solution: The next algorithm yields the solution of Problem 4.8.1.

Algorithm 4.8.2. Define a non-singular matrix N by its $\nabla_{B,C}$-generator of length 1.
Then successively compute:

a) A $\nabla_{A,C}$-generator of length of at most $r+1$ for the matrix $R = MN$ (by applying
   the subroutine MULTIPLY(A, B, C)),

b) A $\nabla_{C,A}$-generator of length of at most $r+1$ for the matrix $R^{-1} = N^{-1}M^{-1}$ (by applying the subroutine INVERT(A, C)),

c) A $\nabla_{B,A}$-generator of length of at most $r+2$ for the matrix $M^{-1} = NR^{-1}$ (by applying the subroutine MULTIPLY(B, C, A)),

d) A $\nabla_{B,A}$-generator of length of at most r for the matrix M^{-1} (by applying the subroutine COMPRESS(B, A)).

Based on the latter algorithm and Theorems 1.5.3, 1.5.4, 4.6.4, and 4.7.2, we obtain

Theorem 4.8.3. *Under Definition 4.7.1 and the assumptions of Theorem 4.7.2, we have* $i_{r,n}(\nabla_{A,B}) \le m_{r,1,n}(\nabla_{A,B}, \nabla_{B,C}) + i_{r,n}(\nabla_{A,C}) + m_{1,r+1}(\nabla_{B,C}, \nabla_{C,A}) + O(r^2 n)$. *Furthermore, if the operators involved in the algorithms that suport the above bound are among the regular or partly regular operators in Example 1.4.1 or Sections 4.4 and 4.5, then we have* $i_{r,n}(\nabla_{A,B}) = i_{r,n}(\nabla_{A,C}) + O(rv_{1,n}(\nabla_{A,B}) + (r+1)v_{1,n}(\nabla_{C,A}) + (2r+1)v_{1,n}(\nabla_{B,C}) + r^2 n)$. *If Assumption 4.7.4 holds for the operators* $\nabla_{A,B}$, $\nabla_{A,C}$, *and* $\nabla_{B,C}$, *then* $i_{r,n}(\nabla_{A,B}) \le i_{r,n}(\nabla_{A,C}) + O(r^2 n + r M(n) \log n)$.

Algorithm 4.8.2 and Theorem 4.8.3 can be extended easily to operators of Stein type (see Theorem 1.3.1) and to various computations with structured matrices other than inversion, such as the computation of the determinants and ranks as well as the bases for the null spaces.

Let us specify some algorithmic transformations of the above kind among the structured matrices of the four classes in Table 1.1/4.1. The simplest are the transformations between Hankel-like and Toeplitz-like matrices in both directions, just by means of their pre- or post-multiplication by the reflection matrix J (see Theorem 2.1.5 and Example 4.4.4), which requires no ops and does not change the L-rank. We further observe that the matrices of all four classes in Table 1.1/4.1 can be transformed into each other at the cost of performing $O(lM(n) \log n)$ ops, l being the length of the input L-generator. Indeed, in addition to the above Toeplitz/Hankel transformations, we may pre- and/or post-multiply the input matrix M by appropriately chosen Vandermonde matrices and/or their transposes and apply Theorems 1.5.4 and 4.7.2 and Algorithm 3.1.5. The choice of appropriate Vandermonde multipliers can be guided by Table 1.4, whereas Table 4.4 shows some upper bounds on the growth of the L-rank of the resulting matrices.

Example 4.8.4. Transformation from a Vandermonde-like to Cauchy-like and Hankel/Toeplitz-like matrices. Given a scalar $f \ne 0$ and an $n \times n$ Vandermonde-like matrix M defined by its $\Delta_{D(s), Z_f}$-generator of length l, fix a vector $\mathbf{t}$ and/or a scalar e and apply Theorem 1.5.4 to compute a $\Delta_{Z_e^T, Z_f}$-generator of length of at most $l+1$ for the matrix $V^T(\mathbf{s}^{-1})M$ as well as a $\Delta_{D(s), D(t)}$-generator of length of at most $l+1$ for the $n \times n$ matrix $MV^T(\mathbf{t})$. In both cases the computation requires $O(lM(n) \log n + lv_{1,n}(\Delta_{D(s), Z_f}))$ ops (see Algorithm 3.1.5 and Theorems 1.5.4, 3.4.1, 4.7.2, and 4.7.3). Similarly we may use the Cauchy matrix $C(\mathbf{p}, \mathbf{s}^{-1})$ as a multiplier to transform the same Vandermonde-like matrix M into another Vandermonde-like matrix $C(\mathbf{p}, \mathbf{s}^{-1})M$ associated with the displacement operator $\Delta_{D(p), Z_f}$, where $\mathbf{p}$ is a vector of our choice (e.g., a scaled vector of roots of 1).

Table 4.4: Upper bounds on the growth of the L-rank in the transformations with Vandermonde multipliers. (T, V, and C denote matrices with the structures of Toeplitz, Vandermonde, and Cauchy types, respectively; the growth is by 2 where both pre- and post-multiplication are required.)

	T	V	C
T		1	2
V	1		1
C	2	1	

The above transformations of the operators and generators enable immediate extension of any succesful algorithm for the inversion of a matrix of any of the four basic classes of structured matrices to the three other classes and similarly for various matrix computations other than inversion. We apply this observation to our superfast algorithms in the next chapters. Meanwhile, here is a similar extension result for fast algorithms.

Theorem 4.8.5. *Let an $n \times n$ Toeplitz-like, Hankel-like, Vandermonde-like, or Cauchy-like matrix M with entries in any fixed field be represented with its displacement generator of length r. Then $O(rn^2)$ ops are sufficient to compute $\det M$, $\operatorname{rank}(M)$, and a shortest displacement generator for a basis for the null space of the matrix M, or a displacement generator of length r for the matrix M^{-1} if the matrix M is non-singular.*

Proof. (Compare the end of the next section.) The fast version of the Gaussian elimination algorithm with complete pivoting supports Theorem 4.8.5 for Cauchy-like matrices M [H95], [GKO95]. (The row and column interchanges preserve the Cauchy-like structure.) Transformations with Vandermonde multipliers (see Table 4.4) enable extension of the algorithm and the computational cost estimates to the three other classes of structured matrices.   □

The transformations of the operator/matrix pair (L, M) based on multiplication of the matrix M can be also applied to regularize

a) the operator L

or

b) the matrix M.

Concerning part a), we may first invert the *preprocessed* structured matrix MP (or PM) associated with a displacement operator L_1 and then compute the inverse $M^{-1} = P(MP)^{-1} = (PM)^{-1}P$ instead of computing the inverse M^{-1} directly where the operator L_1 is regular but L is not.

In Example 5.5.1 and Section 5.6 in the next chapter, we apply similar transformations to accelerate the solution of the tangential Nevanlinna–Pick and matrix Nehari problems and to regularize a matrix M, respectively.

Finally, general Cauchy-like matrices can be transformed by Cauchy multipliers into Cauchy-like matrices associated with scaling operators of a special form such as $D(a\mathbf{w})$ and $D(b\mathbf{w})$ for two scalars a and b and the vector $\mathbf{w}$ of the n-th roots of 1 (of Definition 2.1.1). The transformation techniques remain the same except that Cauchy multipliers replace Vandermonde multipliers. This replacement may have an advantage because we may use the Fast Multipole Algorithm to multiply Cauchy matrices by vectors and because multiplication by vectors of Cauchy-like matrices associated with the operator matrices $D(a\mathbf{w})$, $D(b\mathbf{w})$ is reduced to FFT's and can be performed in $O(n \log n)$ ops. The latter approach was used in Section 3.6.

4.9 Algorithm design based on similarity transformations of operator matrices

In some important cases, Algorithm 4.8.2, Theorem 4.8.3, and the transformations of Table 4.4 rely on Theorem 1.5.4 and can be simplified based on the application of Theorem 4.3.4. In Example 4.4.8 we have applied Theorem 3.11.6 to yield a simple transition (over the real number field) from Toeplitz+Hankel-like to Cauchy-like matrices. Now suppose that the n-th roots of 1 are available (say where the computations are over the field of complex numbers) and combine Theorems 2.6.4 and 4.3.4 to transform the operator matrices from $Z_f = Z_f(\mathbf{e}_1)$ to $D(U_f\mathbf{e}_1)$, for the matrices U_f in (2.6.6), that is, from the matrices of Toeplitz type to those of Vandermonde type and further to Cauchy-like matrices. Then we may *simplify the displacement transformation approach* as follows.

For a Toeplitz-like matrix M, we write (see Definition 2.1.1 and Theorem 2.6.4)

$$\nabla_{Z_1, Z_{-1}}(M) = GH^T,$$

$$D_1 = D(\Omega \mathbf{e}_1) = D(\mathbf{w}_n) = \mathrm{diag}(1, \omega_n, \ldots, \omega_n^{n-1}),$$

$$D_{-1} = D(\omega_{2n}\mathbf{w}_n) = \omega_{2n} \, \mathrm{diag}(1, \omega_n, \ldots, \omega_n^{n-1}),$$

$$D_0 = D((I_n, 0_n)\mathbf{w}_{2n}) = \mathrm{diag}(1, \omega_{2n}, \ldots, \omega_{2n}^{n-1}),$$

and obtain that $Z_1 = Z_1(\mathbf{e}_1) = \Omega^{-1}D(\Omega \mathbf{e}_1)\Omega = \Omega^{-1}D_1\Omega$ (see (2.6.4)), $Z_{-1} = Z_{-1}(\mathbf{e}_1) = U_f^{-1}D(U_f\mathbf{e}_1)U_f = D_0^{-1}\Omega^{-1}D_{-1}\Omega D_0$ (see (2.6.5), (2.6.6) for $f = \omega_{2n}$, $f^n = -1$). Apply Theorem 4.3.4 with

$$A = Z_1, \quad B = Z_{-1}, \quad \widehat{A} = D_1, \quad V = \Omega, \quad \widehat{B} = D_{-1}, \quad W = (\Omega D_0)^{-1}, \qquad (4.9.1)$$

and arrive at the equations

$$\nabla_{D_1, D_{-1}}(\Omega M (\Omega D_0)^{-1}) = \widehat{G}\widehat{H}^T, \quad \widehat{G} = \Omega G, \quad \widehat{H}^T = H^T (\Omega D_0)^{-1}. \qquad (4.9.2)$$

That is, $\widehat{G}, \widehat{H}$ is a short L-generator for the Cauchy-like matrix $\widehat{M} = \Omega M D_0^{-1}\Omega^{-1}$ and for $L = \nabla_{D_1, D_{-1}}$.

The transition between the L-generators (G, H) and $(\widehat{G}, \widehat{H})$ amounts to performing the DFT_n twice and to the diagonal scaling by the matrix D_0^{-1}, that is, to $O(n \log n)$

ops. Within the same bound on the computational cost, a short L-generator for the inverse of a matrix M can be recovered from a short L_1-generator for the inverse of the matrix $\widehat{M}$. This reduces Toeplitz-like linear systems of equations to Cauchy-like ones. The transition between the 4-tuples of matrices G, H, M, M^{-1} and $\widehat{G}$, $\widehat{H}$, $\widehat{M}$, $\widehat{M}^{-1}$ is *stable numerically,* which is a crucial practical improvement for numerical computations versus the transition based on using Vandermonde multipliers.

Similarly, the transitions between Cauchy-like and Vandermonde-like matrices and between Cauchy-like and Toeplitz+Hankel-like matrices based on Theorem 3.11.6 are stable numerically and require $O(n \log n)$ ops. Furthermore, the transition from a real Toeplitz+Hankel-like input matrix to a Cauchy-like matrix can be performed over the field of real numbers.

The same techniques, based on the application of Theorem 4.3.4, can be used to yield numerically stable $O(n \log n)$ ops transition *into the reverse direction:* $C \rightarrow V \rightarrow T$ or $C \rightarrow T+H$, under the following additional restriction on the classes C and V of the matrices with the structures of Cauchy and Vandermonde types, respectively: the scaling operator matrices associated with the latter classes C and V should be of the form $D(U_f e_1)$, D_C, or D_S for some scalar f.

The most acclaimed practical application of the displacement transformation approach was the design of a fast and numerically stable algorithm for Toeplitz and Toeplitz-like linear systems of equations. Indeed, as we have just shown, over the complex number field, such a system can be reduced to a Cauchy-like linear system based on Theorem 4.3.4. The reduction is by means of several FFT's and diagonal scalings. In a similar reduction over the real number field $\mathbb{R}$, FFT's are replaced by the fast sine and cosine transforms (based on the combination of Theorems 3.11.6 and 4.3.4). The reduction can be utilized in the proof of Theorem 4.8.5. The transition to Cauchy-like matrices enables more effective numerical computations. Indeed, both Toeplitz-like and Cauchy-like structures allow fast solution by Gaussian elimination but only a Cauchy-like structure is preserved in the row and column interchanges. The latter property enables a combination of fast computations with pivoting to ensure numerical stability.

4.10   Conclusion

We recalled and substantially extended the techniques required to support the displacement rank approach to computations with structured matrices. We specified its compression and decompression stages, that is, we displayed expressions for various structured matrices via their displacements and showed both numerical and algebraic versions of the compression algorithms. As a consequence, rapid and effectively parallelizable multiplication of Toeplitz/Hankel, Vandermonde, and Cauchy matrices by vectors was extended to much more general classes of structured matrices. As a further consequence, other matrix operations can be extended similarly. Furthermore, we showed how algorithms for structured matrices of one class can be extended to various other classes by means of transformation of the associated displacement operators and generators.

4.11   Notes

Section 4.2   Example 4.2.1 is from [GKO95].

Section 4.3   Theorem 4.3.2 is from [PWa].   Theorem 4.3.4 was used in [H95], [GKO95].   Theorem 4.3.5 is from [PWb].   Theorem 4.3.6 was applied to the inversion of displacement operators in [HR84], [GO92], [W93] in the Toeplitz-like case and in [PRW00], [PRWa], and [PWa] in the general case.

Section 4.4   Example 4.4.8 is from [GKO95]. All other examples and all remarks are from [PWa] and [PWb]. The paper [PWa] treats the more general case of rectangular matrices M and supplies all proofs. The techniques of Examples 4.4.1–4.4.7 are similar to the techniques in [AG89], [GO92], and [GO94], where similar expressions are obtained for some special displacement operators. The techniques in [PWa] applied in Example 4.4.9 and the resulting expressions in this example extend the techniques and an expression in [OS00b] applied to a special subclass of confluent matrices M and associated operators.

Section 4.5   On the inversion of singular displacement operators, see [BP93], [BP94], [GO94], [KO96], [KO97a], [PWa].   The paper [KO97a] uses the nomenclature of "partially-reconstructible matrices" rather than singular operators. (4.5.1) is taken from [GO94], (4.5.2) is [BP94, Theorem 2.11.3], taken from [BP93].   Theorem 4.5.1 is [BP94, Theorem 2.11.4], taken from [BP93].

Section 4.6   Theorem 4.6.2 is from [P92b] (see also [P93]). SVD-based orthogonal displacement generators were extensively studied in [P92b], [P93] and then in [BMa]. The nomenclatures of orthogonal generators and ϵ-ranks are due to [BMa]. Theorem 4.6.4 is from [P92], whose supporting algorithm is now refined and elaborated in more detail. On the most recent practical algorithms for fast $l \times l$ matrix multiplication using $o(l^3)$ flops, see [K99]; for $l > 10,000$ also see [S69], although Kaporin's algorithm in [K99] uses less work space and is substantially more stable numerically. On a parallel implementation of fast matrix multiplication algorithms, see [LP01]. On parallel computational complexity of the compression of displacement generators, see [P92], [P92b]. The SVD approach to compression involves the solution of the real symmetric tridiagonal eigenproblem; nearly optimal sequential and parallel algorithms for this problem are due to [BP91] and [BP98]. On realistic models of parallel computing, see the books [L92], [Q94].

Sections 4.8 and 4.9   The displacement transformation approach to algorithm design was proposed and elaborated upon in [P90]. The approach was first applied to numerical polynomial interpolation and multipoint evaluation [PLST93], [PZHY97], [PACLS98], [PACPS98], and then to the direct and iterative solution algorithms for Toeplitz/Hankel linear systems of equations (via the transformation of the displacement operators associated with their coefficient matrices to the operators of a Cauchy

type) [H95], [GKO95], [H96], [KO96], [HB97], [G97], [HB98], [Oa]. In the conclud-
ing Section 6 of the paper [OS99]/[OS00], the techniques of [P90] were rediscovered
and applied to accelerating algebraic decoding (see Notes to Section 5.8). Further applications of the displacement transformation approach are shown in Chapters 4 and 5.
The most acclaimed application to the direct solution of Toeplitz/Hankel linear systems
of equations relies on the similarity transformation variant of the approach; this is due
to [H95], [GKO95]; some amelioration and implementation details were elaborated in
[G97].

4.12 Exercises

4.1 Complete all omitted proofs of the results in this chapter.

4.2 Specify the constants hidden in the "O" notation in the arithmetic computational
complexity estimates of this chapter.

4.3 Write programs to implement the algorithms of this chapter. Then apply the
algorithms to specific input matrices.

4.4 Extend Table 4.3 by adding the column $\Delta_{A,B}$ and similarly extend equations
(4.2.1)–(4.2.9) (see Theorem 1.3.1).

4.5 (a) Define a Toeplitz+Hankel matrix $T + H$ with $\nabla_{Y_{00}, Y_{11}}(T + H)$-rank 1.

(b) Deduce that the Cauchy matrix $C(\mathbf{s}, \mathbf{t})$ (where $D(\mathbf{s}) = D_S$, $D(\mathbf{t}) = D_C$
for the diagonal matrices D_S and D_C of Theorem 3.11.6) can be multiplied
by a vector in $O(M(n))$ ops for $M(n)$ in (2.4.1), (2.4.2). **Hint:** Apply
some discrete sine and cosine transforms to reduce multiplication of the
above Cauchy matrix $C(\mathbf{s}, \mathbf{t})$ by a vector to the same problem for the matrix
$T + H$ of part (a); the arithmetic cost of the latter problem is $O(M(n))$ as
we estimated in Section 2.4.

(c) Extend the approach of parts (a) and (b) to multiply a real Toeplitz+
Hankel-like matrix by a vector by using $O(M(n))$ ops. Then try to depart
from this approach to use fewer ops (but still of the order of $M(n)$).

(d) Compare the resulting algorithm with the ones based on Theorem 4.5.1.

(e) Based on Example 4.4.8 a), specify bilinear expressions of a matrix via the
generator matrices G and H for the displacement $\nabla_{Y_{00}, Y_{11}}(M)$.

(f) Use the hint in part (b) to express a matrix M given with its $\nabla_{Y_{00}, Y_{11}}$-
generator of length $O(1)$ as a sum $T + H$ of a Toeplitz-like matrix T and
a Hankel-like matrix H both given with their displacement generators of
length $O(1)$.

(g) Estimate from above the $\nabla_{Y_{00}, Y_{11}}$-rank of a matrix $T + H$ provided that
$\text{rank}(\nabla_{Z, Z_1}(T)) = k$ and $\text{rank}(\nabla_{Z, Z_1^T}(H)) = l$.

(h) [BP94, page 219]. Reduce the inversion of the matrices $T + H$ and $T - H$ to the inversion of the matrix $\begin{pmatrix} T & HJ \\ JH & JTJ \end{pmatrix} = Q \operatorname{diag}(T+H, T-H)Q^T$,

$$Q = \sqrt{2} \begin{pmatrix} I & I \\ J & -J \end{pmatrix}.$$

4.6 [PWa].

   (a) Apply Theorem 1.3.1 to extend Theorem 4.3.6 and Corollary 4.3.7 to the case of Sylvester type operators.

   (b) Extend the results of Sections 4.3 and 4.4 to the case of rectangular matrices M.

   (c) Prove that (1.4.2), for an $m \times n$ matrix M, an operator $L = \nabla_{A,B}$, and all positive integers k, implies that

$$M = A^k M B^k + \sum_{i=0}^{k-1} K(A, \mathbf{g}_i, m) K^T (B^T, \mathbf{h}_i, n)$$

   where $K(C, \mathbf{v}, q) = (\mathbf{v}, C\mathbf{v}, \ldots, C^{q-1}\mathbf{v})$ is the $q \times r$ Krylov matrix of Definition 2.1.1 for a vector $\mathbf{v} = (v_i)_{i=0}^{r-1}$.

   (d) Extend Theorem 2.1.8 to the rectangular matrices $K(Z_f, \mathbf{v}, h)$, $K(Z_f^T, \mathbf{v}, h)$ for vectors $\mathbf{v} = (v_i)_{i=0}^{g-1}$, $g \neq h$.

4.7 [HJR88], [GK89], [GKO95], [KO96].

   (a) Extend the expressions of Example 4.4.8 for the inverse of the displacement operator $L = \nabla_{Y_{00}, Y_{11}}$ to invert the operators $\nabla_{Y_{a,b}, Y_{c,d}}$ for various 4-tuples a, b, c, d where $Y_{g,h} = Y_{00} + g\mathbf{e}_0\mathbf{e}_0^T + h\mathbf{e}_{n-1}\mathbf{e}_{n-1}^T$.

   (b) Generalize the latter expressions to the case where $L = \nabla_{Y,Z}$ for tridiagonal matrices Y and Z associated to other discrete sine and cosine transforms.

   (c) Specify the null spaces of the operators L of parts (a) and (b).

4.8 Deduce an upper bound on $\nabla_{Y_{00}, Y_{00}}$-rank of a matrix $M = T + H$ in terms of $\operatorname{rank}(\nabla_{Z_e, Z_f}(T))$ and $\operatorname{rank}(\nabla_{Z_e, Z_f^T}(H))$ (see Examples 4.4.1–4.4.4 and 4.4.8). Deduce a similar bound where Stein type operators are used instead of Sylvester type operators throughout.

4.9 Among operators $L = \Delta_{A,B}$ and $L = \nabla_{A,B}$ for the matrices A, B from a set $\{Z_e, Z_f^T, D(\mathbf{s}), D(\mathbf{t})\}$ and any pairs of scalars e, f and vectors $\mathbf{s}, \mathbf{t}$, show

   (a) singular but internally regular ones

   and

   (b) singular but off-diagonally regular ones.

   (c) Specify the null spaces of these operators.

4.10 For each operator L in Exercise 4.9 and for an L-like matrix M, choose a matrix P and a regular operator L_1 such that PM (or MP) is an L_1-like matrix.

4.11 [HR98]. For an $m \times n$ matrix M, define its *generating polynomial* $m(x, y) = \mathbf{x}^T M \mathbf{y}$ where $\mathbf{x} = (x^i)_{i=0}^{m-1}$, $\mathbf{y} = (y^j)_{j=0}^{n-1}$. Write

$$d_{T,M}(x, y) = (1 - xy)m(x, y),$$
$$d_{H,M}(x, y) = (x - y)m(x, y),$$
$$d_{T+H,M}(x, y) = (1 - xy)(x - y)m(x, y).$$

(a) Prove that $m(x, y) = \sum_{i=1}^{l} g_i(x)h_i(y)$ for some polynomials $g_i(x) = \mathbf{x}^T \mathbf{g}_i$ and $h_i(x) = \mathbf{h}_i^T \mathbf{y}$, $i = 1, \ldots, l$, if and only if $M = \sum_{i=1}^{l} \mathbf{g}_i \mathbf{h}_i^T$.

(b) Call l the *length* of this decomposition. Call a matrix M an *l-Bezoutian of the Toeplitz, the Hankel, or the Toeplitz+Hankel type* provided there exists such a decomposition for the polynomial $d_{T,M}(x, y)$, $d_{H,M}(x, y)$, or $d_{T+H,M}(x, y)$, respectively.

Prove that a non-singular matrix M

(c) is the inverse of a Toeplitz or a Hankel matrix if and only if it is a 2-Bezoutian of the Toeplitz or the Hankel type, respectively,

and

(d) is the inverse of a Toeplitz+Hankel matrix if and only if it is a 4-Bezoutian of the Toeplitz+Hankel type.

4.12 (a) Prove the existence of SVD in (4.6.1)–(4.6.3) for any matrix W and an eigendecomposition for any Hermitian matrix W. Prove the uniqueness of the sets of singular values and eigenvalues.

(b) Extend the concepts of SVD-based numerical orthogonal generators and L-generators, their length minimization property, and Theorem 4.6.2 to the Hermitian case, where all these concepts and properties are based on eigen-decomposition (4.6.5)–(4.6.7).

4.13 For which choices of the operator matrices A and B, the displacement $\nabla_{A,B}(M)$ of a real skew-symmetric matrix M (that is, such that $M^T = -M$) is real symmetric?

4.14 Incorporate pivoting (by using appropriate row and column interchange) to improve numerical stability of the algorithms supporting Theorem 4.6.4.

4.15 Assume bilinear representations of Examples 4.4.1–4.4.4 for Toeplitz-like and Hankel-like matrices M.

(a) Based on these representations and Theorem 2.6.4, multiply the matrices M by a vector in the field of complex numbers by using $a(l)$ DFT_n's and in addition $O(n)$ ops, where $a(l) \leq 3l + 2$.

(b) Suppose that the matrix M of part (a) is real and Toeplitz-like. Is the latter approach faster than the one of Exercise 4.5(b)–(e)?

4.16 Extend Algorithm 4.8.2 and Theorem 4.8.3 to

   (a) operators of Stein type,

   (b) the computation of the determinants, rank, and the null space bases of struc-
       tured matrices.

4.17 [GKO95].   Extend (4.9.1) and (4.9.2) to specify the transformation of
     Toeplitz+Hankel-like matrices into Cauchy-like matrices.

4.18 Specify the subclasses of Cauchy-like and Vandermonde-like matrices that
     allow their simple (low cost) reduction to Toeplitz-like, Hankel-like, and/or
     Toeplitz+Hankel-like matrices by similarity transformations of the operator ma-
     trices based on Theorem 4.3.4.

4.19 Extend the transformations in Sections 4.8 and 4.9 to the case of partly regular
     displacement operators.

Chapter 5

Unified Divide-and-Conquer Algorithm

Summary In this chapter, we describe a superfast divide-and-conquer algorithm for recursive triangular factorization of structured matrices. The algorithm applies over any field of constants. As a by-product, we obtain the rank of an input matrix and a basis for its null space. For a non-singular matrix, we also compute its inverse and determinant. The null space basis and the inverse are represented in compressed form, with their short displacement generators. The presentation is unified over various classes of structured matrices. Treatment of singularities is specified separately for numerical and symbolic implementations of the algorithm. The algorithm enables superfast computations for several fundamental and celebrated problems of computer algebra and numerical rational computations.

Graph for selective reading

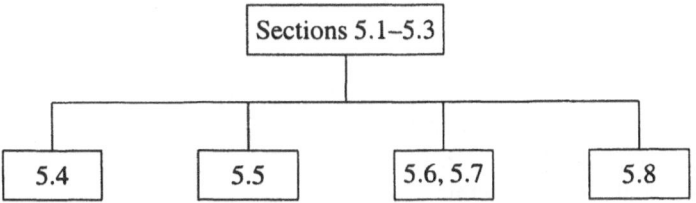

5.1 Introduction. Our subject and results

As pointed out in Section 1.6, the known practical algorithms for the solution of a general non-singular linear system of n equations use the order of n^3 ops (with some practical chances for the reduction of this bound to the order of roughly $n^{2.8}$ [K99]). On the other hand, "fast algorithms" use the order of n^2 ops, whereas "superfast algorithms" use $O(n \log^2 n)$ or $O(n \log^3 n)$ ops provided that the input coefficient matrix is structured.

The first superfast (divide-and-conquer) algorithm proposed by Morf [M80] (see also [M74]) and Bitmead and Anderson [BA80] for the solution of Toeplitz-like and Hankel-like linear systems of n equations used $O(n \log^2 n)$ ops. We refer to this algorithm as the *MBA algorithm*. In principal, one may use Vandermonde multipliers (see Section 4.8) to extend the solution of Toeplitz-like linear systems to Vandermonde-like, Cauchy-like, and various other structured linear systems. The overall asymptotic cost bound of $O(n \log^2 n)$ ops would not change, though the extra steps of the transformation would be a certain complication. On the other hand, one may extend the MBA algorithm to other classes of structured matrices directly with no auxiliary transformation, and it is conceptually interesting to yield such an extension in a unified way for various classes of matrix structures. This is our next subject.

We show a superfast divide-and-conquer algorithm that computes the rank and a basis for the null space of a structured input matrix M as well as its inverse and its determinant if the matrix is non-singular. The algorithm also solves a linear system of equations with this coefficient matrix (if the system is consistent though possibly singular). Furthermore, as a by-product, the algorithm recursively factorizes the matrix M. This factorization of Cauchy-like matrices M immediately implies a superfast solution of the tangential Nevanlinna–Pick and the matrix Nehari problems in Sections 3.8 and 3.9, and the factorization of Vandermonde-like matrices M implies superfast algebraic decoding, whereas only fast (that is, of the order of n^2 ops) solutions for these intensively studied problems have been obtained by other methods so far.

To achieve the superfast level, we compress all auxiliary matrices involved. To make the algorithm work for both singular and non-singular input, we apply regularization techniques of symmetrization (which works numerically for a non-singular input) and multiplicative transformation with randomization (which works over any field for any input).

5.2   Complete recursive triangular factorization (CRTF) of general matrices

The origin of divide-and-conquer algorithms for structured matrices is usually attributed to [M74], [M80], and [BA80] but in the case of general matrices can be traced back at least to [S69], [AHU74]. Let us recall these algorithms. Besides our previous definitions, we use the following:

Definition 5.2.1. $M^{(k)}$ is the $k \times k$ northwestern (leading principal) submatrix of a matrix M. If the submatrices $M^{(k)}$ are non-singular for $k = 1, \ldots, \rho$, where $\rho = \mathrm{rank}(M)$, then the matrix M is said to *have generic rank profile*. If, in addition, the matrix M is non-singular itself, then it is called *strongly non-singular*.

Let an $n \times n$ non-singular matrix M be represented as a 2×2 block matrix,

$$M = \begin{pmatrix} M_{00} & M_{01} \\ M_{10} & M_{11} \end{pmatrix} \qquad (5.2.1)$$

where the $k \times k$ block submatrix $M_{00} = M^{(k)}$ is non-singular. We call this block partition of a matrix M *balanced* if $k = \lceil n/2 \rceil$.

By applying block Gauss–Jordan elimination, we factorize the matrices M and M^{-1} as follows:

$$M = \begin{pmatrix} I & 0 \\ M_{10}M_{00}^{-1} & I \end{pmatrix} \begin{pmatrix} M_{00} & 0 \\ 0 & S \end{pmatrix} \begin{pmatrix} I & M_{00}^{-1}M_{01} \\ 0 & I \end{pmatrix}, \tag{5.2.2}$$

$$M^{-1} = \begin{pmatrix} I & -M_{00}^{-1}M_{01} \\ 0 & I \end{pmatrix} \begin{pmatrix} M_{00}^{-1} & 0 \\ 0 & S^{-1} \end{pmatrix} \begin{pmatrix} I & 0 \\ -M_{10}M_{00}^{-1} & I \end{pmatrix}. \tag{5.2.3}$$

Here the matrix

$$S = S(M_{00}, M) = S^{(k)}(M) = M_{11} - M_{10}M_{00}^{-1}M_{01} \tag{5.2.4}$$

is called the *Schur complement* of the block M_{00} in the matrix M (or the *Gauss transform of* M).

Let us recall some well-known auxiliary results.

Theorem 5.2.2. *If the matrix M is strongly non-singular, then so are the matrices M_{00} in (5.2.1) and S in (5.2.4).*

The next theorem is immediately verified by inspection.

Theorem 5.2.3. *Let the matrices M and M_{00} in (5.2.1) be non-singular. Let us write*
$$M^{-1} = \overline{M} = \begin{pmatrix} \overline{M}_{00} & \overline{M}_{01} \\ \overline{M}_{10} & \overline{M}_{11} \end{pmatrix}. \text{ Then we have } \overline{M}_{11} = S^{-1}, \ \overline{M}_{10} = -S^{-1}M_{10}M_{00}^{-1},$$
$\overline{M}_{01} = -M_{00}^{-1}M_{01}S^{-1}$, *and* $\overline{M}_{00} = M_{00}^{-1} + M_{00}^{-1}M_{01}S^{-1}M_{10}M_{00}^{-1}$ *for the matrices* M_{ij} *in (5.2.1) and S in (5.2.4).*

The next simple theorem shows the transitivity of the Schur complementation and its combination with projecting a matrix into its northwestern blocks.

Theorem 5.2.4. *For some positive integers g, h and k, let the matrices $M_{00} = M^{(k)}$ and $S^{(h)} = (S(M_{00}))^{(h)}$ be non-singular. Then we have $S^{(h)}(S^{(k)}(M)) = S^{(h+k)}(M)$ and $(S^{(k)}(M))^{(g)} = S^{(k)}(M^{(k+g)})$.*

Proof. Apply two steps of the block Gauss–Jordan elimination to obtain successively the Schur complements $S^{(k)}(M)$ and $S^{(h)}(S^{(k)}(M))$. These two steps can be viewed as a single step that produces the Schur complement $S^{(h+k)}(M)$, and we obtain the first equation of Theorem 5.2.4. Apply the block elimination to compute the matrix $S^{(k)}(M)$. Its restriction to the northwestern submatrix $M^{(k+g)}$ produces matrices $S^{(k)}(M^{(k+g)})$ and $(S^{(k)}(M))^{(g)}$, and we arrive at the second equation of the theorem. □

Due to Theorem 5.2.2, we may extend factorizations (5.2.2) and (5.2.3) of a strongly non-singular input matrix M and its inverse to the matrices M_{00} and S and their inverses, and then recursively continue this *descending process* until we arrive at 1×1 matrices. Actual computation starts from there and proceeds bottom up, in the *lifting process*, which begins with the inversion of the 1×1 matrix $M^{(1)}$ and then is extended to the computation of its 1×1 Schur complement in the 2×2 northwestern block $M^{(2)}$, to the inversion of the block $M^{(2)}$ based on its factorization of the form (5.2.3), and so on. The algorithm finally outputs the *complete recursive triangular factorization* of the matrix M; we refer to it as the *CRTF*. We specify the *balanced CRTF* of M (to which we refer as the *BCRTF*) as the one where each recursive step relies on the balanced partition of the input matrices into 2×2 blocks. The BCRTF is naturally associated with a binary tree, whose depth is $\lceil \log_2 n \rceil$ for an $n \times n$ input matrix M.

Algorithm 5.2.5. The BCRTF and the inversion of a strongly non-singular matrix.

INPUT: A field $\mathbb{F}$ and a strongly non-singular matrix M, $M \in \mathbb{F}^{n \times n}$.

OUTPUT: The balanced BCRTF of the matrix M, including the matrix M^{-1}.

COMPUTATION:

1. Apply Algorithm 5.2.5 to the matrix M_{00} in (5.2.1) to compute its balanced BCRTF. Compute the matrix M_{00}^{-1} by performing division if M_{00} is a 1×1 matrix or based on the extension in (5.2.3) to the factorization of the matrix M_{00}^{-1} otherwise.

2. Compute the Schur complement $S = M_{11} - M_{10} M_{00}^{-1} M_{01}$.

3. Apply Algorithm 5.2.5 to the matrix S to compute the balanced BCRTF of S (including S^{-1}).

4. Compute the matrix M^{-1} from (5.2.3).

The correctness of Algorithm 5.2.5 immediately follows from Theorems 5.2.2 and 5.2.4.

By extending Algorithm 5.2.5, we may compute the solution $\mathbf{y} = M^{-1}\mathbf{b}$ to a linear system $M\mathbf{y} = \mathbf{b}$ and $\det M$. Indeed, observe that $\det M = (\det M_{11}) \det S$ and successively compute $\det M_{00}$, $\det S$, and $\det M$ at stages 1, 3, and 4.

In the extension to the case where the matrix M can be singular but has generic rank profile, the **Generalized Algorithm 5.2.5** counts operations of division, that is, the inversion of 1×1 matrices involved. Whenever division by 0 occurs, we have $\rho = \text{rank}(M)$ in the counter, and then continue the computations only where the operations are with the matrices completely defined by the matrix $M^{(\rho)}$ until we compute its CRTF. Of course, this modification may only save some ops, thus *decreasing* their overall number.

Furthermore, assume (5.2.1) with $M_{00} = M^{(\rho)}$, write

$$F = \begin{pmatrix} I_\rho & -M_{00}^{-1} M_{01} \\ 0 & I_{n-\rho} \end{pmatrix}, \quad N = F \begin{pmatrix} 0 \\ I_{n-\rho} \end{pmatrix}, \tag{5.2.5}$$

and easily verify that $MF = \begin{pmatrix} M_{00} & 0 \\ M_{10} & X \end{pmatrix}$, $\text{rank}(MF) = \rho = \text{rank}(M_{00})$. Therefore, $X = 0$, $MN = 0$. Since the $n \times (n - \rho)$ matrix N has full rank, its column vectors form a basis for the null space of the matrix M.

Moreover, the substitution of $\mathbf{y} = F\mathbf{z}$ reduces the solution of a linear system $M\mathbf{y} = \mathbf{b}$ or the determination of its inconsistency to the same task for the system $(MF)\mathbf{z} = \mathbf{b}$, for which the problem is simple because $MN = 0$.

Definition 5.2.6. The *output set* of the Generalized Algorithm 5.2.5 consists of the rank ρ of the input matrix M, the submatrix $M^{(\rho)}$, its inverse (which is just M^{-1} if $\rho = n$), a basis for the null space of the matrix M, a solution $\mathbf{y}$ to the linear system $M\mathbf{y} = \mathbf{b}$ for a given vector $\mathbf{b}$ or "NO SOLUTION" if the system is inconsistent, and the value $\det M$, which is 0 unless $\rho = n$.

The correctness of the Generalized Algorithm 5.2.5 follows from Theorems 5.2.2 and 5.2.4 and (5.2.5).

Theorem 5.2.7. *$F(n) = O(n^{\beta})$ ops are sufficient to compute the output set of the Generalized Algorithm 5.2.5 applied to an $n \times n$ matrix M provided that $\tilde{m}(k) = O(k^{\beta})$ ops for $\beta > 2$ are sufficient to multiply a pair of $k \times k$ matrices.*

Proof. Let $\gamma(k) = k^2$, $\tilde{m}(k)$, and $i(k)$ ops be sufficient to subtract, multiply, and invert $k \times k$ matrices, respectively, assuming strong non-singularity for the inversion. Observe from (5.2.3) and (5.2.4) that

$$i(2k) \leq 2i(k) + 6\tilde{m}(k) + 2\gamma(k). \tag{5.2.6}$$

Obtain similar expression for $i(k), i(k/2), \ldots$. Recursively substitute them on the right-hand side of bound (5.2.6) and deduce that $i(2k) = O(\tilde{m}(k))$. Also observe that the computation by the Generalized Algorithm 5.2.5 is essentially reduced to $O(2^k)$ operations of multiplication and inversion of $(n/2^k) \times (n/2^k)$ matrices for $k = 1, 2, \ldots, \log_2 n$ and deduce the theorem. $\qquad\square$

5.3   Compressed computation of the CRTF of structured matrices

Clearly, the most costly stage of Algorithm 5.2.5 and the Generalized Algorithm 5.2.5 is matrix multiplication. Suppose, however, that the input matrix M is an L-structured matrix of length r, where L is a regular or partly regular operator in Examples 1.4.1, 4.4.1–4.4.9, or Section 4.5 (see Definitions 4.3.8 and 4.5.2), $L = \nabla_{A,B}$ or $L = \Delta_{A,B}$, and the operator matrices A and B are nearly block diagonal (according to the definition in Remark 1.5.7). Then we perform the algorithm more efficiently by operating with short displacement generators instead of the matrices involved in the CRTF of the matrix M. We represent the output matrices $(M^{(\rho)})^{-1}$ and N in a compressed form, by their short displacement generators and the entries on their (possibly empty) irregularity sets.

More specifically, we first follow the descending process of the recursive factorization and apply Theorems 1.5.1, 1.5.3, 1.5.4, 1.5.6, 5.2.3, and 5.2.4 to compute all operator matrices associated either with the BCRTF of the matrices M and M^{-1} if the matrix M is strongly non-singular or with a CRTF of the matrix $(M^{(\rho)})^{-1}$ if the matrix M is singular, of rank $\rho < n$, and has generic rank profile. Theorems 1.5.1, 1.5.3, and 1.5.6 imply that in all products of pairs of structured matrices involved in the recursive algorithm, the associated pairs of displacement operators are compatible (in the sense of Definition 1.5.5), that is, allow application of Theorem 1.5.4.

Furthermore, in our descending process, we estimate the L-ranks of all auxiliary matrices of smaller size involved in the CRTF of the input matrix M, for appropriate operators L. In particular, based on Theorem 1.5.1, 1.5.3, 1.5.4, 1.5.6, and 5.2.2–5.2.4, we bound the L-ranks of all diagonal block matrices in the CRTF by $r + O(1)$ provided that the input matrix M has displacement rank r and that all associated operator matrices are nearly block diagonal. The latter property holds for the operator matrices of the displacement operators in Examples 1.4.1, 4.4.1–4.4.9, and Section 4.5.

For example, assume a non-singular $2k \times 2k$ Cauchy-like input matrix M such that $\text{rank}(\nabla_{D(\mathbf{s}), D(\mathbf{t})}(M)) = r$ for some fixed vectors $\mathbf{s}^T = (\mathbf{s}_0, \mathbf{s}_1)^T$ and $\mathbf{t}^T = (\mathbf{t}_0, \mathbf{t}_1)^T$, where $\mathbf{s}_0, \mathbf{s}_1, \mathbf{t}_0$, and $\mathbf{t}_1$ are subvectors of the vectors $\mathbf{s}$ and $\mathbf{t}$ that have dimension k. Then we immediately deduce from Theorems 1.5.3, 1.5.6, and the equation $\overline{M}_{11} = S^{-1}$ of Theorem 5.2.3 that

$$\text{rank}(\nabla_{D(\mathbf{s}_0), D(\mathbf{t}_0)}(M_{00})) \leq r, \tag{5.3.1}$$

$$\text{rank}(\nabla_{D(\mathbf{t}), D(\mathbf{s})}(M^{-1})) = r, \tag{5.3.2}$$

$$\text{rank}(\nabla_{D(\mathbf{s}_1), D(\mathbf{t}_1)}(S)) = \text{rank}(\nabla_{D(\mathbf{t}_1), D(\mathbf{s}_1)}(S^{-1})) \leq r, \tag{5.3.3}$$

thus bounding by r the displacement ranks of the diagonal blocks in the BCRTF at the first recursive level. The same argument and bounds can be extended to all subsequent levels of the BCRTF based on the application of Theorem 5.2.4 combined with Theorems 1.5.3, 1.5.6, and 5.2.3.

In this example, all operator matrices are diagonal, which slightly simplifies our task. In the case where they are nearly block diagonal, the right-hand sides of bounds (5.3.1) and (5.3.3) may increase but at most by a constant $O(1)$, whereas bound (5.3.2) is preserved. By combining Theorems 1.5.3, 1.5.6, 5.2.3, and 5.2.4, we obtain the same bounds for all diagonal blocks in the CRTF.

In the descending process we also compute the irregularity sets of the displacement operators associated with all matrices in the CRTF. The irregularity sets of the operators in Examples 1.4.1, 4.4.1–4.4.9, and Section 4.5 do not propagate in their projections induced by the projections of the associated matrices into their blocks. That is, regular, off-diagonally regular, and internally regular operators have regular, off-diagonally regular, and internally regular projections, respectively (see Remark 4.5.6), and the irregularity sets propagate in no other steps of the computation of the CRTF.

We combine the results of our analysis of the descending process with Theorems 1.5.1, 1.5.4, 1.5.6, 4.6.2 or 4.6.4, 4.7.2, and 4.7.3 when we follow the lifting process and compute short L-generators of the same matrices of the CRTF whose associated displacement operators L are fixed in the descending process. For an off-diagonally or internally regular operator L, we separately compute the irregularity sets of the matrix entries lying on the diagonal or in the first (or/and the last) column or/and row of the associated matrices of the CRTF. We call this process *compressed computation* of the CRTF because we compute short L-generators of the matrices and their entries on the irregularity sets rather than the matrices themselves, and similarly we represent the matrices F and N in (5.2.5) by a short L-generator of the matrix $-M_{00}^{-1} M_{01}$.

Summarizing the resulting cost estimates and assuming that the input matrices are represented by their generators of length of at most r for the operators in Examples 1.4.1, 4.4.1–4.4.9, and Section 4.5, we arrive at (5.2.6) but this time with $\gamma(k) = O(k)$ and $\bar{m}(k) = O(r^2 v_{1,k}(L))$ (see Definition 4.7.1 and Theorems 4.7.2 and 4.7.3). Recursively, we yield the following result.

Theorem 5.3.1. *Let an $n \times n$ input matrix M in the Generalized Algorithm 5.2.5 have generic rank profile and be given with its regular, off-diagonally regular, or internally regular L-generators of length l, where L denotes an operator in Examples 1.4.1, 4.4.1–4.4.9, or Section 4.5 and $r = \text{rank}(L(M)) \le l \le n$. Let $L = \nabla_{A,B}$ or $L = \Delta_{A,B}$ where the matrices A and B are nearly block diagonal (see Remark 1.5.7). Then compressed computation of the output set of the algorithm requires $O(nl)$ words of memory and $O(l^2 n + v_{1,n}(L)r^2 \log n)$ ops, where the term $l^2 n$ comes from the application of Theorems 4.6.2 or 4.6.4 to the input matrix M.*

Remark 5.3.2. The memory space bound increases to the order of $ln \log n$ words if, besides the output set, short L-generators of all matrices of the CRTF must be also stored (see [PZ00]).

Corollary 5.3.3. *The ops estimate of Theorem 5.3.1 turns into $O((l^2 n + r^2 M(n)) \log n)$ for Toeplitz-like, Hankel-like, Toeplitz+Hankel-like, Vandermonde-like, and Cauchy-like matrices M and their associated operators L (see (2.4.1), (2.4.2)).*

Proof. The bound $v_{1,n}(L) = O(M(n))$ for the classes of Toeplitz-like, Hankel-like, and Toeplitz+Hankel-like matrices and the associated operators L in Examples 4.4.1–4.4.4 and 4.4.8 and Theorem 4.5.1 implies Corollary 5.3.3 for these classes (see Table 1.2). Apply Algorithm 4.8.2 with Vandermonde multipliers V to extend the result to Vandermonde-like and Cauchy-like input matrices. The transition from the output set of the Generalized Algorithm 5.2.5 for the matrices MV or VM to the output set for the matrix M requires $O(r M(n) \log n)$ ops. □

Remark 5.3.4. The (Generalized) Algorithm 5.2.5 allows only limited parallel acceleration yielding at best linear parallel arithmetic time on n processors because the inversion of the northwestern blocks must precede the factorization and/or inversion of their Schur complements. In this respect, more effective are the algorithms of the next chapters.

5.4 Simplified compression of Schur complements

For some operators $L = \nabla_{A,B}$, stage 2 of the computation of Schur complements of the Generalized Algorithm 5.2.5 can be further simplified, although this does not affect the overall asymptotic cost estimates.

Theorem 5.4.1. *Let us be given block representations* $M = \begin{pmatrix} M_{00} & M_{01} \\ M_{10} & M_{11} \end{pmatrix}$, $A = \begin{pmatrix} A_{00} & A_{01} \\ A_{10} & A_{11} \end{pmatrix}$, $B = \begin{pmatrix} B_{00} & B_{01} \\ B_{10} & B_{11} \end{pmatrix}$, $G = \begin{pmatrix} G_0 \\ G_1 \end{pmatrix}$, $H = \begin{pmatrix} H_0 \\ H_1 \end{pmatrix}$ *where the block sizes are compatible. Let the matrix M and its block M_{00} be non-singular, let*

$S = M_{11} - M_{01}M_{00}^{-1}M_{10}$, and let $\nabla_{A,B}(M) = GH^T$. Then $\nabla_{A_{11},B_{11}}(S) = M_{10}M_{00}^{-1}A_{01}S - SB_{10}M_{00}^{-1}M_{01} + (G_1 - M_{10}M_{00}^{-1}G_0)(H_1^T - H_0^TM_{00}^{-1}M_{01})$. In particular, $\nabla_{A_{11},B_{11}}(S) = (G_1 - M_{10}M_{00}^{-1}G_0)(H_1^T - H_0^TM_{00}^{-1}M_{01})$ if $A_{01} = B_{10} = 0$, that is, if the matrices A^T and B are block lower triangular.

Proof. Represent the matrix M^{-1} as a 2×2 block matrix with blocks $\overline{M}_{ij}$ of the same size as M_{ij}, that is, write $M^{-1} = \overline{M} = \begin{pmatrix} \overline{M}_{00} & \overline{M}_{01} \\ \overline{M}_{10} & \overline{M}_{11} \end{pmatrix}$ where $\overline{M}_{11} = S^{-1}$ by Theorem 5.2.3. Recall from Theorem 1.5.3 that $\overline{M}A - B\overline{M} = \overline{M}GH^T\overline{M}$. It follows that $S^{-1}A_{11} - B_{11}S^{-1} + \overline{M}_{10}A_{01} - B_{10}\overline{M}_{01} = (\overline{M}_{10}G_0 + S^{-1}G_1)(H_0^T\overline{M}_{01} + H_1^TS^{-1})$. Therefore, $\nabla_{A_{11},B_{11}}(S) = A_{11}S - SB_{11} = SB_{10}\overline{M}_{01}S - S\overline{M}_{10}A_{01}S + (S\overline{M}_{10}G_0 + G_1)(H_0^T\overline{M}_{01}S + H_1^T)$. Combine the above matrix equations with those from Theorem 5.2.3,
$$\overline{M}_{01} = -M_{00}^{-1}M_{01}S^{-1} \text{ and } \overline{M}_{10} = -S^{-1}M_{10}M_{00}^{-1},$$
and obtain the desired expression for $\nabla_{A_{11},B_{11}}(S)$. □

Remark 5.4.2. The assumption about non-singularity of the matrix M can be relaxed by the confluence argument if the computations are over a field $\mathbb{F}$ containing the field of rational numbers.

Remark 5.4.3. The result of Theorem 5.4.1 can be easily extended to yield explicit expressions for the displacement $\Delta_{A_{11}^{-1},B_{11}}(S)$ and/or $\Delta_{A_{11},B_{11}^{-1}}(S)$ if the blocks A_{11} and/or B_{11} are non-singular (see Theorem 1.3.1).

Remark 5.4.4. Theorem 5.4.1 enables some algorithmic simplifications provided that rank(A_{01}) and rank(B_{10}) are small. The simplifications are greatest where the matrix equations $A_{01} = B_{10} = 0$ hold. This is the case for the operator $\nabla_{A,B}$ with matrices $A \in \{D(s), Z\}$ and $B \in \{D(t), Z^T\}$, which covers displacement operators associated with Hankel-like, Cauchy-like, and Vandermonde-like matrices. In particular, we may use the internally regular operator ∇_{Z,Z^T} to cover the class of Hankel-like matrices, and then we may yield an extension to the Toeplitz-like case by means of pre- or post-multiplication of the input matrix M by the reflection matrix J (see Theorem 2.1.5).

5.5   Regularization via symmetrization

In some applications, the strong non-singularity assumption automatically holds for the input matrix in Algorithm 5.2.5. Over the real and complex number fields, this is the case for positive definite matrices encountered in our study of the Nevanlinna–Pick and Nehari problems in Sections 3.8 and 3.9. Furthermore, any non-singular real or complex input matrix can be transformed into a positive definite one by means of symmetrization. Indeed if M is a non-singular complex or real matrix, then both matrices M^*M and MM^* are Hermitian or real symmetric and positive definite (see Definition 4.6.3) and, therefore, strongly non-singular. Then we may apply Algorithm 5.2.5 to any of these two matrices. From its output set, we immediately recover the matrix $M^{-1} = M^*(MM^*)^{-1} = (M^*M)^{-1}M^*$ and the scalar $|\det M|^2 = \det(M^*M) = \det(MM^*)$.

Let us comment on how symmetrization affects numerical stability and compression/decompression.

The price for achieving strong non-singularity is the squaring of the condition number $\kappa(M)$ of the input matrix M caused by the symmetrization (cf. Definition 6.1.1 on $\kappa(M)$). The condition number of a symmetrized positive definite matrix, however, does not grow any further in the transition to the northwestern (leading principal) submatrices and to their Schur complements [GL96] and consequently to any diagonal block matrix involved in the CRTF. Furthermore, according to an analysis in [B85], the MBA algorithm is weakly stable numerically for Hermitian positive definite Toeplitz-like input but unstable for indefinite Toeplitz-like input. See the Notes on some alternative ways of stabilization.

Now, let a structured matrix M be given with its short $\nabla_{A,B}$-generator. Then Theorem 1.5.2 immediately defines a short ∇_{B^*,A^*}-generator for the Hermitian transpose M^*. Let us specify short displacement generators for the matrices M^*M and MM^*. First assume that the operator matrices A and/or B are real diagonal or, more generally, real symmetric matrices. Then $B^* = B$, $A^* = A$. By applying Theorem 1.5.4, we obtain a desired short L-generator for any of the matrices M^*M or MM^*, where $L = \nabla_{B,B}$ or $L = \nabla_{A,A}$, respectively. First compute short $\nabla_{C,C}$-generators for the matrices M^*M or MM^* and for $C = B$ or $C = A$, respectively, then apply Algorithm 5.2.5 to obtain $\nabla_{C,C}$-generators for the inverse matrices $(M^*M)^{-1}$ or $(MM^*)^{-1}$, respectively, and finally compute a short $\nabla_{B,A}$-generator for the matrix $M^{-1} = (M^*M)^{-1}M^* = M^*(MM^*)^{-1}$ by applying Theorem 1.5.4 again. The auxiliary operators $\nabla_{C,C}$ may become partly regular where the input operators L are regular (for example, $L = \nabla_{D(s),D(t)}$ for two real vectors s and t) but the (Generalized) Algorithm 5.2.5 can be applied in the partly regular case as well.

The above argument fails where $A^* \neq A$ and/or $B^* \neq B$. We, however, first pre- or post-multiply the matrix M by a non-singular structured matrix K represented by its $\nabla_{C,A}$- or $\nabla_{B,E}$-generator of length 1, respectively, where C and E are real symmetric (e.g., real diagonal) matrices (see Algorithm 4.8.2). This reduces the inversion of the input matrix M to the inversion of the matrices KM or MK represented by their short $\nabla_{C,B}$- or $\nabla_{A,E}$-generators, respectively.

Example 5.5.1. Let the input matrix M of Algorithm 5.2.5 be a Hermitian positive definite Cauchy-like matrix given with its short $\nabla_{D(s),D(-s^*)}$-generator of length l, and let $M = W^*W$ for a non-singular matrix W (see Definition 4.6.3). For example, this case must be handled in the matrix approach to the tangential Nevanlinna–Pick Problem 3.8.2. Then Algorithm 5.2.5 can be applied directly to the matrix M, but the computational cost of the solution can be decreased by a logarithmic factor by means of the transition to the *skew-Hankel-like* matrix $\widetilde{M} = V^*(s^*)MV(s^*)$, which clearly remains Hermitian and positive definite because $\widetilde{M} = (WV(s^*))^*WV(s^*)$. In this case, we first compute the matrix $\widetilde{M}$ and then the matrix $M^{-1} = V(s^*)\widetilde{M}^{-1}V^*(s^*)$. The skew-Hankel-like matrix $\widetilde{M}$ can be represented by its L-generator of length of at most $l + 2$ for $L = \nabla_{Z_e^T,-Z_f}$ and any pair of scalars e and f. Our study of decompression in Sections 4.3–4.4 is immediately extended to these operators. In particular, the application of Algorithm 5.2.5 to the matrix $\widetilde{M}$ produces the output set by using as many ops as in the case of an input matrix M given with its $\nabla_{Z_e^T,Z_f}$-generator of length of at most $l + 2$ (see Remark 4.4.12).

5.6 Regularization via multiplicative transformation with randomization

The symmetrization techniques of the previous section may fail for singular matrices (such as $\left(\begin{smallmatrix} 0 & 0 \\ 0 & 1 \end{smallmatrix}\right)$) over any fixed field as well as for non-singular matrices over finite fields. In these cases, we may ensure the generic rank profile property by first applying a multiplicative transformation of the input matrix M into the matrix

$$\widetilde{M} = X M Y \qquad (5.6.1)$$

and then applying the Generalized Algorithm 5.2.5 to the matrix $\widetilde{M}$. Here we use *randomized non-singular Cauchy-like or Toeplitz preconditioner matrices* as the multipliers X and Y. Such preconditioners apply to general matrices M but are most effective where M is a structured matrix. We have $\det M = (\det \widetilde{M})/((\det X)(\det Y))$, $M^{-1} = Y \widetilde{M}^{-1} X$, and $\mathrm{rank}(M) = \mathrm{rank}(\widetilde{M})$. Furthermore, $\widetilde{M}\mathbf{v} = \mathbf{0}$ if and only if $MY\mathbf{v} = \mathbf{0}$. These techniques extend the multiplicative transformation method of Section 4.8. Unlike Vandermonde multipliers of Section 4.8, Toeplitz and Cauchy-like multipliers do not change the structure of the input matrix but just regularize it.

The use of randomization relies on the following basic result.

Theorem 5.6.1. *Let* $p(\mathbf{x}) = p(x_1, \ldots, x_m)$ *be a non-zero m-variate polynomial of total degree* d. *Let* $\mathbf{S}$ *be a finite set of cardinality* $|\mathbf{S}|$ *in the domain of the definition of* $p(\mathbf{x})$. *Let the values* $\hat{x}_1, \ldots, \hat{x}_m$ *be randomly sampled from the set* $\mathbf{S}$, *that is, let them be chosen independently of each other under the uniform probability distribution on the set, and let* $\hat{\mathbf{x}} = (\hat{x}_1, \ldots, \hat{x}_m)^T$. *Then* Probability$(p(\hat{\mathbf{x}}) = 0) \leq d/|\mathbf{S}|$.

Theorem 5.6.2. *Let us be given a matrix* $M \in \mathbb{F}^{n \times n}$ *and n-dimensional vectors* $\mathbf{p} = (p_i)_{i=0}^{n-1}$, $\mathbf{q} = (q_k)_{k=0}^{n-1}$, $\mathbf{u} = (u_i)_{i=0}^{n-1}$, $\mathbf{v} = (v_k)_{k=0}^{n-1}$, $\mathbf{x} = (x_k)_{k=0}^{n-1}$, *and* $\mathbf{y} = (y_k)_{k=0}^{n-1} \in \mathbb{F}^{n \times 1}$, *where each of the two pairs of vectors* $(\mathbf{p}, \mathbf{q})$ *and* $(\mathbf{u}, \mathbf{v})$ *is filled with* $2n$ *distinct scalars,* $x_0 = y_0 = 1$, *and* x_j *and* y_j *are indeterminates for* $j > 0$. *Then the matrix* $\widetilde{M}$ *in (5.6.1) has generic rank profile if* $X = X_\alpha$, $Y = Y_\beta$, $\alpha, \beta \in \{1, 2\}$, $X_1 = \left(\frac{x_k}{p_i - q_k}\right)_{i,k=0}^{n-1}$, $Y_1 = \left(\frac{y_k}{u_i - v_k}\right)_{i,k=0}^{n-1}$, $X_2 = Z^T(\mathbf{x})$, $Y_2 = Z(\mathbf{y})$. *Furthermore, for every* k, $k \leq \rho = \mathrm{rank}(M)$, *the determinant of the* $k \times k$ *leading principal (northwestern) submatrix of the matrix* $\widetilde{M}$ *is a non-vanishing polynomial of degree of at most* $2k$ *in the indeterminates* x_k *and* y_k *for* $k = 1, \ldots, n - 1$.

Proof. For an $n \times n$ matrix W, denote by $W_{\mathbf{I},\mathbf{J}}$ the determinant of the submatrix of W formed by all entries that lie simultaneously in the rows of the matrix M indexed by the set $\mathbf{I}$ and in its columns indexed by the set $\mathbf{J}$. For the sets $\mathbf{I} = \{1, 2, \ldots, i\}$, $\mathbf{J} = \{j_1, j_2, \ldots, j_i\}$, and $\mathbf{K} = \{k_1, k_2, \ldots, k_i\}$, $i = 1, 2, \ldots, \rho$, $\rho = \mathrm{rank}(M)$, we have the Cauchy–Binet formula,

$$\widetilde{M}_{\mathbf{I},\mathbf{I}} = \sum_{\mathbf{J}} \sum_{\mathbf{K}} X_{\mathbf{I},\mathbf{J}} M_{\mathbf{J},\mathbf{K}} Y_{\mathbf{K},\mathbf{I}},$$

where the summation is in all sets $\mathbf{J}$ and $\mathbf{K}$, each made of i distinct indices. Let us prove that

$$\tilde{M}_{\mathbf{I},\mathbf{I}} \neq 0 \text{ for } i = 1, 2, \ldots, \rho. \tag{5.6.2}$$

First assume that $X = X_1$, $Y = Y_1$ and then observe that, for a fixed pair of $\mathbf{J} = \{j_1, j_2, \ldots, j_i\}$ and $\mathbf{K} = \{k_1, k_2, \ldots, k_i\}$, we have $X_{\mathbf{I},\mathbf{J}} = a x_{j_1} \cdots x_{j_i}$ where the monomial $x_{j_1} x_{j_2} \cdots x_{j_i}$ uniquely identifies the set $\mathbf{J}$ and where $a = (C(\mathbf{p}, \mathbf{q}))_{\mathbf{I},\mathbf{J}}$. We have $a \neq 0$ by Corollary 3.6.5. Likewise, $Y_{\mathbf{K},\mathbf{I}} = b y_{k_1} \cdots y_{k_i}$ where the monomial $y_{k_1} \cdots y_{k_i}$ identifies the set $\mathbf{K}$ and where $b = (C(\mathbf{u}, \mathbf{v}))_{\mathbf{K},\mathbf{I}} \neq 0$. Therefore, for distinct pairs $(\mathbf{J}, \mathbf{K})$, the terms $X_{\mathbf{I},\mathbf{J}} M_{\mathbf{J},\mathbf{K}} Y_{\mathbf{K},\mathbf{I}}$ cannot cancel each other. Consequently, $\tilde{M}_{\mathbf{I},\mathbf{I}} \neq 0$ provided that there exists a pair $(\mathbf{J}, \mathbf{K})$ such that $M_{\mathbf{J},\mathbf{K}} \neq 0$. This is true for all $i \leq \rho$ since the matrix M has rank ρ, and we arrive at (5.6.2).

If $X = X_2$, $Y = Y_2$, then we define the lexicographic order for the variables, that is, $1 < x_1 < \ldots < x_{n-1}$ and $1 < y_1 < \ldots < y_{n-1}$, and observe that for each set $\mathbf{J}$ the determinant $X_{\mathbf{I},\mathbf{J}}$ is uniquely defined by its unique lowest order monomial, which appears in no other determinant $X_{\mathbf{I},\mathbf{J}}$ for the same $\mathbf{I}$. A similar property holds for the determinants $Y_{\mathbf{K},\mathbf{I}}$ for all sets $\mathbf{K}$, and then again relations (5.6.2) follow. The same arguments apply to the pairs $X = X_1$, $Y = Y_2$ and $X = X_2$, $Y = Y_1$. $\qquad\square$

Corollary 5.6.3. *Under the assumptions of Theorem 5.6.2, let the values of the $2n - 2$ variables, x_k and y_k, $k = 1, \ldots, n - 1$, be randomly sampled from a fixed finite set $\mathbf{S}$ of cardinality $|\mathbf{S}|$. Then the matrix $\tilde{M} = XMY$ of rank ρ has generic rank profile with a probability of at least $1 - (\rho + 1)\rho/|\mathbf{S}|$.*

Proof. $d_k = \det(\tilde{M}^{(k)})$ is a polynomial in the variables $x_1, y_1, \ldots, x_{n-1}, y_{n-1}$; it has total degree of at most $2k$; for $k \leq \rho$ it does not vanish identically in these variables by Theorem 5.6.2. Therefore, $d = \prod_{k=1}^{\rho} d_k$ is a non-vanishing polynomial of degree of at most $(\rho + 1)\rho$ in the same variables. By Theorem 5.6.1, it may vanish with a probability of at most $\rho(\rho + 1)/|\mathbf{S}|$ under the random sampling of the variables x_k, y_k from the set $\mathbf{S}$. Therefore, the probability that neither $\det(\tilde{M}^{(k)})$ vanishes for $k \leq \rho$ under the random sampling is of at least $1 - \rho(\rho + 1)/|\mathbf{S}|$. $\qquad\square$

Corollary 5.6.3 can be combined with the Generalized Algorithm 5.2.5 in the following algorithm (which should operate with compressed matrices if its input matrix is structured):

Algorithm 5.6.4. Randomized computation of the output set for a general matrix.

INPUT: A positive ϵ, a field $\mathbb{F}$, a pair (α, β) for $\alpha, \beta \in \{1, 2\}$, an $n \times n$ matrix $M \in \mathbb{F}^{n \times n}$, and a vector $\mathbf{b} \in \mathbb{F}^{n \times 1}$.

OUTPUT: FAILURE with a probability of at most ϵ or the output set of the Generalized Algorithm 5.2.5 applied to the matrix M and a vector $\mathbf{b}$.

COMPUTATION:

1. Fix a finite set $\mathbf{S}$ of non-zero elements from the field $\mathbb{F}$ or from its algebraic extension where $|\mathbf{S}| \geq 2n(n+1)/\epsilon$ (see Remarks 5.6.5 and 5.6.6 below). From this set, randomly sample $2n - 2$ elements $x_k, y_k, k = 1, \ldots, n-1$, defining two matrices, $X = X_\alpha$ and $Y = Y_\beta$ of Theorem 5.6.2 and Corollary 5.6.3.

2. Compute the matrix $\widetilde{M} = XMY$.

3. Apply the Generalized Algorithm 5.2.5 to the matrix $\widetilde{M}$, which in particular outputs $\rho = \text{rank}(\widetilde{M}) = \text{rank}(M)$.

4. Compute the matrices F and N in (5.2.5) where the matrix M is replaced by $\widetilde{M}$. Verify whether the matrix $\widetilde{M}N$ (formed by the $n - \rho$ last columns of the matrix $\widetilde{M}F$) is a null matrix. If it is not, output FAILURE, which means that the randomization has failed to ensure the generic rank profile property for the matrix $\widetilde{M}$.

5. Otherwise compute the matrix YN whose columns form a basis for the null space of the matrix M.

6. If the linear system $\widetilde{M}\mathbf{w} = X\mathbf{b}$ has no solution $\mathbf{w}$, then the linear system $M\mathbf{z} = \mathbf{b}$ also has no solution $\mathbf{z}$. In this case output NO SOLUTION. Otherwise compute the solutions $\mathbf{w}$ and $\mathbf{z} = Y\mathbf{w}$.

7. If $\rho = n$, compute $\det M = (\det \widetilde{M})/((\det X)\det Y)$ and the matrix $M^{-1} = Y\widetilde{M}^{-1}X$.

The correctness of Algorithm 5.6.4 is easily verified based on (5.2.5), (5.6.1), and Corollary 5.6.3.

In cases of both general and structured matrices M, the computational cost is clearly dominated by the cost of applying the Generalized Algorithm 5.2.5 and, therefore, is specified in Theorem 5.3.1 and Corollary 5.3.3 provided that we ignore the computational cost of the generation of random parameters.

Remark 5.6.5. Instead of a set $\mathbf{S}$ of cardinality of at least $2n(n+1)/\epsilon$, one may use a set $\mathbf{S}$ of any cardinality $|\mathbf{S}| > 2n(n+1)$ and apply the algorithm repeatedly. In ν applications of the algorithm, the probability of outputting ν FAILUREs is at most $((\rho+1)\rho/|\mathbf{S}|)^\nu$, which is less than ϵ if $\nu \geq (\log((\rho+1)\rho/|\mathbf{S}|))/\log\epsilon$, where $\rho = \text{rank}(M)$, $\rho \leq n$.

Remark 5.6.6. To increase the cardinality $|\mathbf{S}|$ of a set $\mathbf{S}$ chosen from a small field $\mathbb{F}$, one may routinely shift to an algebraic extension of $\mathbb{F}$. The extension of degree $2 + \lceil \log((1+n)n)/\log|\mathbb{F}| \rceil$ has a cardinality of at least $4n(n+1)$, even for the field $\mathbb{F} = GF(2)$.

Remark 5.6.7. We may choose the matrices $X = X_3 = Z_a^T(\mathbf{x})$ and/or $Y = Y_3 = Z_b(\mathbf{y})$, for a and b being among the parameters randomly sampled from the set $\mathbf{S} \cup \{0\}$. Then, clearly, Theorem 5.6.2 can be extended. Yet two other alternative choices of $X_4 = I$, $Y_4 = (t_{i-j})_{i,j=0}^{n-1}$ and $X_5 = I$, $Y_5 = JY_4$ where the scalars t_k, $k =$

$0, \pm 1, \cdots, \pm(n-1)$ are randomly sampled from a fixed finite set $\mathbf{S}$ are due to B. D. Saunders (see [KP91]).

Remark 5.6.8. Recall that for a non-singular real or complex input matrix M, symmetrization may replace randomization. For a singular real or complex matrix M, one may apply randomization to obtain a matrix $\widetilde{M} = XMY$, which is likely to have generic rank profile. The condition number $\kappa(\widetilde{M}) = \text{cond}_2(\widetilde{M})$, however, can be as large as or even larger than $\text{cond}_2(M)$, where $\text{cond}_2(W)$ denotes the ratio $\sigma_1(W)/\sigma_r(W)$ of the largest and smallest singular values of a matrix W of rank r (cf. Definition 6.1.1 on the singular values and condition number).

5.7 Randomization for structured matrices

Now suppose that M is a structured matrix associated with the operator $\nabla_{A,B}$ for two operator matrices A and B in Examples 1.4.1, 4.4.1–4.4.8, or Section 4.5. Then recall Theorem 1.5.4 and choose the structured matrices X and Y in Algorithm 5.6.4 and/or Remark 5.6.7 to simplify both the transition from the matrix M, associated with an operator $\nabla_{A,B}$, to the matrix $\widetilde{M}$ in (5.6.1), associated with an operator $\nabla_{C,E}$, and the subsequent application of the (Generalized) Algorithm 5.2.5 to the latter matrix $\widetilde{M}$. In particular, use the following rules.

Rules 5.7.1.

a) If the operator matrix A is Z_f or Z_f^T for some scalar f, then choose $X = X_2$ in Algorithm 5.6.4 and write $C = Z^T$ or choose $X = X_3$ (cf. Remark 5.6.7) and write $C = Z_a^T$. Apply either of these rules also if $A = Y_{00}$ or $A = Y_{11}$.

b) If $A = D(\mathbf{s})$, then choose $X = X_1$ where $\mathbf{q} = \mathbf{s}$, write $C = D(\mathbf{p})$, and choose the vector $\mathbf{p}$ to simplify subsequent computations with the matrix $\widetilde{M}$, e.g., $\mathbf{p} = (cp^i)_{i=0}^{n-1}$ for a fixed pair of scalars c and p, $cp \neq 0$.

c) If the operator matrix B is Z_f or Z_f^T for some scalar f, then choose $Y = Y_2$ and write $E = Z$ or choose $Y = Y_3$ (cf. Remark 5.6.7) and write $E = Z_b$. Also apply either of these rules if $B = Y_{00}$ or $B = Y_{11}$.

d) If $B = D(\mathbf{t})$, then choose $Y = Y_1$ where $\mathbf{u} = \mathbf{t}$, write $E = D(\mathbf{v})$, and choose the vector $\mathbf{v}$ to simplify subsequent computations with the matrix $\widetilde{M}$, e.g., $\mathbf{v} = (dv^i)_{i=0}^{n-1}$ for a fixed pair of scalars d and v, $dv \neq 0$.

e) If $A = D(\mathbf{s})$, $B = D(\mathbf{t})$, then apply Rules b) and d) and choose the vectors $\mathbf{p}$ and $\mathbf{v}$ having $2n$ distinct components.

f) Wherever the choice of $A \in \{Z_e, Z_e^T\}$ and $B \in \{Z_f, Z_f^T\}$ is recommended for any pair of scalars e and f, the alternative choices $X = I$, $Y = Y_4$ or $X = I$, $Y = Y_5$ in Remark 5.6.7 can be applied instead.

By virtue of Theorems 1.5.4, 4.6.4, and 4.7.2, we have the following results.

Theorem 5.7.2. *Under the notation of Rules 5.7.1, let $r = \mathrm{rank}(\nabla_{A,B}(M))$. Then we have*

1. $\tilde{r} = \mathrm{rank}(\nabla_{C,E}(\widetilde{M})) \leq r + \delta$, *where $\delta \leq 2$ if Rule 5.7.1 (e) is applied, $\delta \leq 3$ if Rules 5.7.1 (b) and either (c) or (f) are applied or if Rules 5.7.1 (a) and either (d) or (f) are applied, and $\delta \leq 4$ if Rules 5.7.1 (a) or (f) and (c) or (f) are applied.*

2. *If the matrix M is given with its $\nabla_{A,B}$-generator of length l, then a $\nabla_{C,E}$-generator of length $\tilde{r}$ for the matrix M can be computed by using $O(l^2 n + v_{1,n}(\nabla_{C,A})r + v_{1,n}(\nabla_{B,E})(r + \gamma) + v_{r,n}(\nabla_{A,B}) + v_{r+\gamma,n}(\nabla_{C,B}))$ ops, where $\gamma \leq 2$.*

Furthermore, due to the special structure of matrices C and E under all Rules 5.7.1, we have $v_{1,n}(\nabla_{C,E}) = O(M(n))$ for $M(n)$ in (2.4.1), (2.4.2), (see Theorem 4.7.3), and thus we obtain the next estimate.

Theorem 5.7.3. *Suppose that Algorithm 5.6.4 has been applied to an $n \times n$ matrix $\widetilde{M}$ whose L-generator of length r for $L = \nabla_{C,E}$ has been obtained according to Rules 5.7.1. Then the algorithm involves $O(r^2 M(n) \log n)$ ops, not counting the ops required for sampling the random parameters.*

Remark 5.7.4. Rules 5.7.1 and Theorems 5.7.2 and 5.7.3 are immediately extended to the case of Stein type operators $L = \Delta_{A,B}$ (see Theorem 1.5.4).

5.8   Applications to computer algebra, algebraic decoding, and numerical rational computations

In Chapters 2 and 3, we reduced various problems of computer algebra and numerical rational interpolation and approximation to computations with structured matrices covered by the Generalized Algorithm 5.2.5, Algorithm 5.6.4, and Rules 5.7.1. Therefore, the reduction immediately implies superfast solution of all these problems at the arithmetic cost bounded according to Theorems 5.3.1, 5.7.2, 5.7.3, and Corollary 5.3.3.

Based on the superfast divide-and-conquer structured matrix computations, we immediately obtain superfast algorithms for the computation of polynomial gcd and lcm, division of polynomials modulo a polynomial, and Problem 3.5.6 of the (m, n) rational interpolation, including the Padé approximation and the Cauchy and Hermite rational interpolation problems as special cases. In Chapters 2 and 3, alternative superfast polynomial algorithms for the same computational problems are shown but the customary techniques for numerical stabilization are more readily available for the matrix versions of these algorithms.

Among numerous problems that can be reduced to computations with structured matrices, we also single out the problems of *algebraic decoding*, which in the paper [OS99]/[OS00] have been reduced to computing in finite fields of a vector in the null space of a Vandermonde-like matrix. The algorithms of this chapter immediately produce a superfast solution of the latter problem and, consequently, of the algebraic decoding problems, versus the fast solution algorithms in [OS99]/[OS00].

Rational interpolation and approximation is another prominent area of application of our superfast algorithms. [BGR90] reduced the solution of Problems 3.8.1–3.8.5 of the Nevanlinna–Pick interpolation to inverting a Cauchy-like or confluent Cauchy-like Hermitian matrix R satisfying (3.8.6),

$$AR + RA^* = BDB^*,$$

where we are given the matrices A, B, D, and R in (3.8.7)–(3.8.12). We may restrict the study to the case where the matrix R is positive definite and, therefore, strongly non-singular (recall the solvability conditions of Sections 3.8 and 3.9). Algorithm 5.2.5 and Theorem 5.3.1 combined imply a superfast solution using $O((M + N) \log^3 (M + N))$ ops. Moreover, the *transformation of Example 5.5.1 from a Cauchy-like matrix R to a skew-Hankel-like matrix $\widetilde{R}$ enables acceleration to yield the overall solution in $O((M + N) \log^2 (M + N))$ ops.*

Remark 5.8.1. The comments of this section on Problems 3.8.1–3.8.5 also apply to the matrix Nehari Problem 3.9.1, whose solution reduces to the inversion of the Hermitian and positive definite Cauchy-like matrices P and $Q - P^{-1}$ for the matrices P and Q in (3.9.1)–(3.9.3).

Remark 5.8.2. Our techniques yield a superfast solution of Problem 3.8.3 of the tangential confluent Nevanlinna–Pick problem in the cases where $n = 1$ (for any m_k) as well as where $m_k = 1$ (or $m_k = O(1)$) for all k, but our algorithm of Remark 4.4.10 (c) only leads to a fast solution in the general case where no restriction is imposed on the integer parameters n and $m_1, \ldots, m_n$.

For the experts on numerical rational interpolation and approximation, it could be interesting that the solution by our divide-and-conquer algorithm is closely related to a well-known variant of the *State-Space solution* of Problems 3.8.1–3.8.5, that is, to a variant based on the following *cascade decomposition* of the function $W(z) = I_{M+N} - B^*(zI + A^*)^{-1} R^{-1} BD$ in (3.8.3):

$$W(z) = \Theta_1(z) \cdots \Theta_n(z) \text{ with } \Theta_k(z) = I_{M+N} - \mathbf{b}_k^* \frac{1}{z - z_k} \frac{1}{d_k} \mathbf{b}_k^T. \qquad (5.8.1)$$

Here the matrices $\Theta_k(z) \in \mathbb{C}^{(M+N) \times (M+N)}$ are first order factors, each having just one pole and one zero and having also format (3.8.3), whereas $\mathbf{b}_k$ are vectors, $\mathbf{b}_k \in \mathbb{C}^{M+N}$, $k = 1, \ldots, n$. Factorization (5.8.1) is a convenient way of representing the solution both for the realization of the factors $\Theta_k(z)$ and the global transfer function $W(z)$ on electronic devices and for simple evaluation of the function $W(z)$, in $O((M + N)^2 n)$ ops.

The divide-and-conquer Algorithm 5.2.5 can be naturally extended to yield a super-fast computation of the cascade decomposition. This is done based on the BCRTF of the matrix R in (3.8.3), (3.8.6) (this factorization can be computed by Algorithm 5.2.5) and on the respective recursive 2×1 block decomposition of the matrices B and BD. The matrix function $W(z)$ is decomposed into the product $W(z) = W_0(z) W_1(z)$, and then the factors $W_0(z)$ and $W_1(z)$ are recursively decomposed in a similar way until the complete cascade decomposition (5.8.1) is computed. Here we have (see (5.2.4))

$$W_0(z) = I - B_0^*(zI + A_{00}^*)^{-1} R_{00}^{-1} B_0 D_{00},$$
$$W_1(z) = I - S_R^*(B_0, B)(zI + A_{11}^*)S^{-1}(R_{00}, R)S_R(B_0 D_{00}, BD),$$
$$S(R_{00}, R) = R_{11} - R_{10} R_{00}^{-1} R_{01},$$
$$S_R^*(B_0, B) = B_1^* - B_0^* R_{00}^{-1} R_{01},$$
$$S_R(B_0 D_0, BD) = B_0 D_{00} - R_{10} R_{00}^{-1} B_1 D_{11},$$

and A_{00}, A_{11}, B_0, B_1, D_{00}, D_{11}, R_{00}, R_{01}, R_{10}, and R_{11} are the respective blocks of the 2×2 block matrices A, D, and R and of the 2×1 matrix B. The matrices R_{00}^{-1}, $S^{-1}(R_{00}, R)$, $S_R(B_0, B)$, and $S_R(B_0 D_{00}, BD)$ are computed as by-products of application of Algorithm 5.2.5 to the inversion of the matrix R, and so are all matrices obtained similarly at the subsequent steps of recursive factorization of the matrix $W(z)$. It follows that the same asymptotic cost bounds of Theorem 5.3.1 apply.

5.9   Conclusion

We described in some detail the unified superfast divide-and-conquer algorithm for various fundamental computations with structured matrices. The algorithm relies on recursive triangular factorization of structured matrices and covers computing the rank and a basis for the null space of a structured matrix and the solution of a consistent structured linear system of equations. For a non-singular matrix, also its determinant and inverse are computed. The input, output, and all auxiliary matrices are represented in compressed form — via their displacements and (possibly empty) irregularity sets. The algorithm can be applied over any field of constants. We presented it in a unified way over various classes of structured matrices. The description was more complete than in earlier literature. We described also numerical and algebraic approaches to regularizing the computation (by means of symmetrization and multiplicative transformation with randomization, respectively) and gave our comments on applications to a superfast solution of the problems of computer algebra and numerical rational interpolation and approximation of Chapters 2 and 3.

5.10   Notes

Divide-and-conquer factorization algorithm for general matrices can be traced back to [S69], for Toeplitz-like matrices to [M74] (a fast version) and [M80], [BA80] (the superfast MBA algorithm). In [K94] and [K95], the MBA algorithm was extended to the singular Toeplitz-like input and the computations over any field. The extension used the order of $n \log n$ random parameters versus the order of n in this chapter. The paper [PZ00] (submitted to *Linear Algebra and Its Applications* in 1996 and published with a minor revision in 2000) was a detailed extension of the MBA algorithm to the singular and non-singular Cauchy-like input over any field. An outline of the algorithm of [PZ00] was included in the paper [PACPS98]. The algorithm was presented in the form readily extendable to other matrix structures. For Vandermonde-like matrices, both singular and non-singular, the extension was worked out in the paper [PZACP99]. The first outline of the unified version of the MBA algorithm appeared in the proceedings paper [OP98]; it occupied one of the 10 pages of the entire paper, mostly devoted to the rational interpolation and approximation problems of Sections 3.8 and 3.9 and the application of the MBA algorithm to their solution following an approach proposed by Sachnovich in 1976. Similar outline was published in the Fall of 1999 in [OS99a]. Unlike the earlier papers [PZ00] and [PZACP99] on the Cauchy-like case, the outlines in [OP98] and [OS99a] addressed neither regularization and compression problems nor the computations over any field except for reproducing a result of [GO94b] on the compression of Schur complements. The present chapter follows the papers [P99], [P00], which presented the unified algorithm in some detail, and further elaborates upon the compression and the regularization techniques. The presentation is more complete and comprehensive than in other available expositions.

The divide-and-conquer algorithm of this chapter follows the MBA pattern of computing a recursive triangular factorization of a structured matrix. The papers [BP91] and [BP98] show an important distinct application of the divide-and-conquer techniques to a *superfast solution of the eigenproblem for a real symmetric tridiagonal input matrix*. The first superfast solution of the problem is due to the paper [BP91]. The journal version [BP98] was submitted in 1990 and published with no revision. Various details of implementing this approach were covered in [BP92]. A later work [GE95] also presented a superfast divide-and-conquer algorithm for the same problem.

Section 5.2 Algorithm 5.2.5 and Theorem 5.2.7 are essentially from [S69]. Theorems 5.2.2 and 5.2.4 are well known (see, e.g., [BP94], [GL96]).

Section 5.3 The presentation follows the papers [P99], [P00].

Section 5.4 Theorem 5.4.1 is from [P99], [P00]. It is a simple extension of a result from [GO94b].

Section 5.5 The presentation extends [P99], [P00]; Example 5.5.1 and the class of skew-Hankel-like matrices are new. Alternative means of stabilization of the superfast divide-and-conquer algorithms are given by the transformation approach of [P90], generalized to include a transformation of the problem to rational interpolation [VK98], [VHKa], and the somewhat related techniques of transition to the generalized Schur approach [KS99].

Section 5.6 Theorem 5.6.1 is due to [DL78]. Its proof also can be found in [Z79], [S80], or [BP94, pages 44–45]. Unstructured preconditioners were used already in [BGH82]. The Toeplitz preconditioners X_2, Y_2 are from [KS91]; the Cauchy-like preconditioners X_1, Y_1 are from [P99], [P00]; they refine those of [PZ00]. See other Toeplitz preconditioners in [KP91], [BP94], preconditioners for sparse matrices in [CEKSTV00]. Theorem 5.6.2 and Corollary 5.6.3 combine the results of [KS91], [PZ00], and [P99].

Section 5.7   The presentation extends [P99], [P00].

Section 5.8   The paper [OS99] proposed novel techniques for the reduction of algebraic decoding to computations with Vandermonde-like matrices. On the other hand, the acceleration of the algebraic decoding to the superfast level was most desirable in the environment of ACM STOC'99 where estimates for asymptotic complexity of computations were the primary issue and where the paper [OS99] was presented. The acceleration, however, was not achieved at that time. The missing ingredients of the techniques for computations over finite fields for a singular Vandermonde-like input were supplied in the papers [P99], [PZACP99], [P00].

The presentation in this section follows the paper [OP98] regarding application of superfast matrix algorithms to rational interpolation and approximation and follows [BGR90], [BGR90a], and [GO94b] regarding the reduction of rational interpolation and approximation to structured matrix computation.

A superfast solution of the tangential Nevanlinna–Pick and matrix Nehari problems using $O((M + N)\log^2(M + N))$ flops was immediately implied by combining their reduction to Cauchy-like matrix inversion, due to [BGR90] (see also [BGR90a] and [GO94b]), with the transformation of the latter problem into the problem of Toeplitz-like matrix inversion proposed in [P90] and using also $O((M+N)\log^2(M+N))$ flops. The subsequent solution of the latter problem by the MBA algorithm is within the same cost bound. Positive definiteness was not preserved in the transformations of [P90], and additional symmetrization was required, but this did not affect the overall asymptotic cost of the computations. Until [OP98], however, the existence of a superfast solution was very little (if at all) known to the research community of rational interpolation and approximation.

A superfast solution of the confluent Problem 3.8.3 was claimed in the three proceeding papers [OS99a], [OS00a] and [OS00b]. This claim is the main point of distinction of the three papers from [OP98].

In the paper [H95], specifics of Cauchy-like matrices and their association with rational interpolation were used to yield superfast Cauchy-like inversion algorithms. Coupled with the transformation approach in [P90], they may immediately serve as a basis for the superfast matrix algorithms for Toeplitz-like and Vandermonde-like matrices, thereby completing the correlation between matrix structure and interpolation. This approach was extended in [VK98], [VHKa] to practical algorithms for Toeplitz inversion. The rational interpolation approach, however, does not lead to unifying computations with structured matrices, whereas, on the contrary, reduction of rational interpolation and approximation problems in Sections 3.8 and 3.9 to structured matrix computation enables the unifying treatment of all these problems.

5.11   Exercises

5.1 Complete all omitted proofs of the results claimed in this chapter.

5.2 Specify the constants hidden in the "O" notation in the arithmetic computational complexity estimates of this chapter.

5.3 Write programs to implement the algorithms of this chapter. Then apply the algorithms to some specified structured input matrices.

5.4 Verify that no division by zero may occur in Gaussian elimination with no pivoting applied to a strongly non-singular matrix M.

5.5 [B68], [P00b]. Suppose that an $n \times n$ strongly non-singular matrix $M = (m_{ij})$ is filled with integers. Modify Algorithm 5.2.5 to minimize the absolute values of the denominators in all rational values computed by the algorithm. Estimate from above the magnitudes of the denominators in terms of n and $\max_{i,j} |m_{ij}|$.

5.6 Verify that the Schur complement $S^{(k)}(M)$ for a strongly non-singular $n \times n$ matrix M and a positive integer k, $k < n$, is the matrix obtained in k steps of Gaussian elimination applied to the matrix M.

5.7 (a) Apply Algorithm 5.2.5 to a Toeplitz+Hankel-like matrix, a Cauchy-like matrix, and a Vandermonde-like matrix. Estimate the displacement ranks of all computed Schur complements.

   (b) Specify the irregularity sets for all partly regular displacement operators in Section 4.5 associated with the Schur complements of part (a).

   (c) For the partly regular displacement operators associated with the Schur complements of parts (a) and (b), consider the irregularity sets of all matrices involved in the CRTF. Estimate that the overall cardinality is $O(n)$ for all these sets processed at every level of a binary tree associated with the CRTF.

5.8 Estimate the number of ops involved in Algorithm 5.2.5. Assume a strongly non-singular Toeplitz-like input matrix M given with its short L-generator. Show the estimates in terms of the matrix size $n \times n$ and length l of the generator $L(M)$. Then do the same for a Cauchy-like input matrix. In both cases first assume that the compression of the matrices of the BCRTF is limited to application of

   (a) Theorem 5.4.1 only (without using Theorems 4.6.2 and 4.6.4),

   (b) Theorem 4.6.2 only,

   (c) Theorem 4.6.4 only.

   (d) Finally, show upper estimates in the cases where only the latter two theorems are allowed to be applied and where all three theorems can be applied.

   In which of the cases (a)–(d) is the resulting algorithm superfast?

5.9 (See Remark 5.3.4.) Estimate parallel arithmetic time for algorithms of this section. Allow n processors and computations in the complex number field $\mathbb{C}$ where the input matrix M has size $n \times n$ and FFT uses $O(\log n)$ arithmetic time with n processors [L92], [BP94], [Q94].

5.10 To invert an $n \times n$ positive definite matrix M given with its short $\nabla_{D(s),D(s)}$-generator and with its diagonal, one may first apply Algorithm 5.2.5 to compute a shortest $\nabla_{D(s),D(t)}$-generator of the matrix $MC(\mathbf{s}, \mathbf{t})$ for a fixed vector $\mathbf{t}$ such that the $2n$ components of the vectors $\mathbf{s}$ and $\mathbf{t}$ are all distinct, and then compute shortest generators for the inverses $Y = (MC(\mathbf{s}, \mathbf{t}))^{-1}$ and $M^{-1} = C(\mathbf{s}, \mathbf{t})Y$. Compare this variation with Algorithm 5.2.5 applied directly to the matrix M. Which solution is faster and which has better numerical stability?

5.11 (V. Olshevsky). Consider a modification of the proof of Corollary 5.6.3 where we first proceed as in the original proof to deduce the upper bound $2k/|S|$ on the probability that the $\det(M^{(k)})$ vanishes for $k = 1, \ldots, \rho$ and then, as an implication, yield the lower bound $\prod_{k=1}^{\rho}(1 - 2k/|S|) \geq 1 - (\rho + 1)\rho/M$ on the probability that neither of the $\det(M^{(k)})$ vanishes. Show a flaw in this modification of the proof.

5.12 Try to find other structured preconditioners besides the matrices X_k, Y_k in Section 5.6 for $k = 1, 2, 3, 4, 5$. What about X_3 and Y_3 with fixed non-random scalars a and b? Estimate the probability that in this case the matrix $X_3 M Y_3$ has generic rank profile?

5.13 Assume randomization of Remark 5.6.8 and subsequent symmetrization for a non-singular but ill conditioned general input matrix M. Comment on the expected impact of these operations on numerical conditioning of the problem of recursive factorization of the resulting matrix and on numerical stability of the solution by means of the Generalized Algorithm 5.2.5. How will your comment change if M is a Toeplitz-like matrix [B85]? A Vandermonde-like matrix? A Cauchy-like matrix?

5.14 How will the complexity estimates of Theorems 5.7.2 and 5.7.3 change if the sets of the components cp^i and/or dv^i, $i = 0, 1, \ldots, n - 1$, of the vectors $\mathbf{p}$ and/or $\mathbf{v}$ are:

(a) changed into $c_2 p^{2i} + c_1 p^i + c_0$ and/or $d_2 v^{2i} + d_1 v^i + d_0$, respectively, for some fixed scalars $c_0, c_1, c_2, d_0, d_1, d_2$, where $c_1 d_1 \neq 0$ or $c_2 d_2 \neq 0$, $i = 0, 1, \ldots, n - 1$, or

(b) replaced by (scaled and shifted) sets of Chebyshev points, that is, of the zeros of Chebyshev polynomials?

5.15 Supply the details for the extension to Stein type operators suggested in Remark 5.7.4.

5.16 Give a list of applications of Algorithms 5.2.5, the Generalized Algorithm 5.2.5, and Algorithm 5.6.4 to the computer algebra problems of Chapters 2 and 3. Compare the numbers of ops involved in the matrix and polynomial versions of the superfast solution of each problem and show for which of the problems the superfast matrix solution requires randomization.

5.17 (a) Specify application of Algorithm 5.2.5 to Problems 3.8.1–3.8.5 and 3.9.1. Elaborate upon the details of the transformation to a skew-Hankel-like matrix $\widetilde{M}$ described in Example 5.5.1 and of the subsequent application of Algorithm 5.2.5 to this matrix $\widetilde{\widetilde{M}}$.

(b) If you do not know whether a problem of part (a) is solvable for a given input, how would you modify the algorithm? Would you use symmetrization or randomization in this case?

(c) Answer the same questions as in part (b) provided the complete cascade decomposition of the global transfer matrix function $W(z)$ is required as the output.

5.18 Implement the algorithms of this chapter:

(a) numerically, with rounding to a finite precision and

(b) symbolically, with infinite precision.

(c) Perform experimental tests for both implementation variants.

Chapter 6

Newton-Structured Numerical Iteration

Summary   Newton's iteration rapidly refines a crude initial approximation to the inverse or the Moore–Penrose generalized inverse of a structured matrix. The algorithm can be applied to a general matrix but is dramatically accelerated and becomes superfast where the well conditioned matrix is structured. Newton's iteration is strongly stable numerically and is reduced essentially to performing matrix multiplication twice per an iterative step. Multiplication is inexpensive, easy to implement on computers, and effectively parallelizable for structured matrices represented in compressed form, but the structure rapidly deteriorates in the process of the iteration. In this chapter we cover the following main subjects.

  a) We develop two general techniques for preserving both structure and superlinear convergence. One of the techniques is purely numerical, SVD based. It allows variation of the compression level, which enables a trade-off between the structure and the convergence rate or, equivalently, between the levels of compression and the approximation errors. Another technique can be applied over any field, although in this chapter we assume numerical application.

  b) For our two resulting versions of the Newton-Structured Iteration, we estimate their convergence rates and computational cost in flops.

  c) This analysis is ultimately reduced to estimating the norms of the inverse displacement operators, and we show five approaches to obtaining these estimates.

  d) We describe some policies for convergence acceleration and choosing an initial approximation to the inverse matrix.

  e) We describe homotopic processes with Newton's iteration as their basic subroutine (for general and structured matrices). The homotopic approach ensures superfast numerical inversion of well conditioned Toeplitz-like and other structured matrices, which can be implemented in polylogarithmic parallel arithmetic time on n processors.

f) Non-homotopic versions of the Newton-Structured Iteration do not have theoretical support of their superfast performance, but this may partly be the result
   of the overly pessimistic nature of the available analysis techniques. Numerical
   experiments should ultimately determine the relative efficiency of the homotopic
   and non-homotopic approaches for practical computation and help optimize their
   variations, in particular in choosing appropriate policies for choosing the step
   sizes in the homotopic processes and for compression and scaling the iterates.
   We show the results of some experimental numerical computations with Toeplitz
   matrices.

The study in this chapter brings the reader to some open problems in the ongoing
research. More than in other chapters, the presentation omits proofs, derivations, and
other technical details, which we replace by references to bibliography.

Graph for selective reading

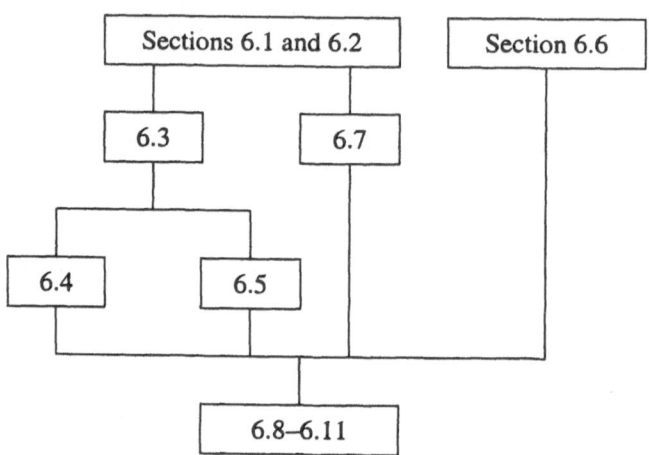

6.1 Some definitions and preliminaries

Definition 6.1.1. $\|v\|_1 = \sum_i |v_i|$, $\|v\|_2 = (\sum_i |v_i|^2)^{1/2}$, $\|v\|_\infty = \max_i |v_i|$ for a vec-
tor $v = (v_i)_i$. $\|\cdot\|$ covers the norms $\|\cdot\|_l$ for $l = 1, 2, \infty$. $\|M\| = \sup_{v \neq 0} \|Mv\|/\|v\|$
is a fixed *operator norm* or *consistent norm* of a matrix M. In particular $\|M\|_l$, the
l-norm of M is the operator norm of M induced by the vector norm $\|v\|_l$, $l = 1, 2, \infty$.
$\kappa(M) = \text{cond}_2(M) = \|M\|_2 \|M^+\|_2 = \sigma_1(M)/\sigma_r(M)$ where $\sigma_i(M)$ is the i-th singular value of the matrix M, M^+ is its Moore–Penrose generalized (pseudo) inverse (see
Definition 4.6.1), and $r = \text{rank}(M)$. $\|M\|_F = (\text{trace } (M^*M))^{1/2} = (\sum_{i,j} |m_{i,j}|^2)^{1/2}$
is the *Frobenius norm* of a matrix $M = (m_{i,j})$.

Theorem 6.1.2. *For every $m \times n$ matrix $M = (m_{i,j})$, we have the following relations:*

a) $\|M\|_2 = \|M^T\|_2 = \|M^*\|_2 = \sigma_1(M)$, $\|M\|_F^2 = \sum_{i=1}^{\rho} \sigma_i^2(M)$, *where $\rho = $*
   rank(M), *and consequently,* $\|M\|_2 \leq \|M\|_F \leq \|M\|_2 \sqrt{\rho}$,

b) $\|M\|_1 = \|M^T\|_\infty = \|M^*\|_\infty = \max_j \sum_i |m_{i,j}|,$

c) $\|M\|_1/\sqrt{m} \le \|M\|_2 \le \|M\|_1 \sqrt{n}, \ \|M\|_2^2 \le \|M\|_1 \|M\|_\infty,$

d) $\|MN\|_l \le \|M\|_l \|N\|_l$ *for any pair of $m \times n$ matrix M and $n \times p$ matrix N and for $l = 1, 2, \infty$.*

Theorem 6.1.3. *Define scalars $r, \sigma_1, \ldots, \sigma_r$ by (4.6.1)–(4.6.3). Write $\sigma_q = 0$ for $q > r$. Then*

a) $\sigma_{q+1} = \displaystyle\min_{V:\text{rank}(V)\le q} \|W - V\|_2,$

b) $\displaystyle\min_{V:\text{rank}(V)\le q} \|W - V\|_F^2 = \sum_{i=q+1}^{r} \sigma_i^2.$

Definition 6.1.4. For a regular or partly regular operator L, write

$$\nu = \nu_{r,l}(L) = \sup_M (\|L(M)\|_l / \|M\|_l),$$

$$\nu^- = \nu_{r,l}(L^{-1}) = \sup_M (\|M\|_l / \|L(M)\|_l),$$

where l stands for $1, 2, \infty$, or F, and the supremum is over all matrices M having positive L-rank of at most r and vanishing in all their entries of the (possibly empty) associated irregularity set.

In this chapter we study numerical algorithms. Our previous results are assumed over the fields of complex or real numbers; we assume bound (2.4.2) and measure arithmetic computational cost in flops, parallel arithmetic time, and the number of arithmetic processors operating with real or complex numbers.

6.2 Newton's iteration for root-finding and matrix inversion

Newton's iteration

$$x_{i+1} = x_i - f(x_i)/f'(x_i), \quad i = 0, 1, \ldots, \tag{6.2.1}$$

rapidly improves of a crude initial approximation $x = x_0$ to the solution $x = r$ of an equation $f(x) = 0$, provided that $f(x)$ is a nearly linear function on the open line interval covering points r and x_0. (6.2.1) is obtained by truncating all terms of the orders of at least two in Taylor's expansion of the function $f(x)$ at $x = r$.

Let us show quadratic convergence of iteration (6.2.1) to the root r. Recall that

$$f(r) = f(x_i) + (r - x_i)f'(x_i) + \frac{(r - x_i)^2}{2} f''(\alpha)$$

for some $\alpha \in [r, x_i]$ provided that the function $f''(x)$ is defined and is continuous on the closed interval $[r, x_i]$. Substitute $f(r) = 0$ and combine the resulting equation with (6.2.1). Obtain that

$$x_{i+1} - r = (x_i - r)^2 f''(\alpha)/(2f'(x_i)), \tag{6.2.2}$$

which implies quadratic convergence.

Let us show a sample application where $f(x) = (1/x) - s$ for a fixed integer s.

Problem 6.2.1. Integer division. Given four integers m, n, S, and T, $2^{n-1} \leq T < 2^n \leq S < 2^m$, compute the unique pair of integers Q and R such that

$$S = QT + R, \quad 0 \leq R < T. \tag{6.2.3}$$

Solution: The problem is reduced to computing a rational value w_k that approximates $s^{-1} = 2^n T^{-1}$ within 2^{n-m-1} so that $|Q_k - S/T| < 2^{-m-1}S < 1/2$ for $Q_k = 2^{-n} w_k S$, and, therefore, $|Q_k T - S| < 0.5T$. As soon as w_k is available, the integer Q is obtained by rounding Q_k to the closest integer, and the integer R is recovered by using (6.2.3). The value w_k is computed by means of Newton's iteration, $w_0 = 2$, $w_{j+1} = \lceil 2w_j - w_j^2 s \rceil_j$, $j = 0, 1, \ldots, k - 1$, where $\lceil x \rceil_j$ denotes the value x rounded up to $2^{j+1} + 1$ bits. It can be immediately verified that $0 \leq w_j - s^{-1} < 2^{1-2^j}$ for $j = 0$. Then we extend these bounds inductively to all j based on the implication of the equation $1/s - w_{j+1} = s(1/s - w_j)^2$ by the equation $w_{j+1} = 2w_j - w_j^2 s$. Therefore, we arrive at the desired value w_k for $k = \lceil \log_2(m - n + 2) \rceil$. The solution uses $O(\mu(m))$ bit-operations for $\mu(m)$ in Exercise 2.9. $\qquad \square$

Newton's iteration (6.2.1) can be extended to the case where x, x_i, and $f(x_i)$ are vectors or matrices. In particular, let $f(X) = M - X^{-1}$ for two matrices M and X. In this case Newton's iteration rapidly improves a crude initial approximation X_0 to the inverse of a non-singular $n \times n$ matrix M:

$$X_{i+1} = X_i(2I - MX_i), \quad i = 0, 1, \ldots. \tag{6.2.4}$$

Indeed, let us define the error and residual matrices:

$$E_i = M^{-1} - X_i, \qquad\qquad e_i = \|E_i\|, \quad e_{l,i} = \|E_i\|_l, \tag{6.2.5}$$

$$R_i = ME_i = I - MX_i, \qquad \rho_i = \|R_i\|, \quad \rho_{l,i} = \|R_i\|_l, \tag{6.2.6}$$

for all i and $l = 1, 2, \infty$. Then (6.2.4) implies that

$$R_i = R_{i-1}^2 = R_0^{2^i}, \quad \rho_i \leq \rho_0^{2^i}, \quad \rho_{l,i} \leq \rho_{l,0}^{2^i}, \tag{6.2.7}$$

$$ME_i = (ME_{i-1})^2 = (ME_0)^{2^i}, \quad e_i = e_0^{2^i} \|M\|^{2^i - 1}, \quad e_{i,l} = e_{0,l}^{2^i} \|M\|_l^{2^i - 1}. \tag{6.2.8}$$

(6.2.7) and (6.2.8) show quadratic convergence of the approximations X_i to the inverse matrix M^{-1} provided that $\rho_{l,0} < 1$. The strong numerical stability of Newton's iteration (6.2.4) and extensions is proved in [PS91]. Each step (6.2.4) amounts essentially to performing matrix multiplication twice and, therefore, allows effective parallel implementation [Q94], [LP01].

The *residual correction algorithms*

$$X_i = X_{i-1}(I + R_{i-1} + \ldots + R_{i-1}^{p-1}), \quad i = 0, 1, \ldots, \tag{6.2.9}$$

are defined for any integer $p \geq 2$ and turn into process (6.2.4) for $p = 2$. It is possible to extend the above estimates and our subsequent study to these processes. In particular (6.2.9) immediately implies that

$$R_i = R_{i-1}^p = R_0^{p^i}, \quad i = 0, 1, \ldots. \tag{6.2.10}$$

6.3 Newton-Structured Iteration

For a general input matrix M, the computational cost of iteration (6.2.4) and its known modifications is quite high because matrix products must be computed at every step. For the structured matrices covered in Examples 1.4.1, 4.4.1–4.4.9, and Section 4.5, matrix multiplication has a lower computational cost of $O(rr_1 n \log^d n)$ flops for $d \leq 2$ provided that we operate with displacement generators of length r and r_1 for the two input matrices (see (2.4.2), Definition 4.7.1, and Theorems 4.7.2 and 4.7.3). The length of the displacement generators, however, is generally tripled in every step (6.2.4) (see Theorem 6.3.2). Thus we modify Newton's iteration in the case where M and X_0 are structured matrices. To simplify the description, we assume throughout that all matrices involved are linearly expressed via their displacement generators, that is, we deal with regular operators (see Remark 6.3.4).

Algorithm 6.3.1. Unified Newton-Structured Iteration.

INPUT: A positive integer r, two matrices A and B defining a regular or partly regular operator $\nabla_{A,B}$, $\nabla_{A,B}$-generators of length (of at most) r for an $n \times n$ non-singular structured matrix M having $\nabla_{A,B}$-rank r and for a sufficiently close initial approximation Y_0 to the inverse matrix M^{-1}, an upper bound q on the number of iteration steps, and a Compression Subroutine $\mathbf{R}$ for the transition from a $\nabla_{B,A}$-generator of length of at most $3r$ for an $n \times n$ matrix approximating M^{-1} to a $\nabla_{B,A}$-generator of length of at most r for another approximation to M^{-1}.

OUTPUT: A $\nabla_{B,A}$-generator of length of at most r for a matrix Y_q approximating M^{-1}.

COMPUTATION: For every i, $i = 0, 1, \ldots, q-1$, compute a $\nabla_{B,A}$-generator of length of at most $3r$ for the matrix

$$X_{i+1} = Y_i(2I - MY_i) \tag{6.3.1}$$

and then apply the Compression Subroutine $\mathbf{R}$ to the matrix X_{i+1} to compute a $\nabla_{B,A}$-generator of length r for the matrix Y_{i+1}.

Theorem 6.3.2. *At the i-th step (6.3.1) of Algorithm 6.3.1, a $\nabla_{B,A}$-generator of length of at most $3r$ for the matrix X_{i+1} can be computed by using $O(r v_{r,n}(\nabla_{B,A}))$ flops for $v_{r,n}(\nabla_{B,A})$ in Definition 4.7.1.*

Proof. Write $Y = Y_i$, $X = X_{i+1} = 2Y - YMY$, $\nabla_{A,B}(M) = GH^T$, $\nabla_{B,A}(X) = G_X H_X^T$, and $\nabla_{B,A}(Y) = G_Y H_Y^T$. Then obtain that

$$\nabla_{B,A}(X) = \nabla_{B,A}(2Y - YMY)$$
$$= \nabla_{B,A}(Y)(2I - MY) - Y\nabla_{A,B}(M)Y - YM\nabla_{B,A}(Y),$$

$$G_X^T = \left(G_Y^T, (YG)^T, (YMG_Y)^T \right),$$

$$H_X^T = \left(H_Y^T(2I - MY), -H^T Y, -H_Y^T \right). \qquad \square$$

Remark 6.3.3. If $\nabla_{B,A}$ is an operator in Examples 1.4.1, 4.4.1–4.4.9, or Section 4.5, then in the statement of Theorem 6.3.2 we may substitute $v_{r,n}(\nabla_{B,A}) = O(rn \log^d n)$, for $d \le 2$, due to Theorems 4.7.2 and 4.7.3. Furthermore, in this case, by virtue of Theorem 4.7.6, we may perform the i-th step (6.3.1) by using parallel arithmetic time $O(\log^c n)$ and $O(r^2 n \log^{d-c} n)$ processors, where $c \le 2$. If the operator $\nabla_{B,A}$ satisfies the assumptions of part a) of Theorem 4.7.3, then $d \le 1$ in the above estimates.

Remark 6.3.4. We explicitly analyze the iteration for non-singular operators of Sylvester type. Theorem 1.3.1 enables immediate extension to the Stein type operators. We may avoid using singular displacement operators (see the recipes of Remark 4.5.5), but it is not hard to extend the Newton-Structured Iteration to the case of partly regular operators by maintaining the computation of the entries of the matrices on the associated irregularity sets (for example, see (6.5.3)).

6.4 Compression of the displacements by the truncation of their smallest singular values

To complete the description of Algorithm 6.3.1, we specify two variants of Compression Subroutine **R**, which we denote Subroutines **R1** (in this section) and **R2** (in the next section).

Subroutine R1. Compute the SVD based orthogonal generator of the displacement $W = L(X_i)$ (see (4.6.1)–(4.6.4)) and decrease the length of the L-generator to r by zeroing (truncating) all singular values σ_i for $i \ge r+1$. Output the resulting orthogonal L-generator of the matrix Y_i to define a unique matrix Y_i.

Example 6.4.1. Let $L = \nabla_{Z_e, Z_f}$.

a) $e \ne f$, $(e-f)X_i = \sum_{j=1}^{l} Z_e(\mathbf{g}_j) Z_f(J\mathbf{h}_j)$ (see Example 4.4.2) where the pair $G = (\mathbf{g}_j)_{j=1}^{l}$, $H = (\mathbf{h}_j)_{j=1}^{l}$ is the orthogonal, SVD based generator for the displacement $L(X_i)$. Let $\text{rank}(L(M^{-1})) = \text{rank}(\nabla_{Z_f, Z_e}(M)) = r \le l$. Then Subroutine **R1** applied to the matrix X_i outputs the matrix

$$Y_i = (e-f)^{-1} \sum_{j=1}^{r} Z_e(\mathbf{g}_j) Z_f(J\mathbf{h}_j).$$

b) Under the same assumptions as in part a) except that $e = f = 0$, $L = \nabla_{Z,Z}$, we have $X_i = Z(X_i e_0) - \sum_{j=1}^{l} Z(\mathbf{g}_j) Z^T(Z\mathbf{h}_j)$ (see (4.5.2)). In this case $Y_i = Z(X_i e_0) - \sum_{j=1}^{r} Z(\mathbf{g}_j) Z^T(Z\mathbf{h}_j)$. □

Subroutine **R1** is supposed to be implemented numerically. In Section 6.8, we consider its variations allowing various levels of compression.

By combining Theorems 6.3.2, 6.1.2, and 4.6.2, we obtain

Theorem 6.4.2. *It is sufficient to use*

$$O(r v_{r,n}(L) + r^2 n + (r \log r) \log \log(\|\nabla_{A,B}(M)\|_2/\epsilon))$$

flops in a single step (6.3.1) of Algorithm 6.3.1 combined with Subroutine **R1** *where* $L = \nabla_{B,A}$ *and* ϵ *is the tolerance to the errors of computing the SVD (see Section 4.6.1). If*

$$\log(\|L(M)\|_2/\epsilon) = 2^{O(nr/\log r)}, \tag{6.4.1}$$

then the cost bound turns into $O(r v_{r,n}(L) + r^2 n)$.

How much does the compression based on Subroutine **R1** affect the approximation to the inverse M^{-1} by the matrix X_i? This can be estimated based on Theorem 6.1.3. Indeed, by Theorem 6.1.3 a), the 2-norm $\|L(Y_i) - L(X_i)\|_2$ is within ϵ from its minimum over all matrices Y_i such that $\text{rank}(L(Y_i)) \le r$. It follows that $\|L(Y_i) - L(X_i)\|_2 \le \epsilon + \|L(X^{-1}) - L(X_i)\|_2$. Therefore, $Y_i \approx M^{-1}$ because $X_i \approx M^{-1}$.

Formally, we have

$$\|M^{-1} - Y_i\|_2 \le (1 + (\|A\|_2 + \|B\|_2)\nu^-)e_{2,i} \tag{6.4.2}$$

for a non-singular operator $\nabla_{B,A}$, $\nu^- = v_{r,2}(\nabla_{B,A}^{-1})$ in Definition 6.1.4 and $e_{2,i}$ in (6.2.5). By analyzing and summarizing our estimates, we obtain the following corollary, where we write

$$\hat{e}_{2,i} = \|M^{-1} - Y_i\|_2, \tag{6.4.3}$$

$$\hat{\rho}_{2,i} = \|I - MY_i\|_2, \tag{6.4.4}$$

$$\mu = (1 + (\|A\|_2 + \|B\|_2)\nu^-)\kappa(M) \tag{6.4.5}$$

for $\nu^- = v_{r,2}(\nabla_{B,A}^{-1})$ in Definition 6.1.4 and $\kappa(M)$ in Definition 6.1.1.

Corollary 6.4.3. *Let Algorithm 6.3.1 be applied for a non-singular operator* $\nabla_{B,A}$ *and combined with Subroutine* **R1**. *Then we have*

$$\mu \hat{\rho}_{2,i} \le (\mu \hat{\rho}_{2,0})^{2^i}, \quad i = 1, 2, \ldots,$$

$$\mu \hat{e}_{2,i} = \mu \|M^{-1} - Y_i\|_2 \le (\mu \hat{\rho}_{2,0})^{2^i} \|M^{-1}\|_2, \quad i = 1, 2, \ldots.$$

Corollary 6.4.4. *Under the assumptions of Corollary 6.4.3, let positive* δ *and* θ *be less than 1 and let*

$$\mu \hat{\rho}_{2,0} \le \theta < 1. \tag{6.4.6}$$

Then we have

$$\mu \hat{\rho}_{2,i} \le \theta^{2^i}, \quad \mu \hat{e}_{2,i} \le \theta^{2^i} \|M^{-1}\|_2, \quad i = 1, 2, \ldots, \tag{6.4.7}$$

so the bounds $\hat{\rho}_{2,q} \le \delta/\mu$ *and* $\hat{e}_{2,q} \le \delta\kappa(M)/\mu$ *on the residual and error norms can be ensured in* $q = \lceil \log((\log \delta)/\log \theta) \rceil$ *steps of Algorithm 6.3.1 with Subroutine* **R1**. *Under (6.4.1),* $O((r v_{r,n}(\nabla_{B,A}) + r^2 n)q)$ *flops are used in all these steps, which is* $O(q n r^2 \log^d n)$ *flops if* $\nabla_{A,B}$ *and* $\nabla_{B,A}$ *are regular operators in Examples 1.4.1 and 4.4.1–4.4.9 and if* $v_{r,n}(L) = O(nr \log^d n)$ *for* $L = \nabla_{A,B}$ *and* $L = \nabla_{B,A}$; *furthermore, we have* $d \le 2$ *under Assumption 4.7.4.*

The estimates of Corollaries 6.4.3 and 6.4.4 have been obtained provided that the initial residual norm $\hat{\rho}_{2,0}$ is bounded in terms of the norm $\nu^- = \nu_{2,r}(\nabla_{A,B}^{-1})$ (see (6.4.5) and (6.4.6)) and similarly with the alternative Subroutine **R2** of the next section (see (6.5.5) and (6.5.6)). The estimates for the norm ν^- are supplied in Section 6.6, and some techniques for choosing an initial matrix X_0 are described in Sections 6.7 and 6.9. On further acceleration of the convergence of Newton's iteration by scaling, see Section 6.7. In Section 6.8, we comment on variations of the truncation policy where we truncate fewer singular values to trade stronger compression for closer approximation (particularly at the initial stages of Newton iteration where the convergence is fragile and can be easily destroyed by compression).

6.5 Compression of displacement generators of approximate inverses by substitution

To derive an alternative variant of the Subroutine **R**, denoted by **R2**, we substitute $X_{i+1} = Y_i(2I - MY_i) \approx M^{-1}$ for M^{-1} in the expression in Theorem 1.5.3 for $L(M^{-1})$.

Subroutine R2.   Let $\nabla_{A,B}(M) = GH^T$, for two given matrices G and H. Compute and output the two matrices

$$G_{i+1} = Y_i(-2I + MY_i)G, \tag{6.5.1}$$

$$H_{i+1}^T = H^T Y_i(2I - MY_i). \tag{6.5.2}$$

The subroutine can be applied numerically and algebraically (over any field). In this chapter we assume its numerical implementation.

The computation by the Subroutine **R2** uses $O(r\nu_{r,n}(\nabla_{B,A}))$ flops and outputs $\nabla_{B,A}$-generator G_{i+1}, H_{i+1} of length r for a matrix Y_{i+1} satisfying the equation $\nabla_{B,A}(Y_{i+1}) = G_{i+1}H_{i+1}^T$. We assume regular operators $L = \nabla_{B,A}$; then the matrix Y_{i+1} is uniquely defined by its L-generator G_{i+1}, H_{i+1}, and (6.5.1) and (6.5.2) completely define Newton's process without involving the matrix X_{i+1} in (6.3.1). For partly regular operators L, the same algorithm applies, but in addition one should compute the entries of the matrix Y_{i+1} on the associated irregularity set. For instance, if the set has been made of the first column, that is, if $M_{I.S.}(L) = Me_0$, then in addition to the matrices G_{i+1} and H_{i+1} in (6.5.1) and (6.5.2), the i-th Newton step also computes the vector

$$Y_{i+1}e_0 = Y_i(2I - M_iY_i)e_0. \tag{6.5.3}$$

Then again (6.5.1)–(6.5.3) completely define Newton's process, without involving the matrix X_{i+1}.

We may equivalently rewrite (6.5.1) and (6.5.2) as

$$G_{i+1} = -X_{i+1}G, \quad H_{i+1}^T = H^T X_{i+1}. \tag{6.5.4}$$

Subroutine **R2** achieves *compression* because $\mathrm{rank}(G_{i+1}H_{i+1}^T) \leq \mathrm{rank}(GH^T)$; furthermore, it preserves *approximation*. Indeed, since $X_{i+1} \approx M^{-1}$, we deduce from

Theorem 1.5.3 that

$$G_{i+1} H_{i+1}^T \approx -M^{-1} G H^T M^{-1} = \nabla_{B,A}(M^{-1}).$$

The *convergence rate* of the Newton-Structured Iteration with Subroutine **R2** is estimated in the next theorem.

Theorem 6.5.1. *Assume the Newton-Structured Iteration of Algorithm 6.3.1 combined with Subroutine* **R2** *for a non-singular operator L. Write* $\hat{e}_i = \|M^{-1} - Y_i\|$ *and* $C_i = \nu^- \|G H^T\|(e_i + 2\|M^{-1}\|)$, *for* e_i *in (6.2.5) and* ν^- *in Definition 6.1.4. Then* $\hat{e}_i \le C_i e_i$, $e_i \le (C_{i-1} e_{i-1})^2 \|M\|$, $i = 1, 2, \ldots$.

Theorem 6.5.1 is proved in [PRWa] using the bound $e_{i+1} \le \|M\| \hat{e}_i^2$. This bound follows because, by (6.3.1), we have $M^{-1} - X_{i+1} = (M^{-1} - Y_i) M (M^{-1} - Y_i)$.

Based on Theorem 6.5.1, we estimate the computational cost of performing Algorithm 6.3.1 with Subroutine **R2**. We arrive at the following estimates.

Theorem 6.5.2. *Under the assumptions of Theorem 6.5.1, let* $\|I - M Y_0\| = \hat{\rho}_0 \le 1$ *and let* $e_i \le \|M^{-1}\|$ *for all i. Write*

$$C = 3\nu^- \|G H^T\| \; \|Y_0\|/(1 - \hat{\rho}_0), \quad \tilde{\mu} = C^2 \|M\|, \qquad (6.5.5)$$

and $\tilde{e} = \hat{\rho}_0 \|Y_0\|/(1 - \hat{\rho}_0)$. *Let*

$$\tilde{\mu} e_1 \le (C\tilde{e}\|M\|)^2 \le \theta^2. \qquad (6.5.6)$$

Then $\tilde{\mu} e_{i+1} \le (\tilde{\mu} e_i)^2 \le (C\tilde{e}\|M\|)^{2^{i+1}} \le \theta^{2^{i+1}}$, $\hat{e}_{i+1} \le C e_{i+1}$, $i = 0, 1, \ldots$.

Corollary 6.5.3. *Given positive* $\hat{\epsilon}$ *and* θ, *both less than 1, write* $\hat{\epsilon} = \epsilon/\tilde{\mu}$, $k = \lceil \log((\log \epsilon)/\log \theta) \rceil$. *Then, under the assumptions of Theorem 6.5.2, we have* $e_k \le \hat{\epsilon}$, $\hat{e}_k \le C\hat{\epsilon}$. *The latter bounds are reached in* $O(r \upsilon_{r,n}(\nabla_{B,A}))$ *flops. This turns into* $O(kr^2 n \log^d n)$ *flops if* $\nabla_{A,B}$ *and* $\Delta_{A,B}$ *are regular operators in Examples 1.4.1 and 4.4.1–4.4.9 and if* $\upsilon_{r,n}(\nabla_{B,A}) = O(rn \log^d n)$; *furthermore,* $d \le 2$ *under Assumption 4.7.4.*

Remark 6.5.4. The computational cost of performing Subroutine **R1** is dominated by the computation of the SVD of the displacement. This operation allows its low cost parallel implementation for smaller values r (see Remark 4.6.8). Otherwise Algorithm 6.3.1 (with both Subroutines **R1** and **R2**) essentially reduces to multiplication of structured matrices performed $2q$ or $2k$ times for q of Corollary 6.4.4 and k of Corollary 6.5.3. The parallel arithmetic cost of this computation is estimated in Remark 6.3.3.

We conclude this section by specifying three variants of **Newton–Toeplitz Iteration,** that is, the Newton-Structured Iteration with Subroutine **R2** applied to a Toeplitz input matrix M.

Based on the two formulae of Exercise 2.24 (b), (c) for the inverse T^{-1} of a Toeplitz matrix $T = (t_{i-j})_{i,j=0}^{n-1}$, we define two variants of the Newton-Toeplitz Iteration as follows:

$$Y_{i+1} = Z(\mathbf{p}_i) Z^T(\mathbf{u}_i) - Z(\mathbf{x}_i) Z^T(\mathbf{q}_i) \qquad (6.5.7)$$

where $\mathbf{p}_i = Y_i(2I - TY_i)\tilde{\mathbf{t}}$, $\mathbf{x}_i = Y_i(2I - TY_i)\mathbf{e}_0$, $\tilde{\mathbf{t}} = (t_{i-n})_{i=0}^{n-1}$ for any fixed scalar t_{-n}, $\mathbf{q}_i = ZJ\mathbf{p}_i - \mathbf{e}_0$, $\mathbf{u}_i = ZJ\mathbf{x}_i$, $i = 0, 1, \dots$, and

$$Y_{i+1} = \frac{1}{1 - bf} \left(Z_f(\mathbf{x}_i) Z_{1/b}(\mathbf{v}_i) - Z_f(\mathbf{v} - (1 - bf)\mathbf{e}_0) Z_{1/b}(\mathbf{x}_i) \right) \qquad (6.5.8)$$

where $\mathbf{v}_i = Y_i(2I - TY_i)\tilde{\mathbf{v}}$ and $\mathbf{x}_i = Y_i(2I - TY_i)\mathbf{e}_0$ are defined similarly to the vectors $\mathbf{p}_i$ and $\mathbf{x}_i$ in (6.5.7), $\tilde{\mathbf{v}} = (a, t_1 - bt_{1-n}, t_2 - bt_{2-n}, \dots, t_{n-1} - bt_{-1})^T$, and a, b and f are three fixed scalars, $b \neq 0$, $f \neq 1/b$.

Furthermore, for a positive definite Toeplitz matrix $T = (t_{i-j})$, its inverse and close approximations Y_i to the inverse have positive northwestern entries $(T^{-1})_{00}$ and $(Y_i)_{00}$, respectively, which are no less than the 2-norms of the matrices T^{-1} and Y_i, respectively [GL96]. Then we define the Newton–Gohberg–Semencul Iteration based on (2.11.4):

$$(Y_{i+1})_{00} Y_{i+1} = Z(\mathbf{x}_i) Z^T(J\mathbf{y}_i) - Z(Z\mathbf{y}_i) Z^T(ZJ\mathbf{x}_i) \qquad (6.5.9)$$

where $\mathbf{x}_i = Y_i\mathbf{e}_0$, $\mathbf{y}_i = Y_i\mathbf{e}_{n-1}$, and $(Y_{i+1})_{00}$ equals the first coordinate of the vector $\mathbf{x}_i$.

We refer the reader to the Notes to Section 1.5 on other Toeplitz inversion formulae, which immediately define further modifications of the Newton–Toeplitz Iteration.

6.6   Bounds on the norms of the inverse operator

6.6.1   Introductory comments

We have to estimate the operator norm ν^- of Definition 6.1.4 to complete the estimates of Corollaries 6.4.3 and 6.5.3. This is also a natural subject in the numerical study of the inverse displacement operators L^{-1} and the DECOMPRESS stage of the displacement rank approach.

In the next four subsections, we recall five approaches to obtaining these estimates. All approaches use distinct techniques of independent interest. We refer the reader to the original papers [PRWa] and [PWa] for proofs, further details, and an additional SVD based approach. Before recalling the five approaches, let us briefly comment on their predecessor from the papers [P92b] and [P93], where we proved in the Toeplitz-like case that

$$\|M^{-1} - Y_i\|_2 \leq (1 + 2(l_i - r)n)\|M^{-1} - X_i\|_2$$

under Subroutine **R1** for $r = \text{rank}(L(M^{-1}))$, $l_i = \text{rank}(L(Y_i)) \leq 3r$, and $L = \Delta_{Z^T, Z}$. The factor n above has come from double application of Theorem 6.1.2 c) where $m = n$. The theorem is applied to support the transition between 1-norms and 2-norms of matrices. Each application contributes at worst the factor of $\sqrt{n}$, which tends to be smaller on the average input. Likewise, the factor $l_i - r$ is attained only where $\sigma = \sigma_j(L(X_i))$, $j = 1, \dots, l_i$, that is, where all singular values of the matrix $L(X_i)$ are exactly the same, which is a rather rare case. Most important, the upper bounds would not be reached unless the truncation of the singular values moves the computed approximation X_i in the direction precisely opposite to the direction to the inverse, but again, this is a rather exceptional case. Thus the above upper bound on the norm

$\|M^{-1} - Y_i\|$ is overly pessimistic on the "average input," whatever this term realistically means. Similar comments apply to the upper estimates of the next subsections.

6.6.2 Bounds via powering the operator matrices

Corollary 4.3.7 implies the following result:

Corollary 6.6.1. *Let* $\Delta = \Delta_{A,B}$ *be a non-singular operator of Stein type where* $A^k = fI$ *(or* $B^k = fI$ *) for some scalar* f *and positive integer* k. *Then we have*

$$v_{r,l}(\Delta^{-1}) \leq \|(I - fB^k)^{-1}\|_l \sum_{i=0}^{k-1} \|A\|_l^i \|B\|_l^i$$

(or, respectively, $v_{r,l}(\Delta^{-1}) \leq \|(I - fA^k)^{-1}\|_l \sum_{i=0}^{k-1} \|A\|_l^i \|B\|_l^i)$ *where* $l = 1, 2, \infty$.

Based on Corollary 6.6.1, we estimate the inverse operator norm $v_{r,l}(\Delta_{A,B}^{-1})$ where $A \in \{Z_e, Z_e^T\}$, $B \in \{Z_f, Z_f^T, D(\mathbf{t})\}$ or $A = D(\mathbf{t})$, $B \in \{Z_f, Z_f^T\}$ for some scalars e and f and vector $\mathbf{t}$. In particular, for operators L associated with Toeplitz-like and Hankel-like matrices we obtain that $v_{r,l}(\Delta_{A,B}^{-1}) \leq an$ for $l = 1, \infty$ and any $r \geq 1$ where $A, B \in \{Z_e, Z_f, Z_e^T, Z_f^T\}$, $|e| \leq 1$, $|f| \leq 1$, $a = 1/\max\{|1 - e|, |1 - f|\}$; $a = 1/2$ where $e = -1$ or $f = -1$.

6.6.3 The spectral approach

The technique of the previous subsection can be generalized based on the next theorem.

Theorem 6.6.2. *Let* $\Delta = \Delta_{A,B}$ *be a non-singular operator of Stein type with* $n \times n$ *operator matrices* A *and* B. *Let* $\lambda_1, \ldots, \lambda_n$ *be the eigenvalues of the matrix* A *and let neither of them be an eigenvalue of the matrix* B. *Write* $A_{\lambda_i} = A - \lambda_i I$, $B_{\lambda_i} = I - \lambda_i B$, $i = 1, \ldots, n$; $\prod_{i=1}^{0} A_{\lambda_i} = I$, $\prod_{i=1}^{0} \|A_{\lambda_i}\| = 1$. *Then we have*

$$M = \sum_{k=1}^{n} \left(\prod_{i=1}^{k-1} A_{\lambda_i} \right) \Delta(M) \left(\prod_{i=1}^{k} B_{\lambda_i}^{-1} \right) B^{k-1},$$

and consequently,

$$v_{r,l}(\Delta_{A,B}^{-1}) \leq \sum_{k=1}^{n} \|B^{k-1}\|_l \left(\prod_{i=1}^{k-1} \|A_{\lambda_i}\|_l \right) \left(\prod_{j=1}^{k} \|B_{\lambda_j}^{-1}\|_l \right), \quad l = 1, 2, \infty.$$

Theorem 6.6.2 enables us to estimate the norm of the inverse operators associated with Cauchy-like and Toeplitz+Hankel-like matrices.

6.6.4   The bilinear expression approach

In Table 6.1 we estimate the norm $\|L^{-1}\|$ based on the bilinear expressions for a structured matrix M via its L-generator (G, H) (cf. Examples 1.4.1, 4.4.1–4.4.9, and Section 4.5). In particular, for $A \in \{Z_e, Z_e^T\}$ and $B \in \{Z_f, Z_f^T\}$, $e = 0$, $f = 1$ or $e = 1$, $f = 0$, we obtain that $\|\nabla_{A,B}^{-1}\| \le n$, $\|\Delta_{A,B}^{-1}\| \le n$.

Table 6.1: Upper bounds on the 2-norms of the inverse displacement operators. We use parameters $\tilde{e} = \max(1, |e|)$, $\tilde{f} = \max(1, |f|)$, $\tilde{v}_j = \max(1, |v_j|)$. **The norm bounds in the last line (for $A = D(s)$, $B = D(t)$) are restricted to the matrices having displacement rank of at most r.**

$A \in \mathbb{C}^{n \times n}$	$B \in \mathbb{C}^{n \times n}$	$\|\nabla_{A,B}^{-1}\|_2$	$\|\Delta_{A,B}^{-1}\|_2$
Z_e or Z_e^T	Z_f or Z_f^T	$\le n \dfrac{\tilde{e}\tilde{f}}{|e-f|}$	$\le n \dfrac{\tilde{e}\tilde{f}}{|1-ef|}$
Z_e or Z_e^T	$D(v)$	$\le \max\limits_j \dfrac{1}{|e-v_j^n|} \dfrac{1-\tilde{v}_j^n}{1-\tilde{v}_j}$	$\le \max\limits_j \dfrac{e}{|1-ev_j^n|} \dfrac{1-\tilde{v}_j^n}{1-\tilde{v}_j}$
$D(s)$	$D(t)$	$\le r \|(\frac{1}{s_i-t_j})_{i,j}\|_2$	$\le r \|(\frac{1}{1-s_i t_j})_{i,j}\|_2$

6.6.5   Estimation using the Frobenius norm

Theorem 6.1.2 a) implies the following corollary, which closely relates the norms $\|L^{-1}\|_2$ and $\|L^{-1}\|_F$ to each other.

Corollary 6.6.3. *Let a linear operator L be defined on the space of $n \times n$ matrices having positive L-rank of at most r. Then $1/\sqrt{r} \le \|L^{-1}\|_F / \|L^{-1}\|_2 \le \sqrt{n}$, that is, $v_{r,F}(L^{-1})/\sqrt{n} \le v_{r,2}(L^{-1}) \le \sqrt{r} v_{r,F}(L^{-1})$, $1/\sqrt{n} \le \|L\|_F / \|L\|_2 \le \sqrt{r}$.*

Let us next estimate the Frobenius norms $\|L^{-1}\|_F$.

Theorem 6.6.4. *For non-singular operators $\nabla_{A,B}$ and $\Delta_{A,B}$, we have*

$$\|\nabla_{A,B}^{-1}\|_F = \|(I_n \otimes A - B^T \otimes I_m)^{-1}\| \ge \max_{i,j} |\lambda_i(A) - \lambda_j(B)|^{-1},$$

$$\|\Delta_{A,B}^{-1}\|_F = \|(I_{mn} - B^T \otimes A)^{-1}\| \ge \max_{i,j} |1 - \lambda_i(A)\lambda_j(B)|^{-1},$$

where $\otimes$ is the Kronecker product.

Theorems 4.3.5 and 6.6.2 together imply the next result.

Corollary 6.6.5. *Let $\widehat{A} = VAV^{-1}$, $\widehat{B} = W^{-1}BW$ for some non-singular matrices V and W. Then we have*

$$\|\nabla_{A,B}^{-1}\|_2 \le C\|\nabla_{\widehat{A},\widehat{B}}^{-1}\|_2, \quad \|\nabla_{A,B}^{-1}\|_F \le C\|\nabla_{\widehat{A},\widehat{B}}^{-1}\|_F,$$

$$\|\Delta_{A,B}^{-1}\|_2 \le C\|\Delta_{\widehat{A},\widehat{B}}^{-1}\|_2, \quad \|\Delta_{A,B}^{-1}\|_F \le C\|\Delta_{\widehat{A},\widehat{B}}^{-1}\|_F,$$

where $C = \|V\|_2 \|V^{-1}\|_2 \|W\|_2 \|W^{-1}\|_2$.

Table 6.2: Upper bounds on the Frobenius norms of the inverse displacement operators

$A \in \mathbb{C}^{n \times n}$	$B \in \mathbb{C}^{n \times n}$	$\lVert \nabla_{A,B}^{-1} \rVert_F$
Z_e or Z_e^T	Z_f or Z_f^T	$\leq \frac{n}{\lvert e-f \rvert} \max(\lvert e \rvert + \lvert f \rvert, 1 + \lvert ef \rvert)$
Z_e or Z_e^T	$D(v)$	$\leq \max_j \frac{1}{\lvert e-v_j^n \rvert}(\lvert v_j \rvert^{n-1} + \frac{1-\lvert v_j \rvert^{n-1}}{1-\lvert v_j \rvert} \max(1, \lvert e \rvert))$

$A \in \mathbb{C}^{n \times n}$	$B \in \mathbb{C}^{n \times n}$	$\lVert \Delta_{A,B}^{-1} \rVert_F$
Z_e or Z_e^T	Z_f or Z_f^T	$\leq \frac{n}{\lvert 1-ef \rvert} \max(1 + \lvert e \rvert, 1 + \lvert f \rvert, 1 + \lvert ef \rvert)$
Z_e or Z_e^T	$D(v)$	$\leq \max_j \frac{1}{\lvert 1-ev_j^n \rvert}(1 + \frac{\lvert v_j \rvert - \lvert v_j \rvert^n}{1-\lvert v_j \rvert} \max(1, \lvert e \rvert))$

Based on the latter results, we obtain the estimates of Tables 6.2 and 6.3 for the norms $\lVert L^{-1} \rVert_F$ for several most used operators L.

Table 6.3: Lower and upper bounds for the Frobenius norms of the inverse displacement operators. Parameters $c_1, \ldots, c_4$ satisfy $1 \leq c_1, c_3 \leq d_1 d_2, 1 \leq c_2, c_4 \leq d_1$ where $d_1 = \max(\lvert e \rvert^{\frac{n-1}{n}}, \lvert e \rvert^{\frac{1-n}{n}}), d_2 = \max(\lvert f \rvert^{\frac{n-1}{n}}, \lvert f \rvert^{\frac{1-n}{n}})$

$A \in \mathbb{C}^{n \times n}$	$B \in \mathbb{C}^{n \times n}$	$\lVert \nabla_{A,B}^{-1} \rVert_F$	$\lVert \Delta_{A,B}^{-1} \rVert_F$
Z_e or Z_e^T	Z_f or Z_f^T	$c_1 \max_j \lvert e^{\frac{1}{n}} - f^{\frac{1}{n}} \omega_n^j \rvert^{-1}$	$c_2 \max_j \lvert 1 - e^{\frac{1}{n}} f^{\frac{1}{n}} \omega_n^j \rvert^{-1}$
Z_e or Z_e^T	$D(v)$	$c_3 \max_{i,j} \lvert e^{\frac{1}{n}} \omega_n^i - v_j \rvert^{-1}$	$c_4 \max_{i,j} \lvert 1 - e^{\frac{1}{n}} \omega_n^i v_j \rvert^{-1}$
$D(s)$	$D(t)$	$\max_{i,j} \lvert s_i - t_j \rvert^{-1}$	$\max_{i,j} \lvert 1 - s_i t_j \rvert^{-1}$

In particular, let A denote Z_e or Z_e^T and let B denote Z_f or Z_f^T. Then the first lines of Tables 6.1 and 6.2 for $e = 0$, $f = 1$ as well as for $e = 1$, $f = 0$ imply the bounds $\lVert \nabla_{Z_e,Z_f}^{-1} \rVert_2 \leq n$, $\lVert \Delta_{Z_e,Z_f}^{-1} \rVert_2 \leq n$, $\lVert \nabla_{Z_e,Z_f}^{-1} \rVert_F \leq n$, $\lVert \Delta_{Z_e,Z_f}^{-1} \rVert_F \leq 2n$, respectively. Furthermore, the first line of Table 6.3 implies that $\lVert \nabla_{Z_e,Z_f}^{-1} \rVert_F = 1/(2 \sin \frac{\pi}{2n})$ where $\lvert e \rvert = 1$, $f = -e$, and $\lVert \Delta_{Z_e,Z_f}^{-1} \rVert_F = 1/(2 \sin \frac{\pi}{2n})$ where $\lvert e \rvert = 1$, $ef = -1$.

6.7 Initialization, analysis, and modifications of Newton's iteration for a general matrix

When we defined Newton's iteration (6.2.4) and the Newton-Structured Iteration (Algorithm 3.3.1), we assumed that a close initial approximation X_0 to the inverse of the input matrix M is available. In some cases this is a realistic assumption, for instance, where we update a dynamic process and use the data from the preceding period of a dynamic model or where the approximation X_0 can be supplied either by the preconditioned conjugate gradient method, which has linear convergence, or by a direct method performed with rounding errors.

Now suppose that no initial approximation X_0 is given from outside and let us *generate the approximation based on the SVD of M*. First consider Newton's iteration (6.2.4) for a general $n \times n$ matrix M. Write the SVDs of the matrices M, M^* and M^*M:

$$M = U\Sigma V^*, \quad M^* = V\Sigma U^*, \quad M^*M = V\Sigma^2 V^*, \quad U^*U = I, \quad V^*V = I,$$
$$\Sigma = \mathrm{diag}((\sigma_j)_{j=1}^n), \quad \sigma_+ \geq \sigma_1 \geq \ldots \geq \sigma_n \geq \sigma_- > 0$$

for two fixed constants σ_+ and σ_- (cf. Definition 4.6.1). Then the choice of

$$X_0 = a_0 M^*, \quad a_0 = 2/(\sigma_+ + \sigma_-), \tag{6.7.1}$$

symmetrizes the matrices $X_i M$ and $I - X_i M$ for all i.

Let us estimate the *convergence rate* of the iteration with this choice of X_0 based on the observation that the matrices M, M^{-1}, M^+, and X_i for all i share all their singular vectors. First deduce that

$$X_i = V\Sigma_i U^*, \quad X_i M = V S_i V^*,$$

where

$$\Sigma_i = \mathrm{diag}(\sigma_j^{(i)})_{j=1}^n, \quad S_i = \mathrm{diag}(s_j^{(i)})_{j=1}^n,$$
$$\sigma_j^{(i)} = \sigma_j^{(i-1)}(2 - \sigma_j \sigma_j^{(i-1)}), \quad s_j^{(0)} = a_0\sigma_j^2, \quad s_j^{(i)} = \sigma_j^{(i)}\sigma_j,$$
$$j = 1, \ldots, n, \quad i = 1, 2, \ldots.$$

Consequently, we have

$$1 - s_j^{(i)} = (1 - s_j^{(i-1)})^2 = (1 - s_j^{(0)})^{2^i} = (1 - a_0\sigma_j^2)^{2^i}, \quad \text{for all } j \text{ and } i. \tag{6.7.2}$$

Observe that

$$1 - a_0\sigma_j^2 = 1 - 2\sigma_j^2/(\sigma_+^2 + \sigma_-^2), \quad j = 1, \ldots, n,$$
$$1 - a_0\sigma_+^2 \leq 1 - a_0\sigma_1^2 \leq 1 - a_0\sigma_j^2 \leq 1 - a_0\sigma_n^2 \leq 1 - a_0\sigma_-^2, \quad j = 1, \ldots, n.$$

For $\sigma_+ = \sigma_1, \sigma_- = \sigma_n, 2/a_0 = \sigma_+^2 + \sigma_-^2$, we have

$$-(1 - a_0\sigma_+^2) = 1 - a_0\sigma_-^2 = \frac{\sigma_+^2 - \sigma_-^2}{\sigma_+^2 + \sigma_-^2} = \frac{\kappa_+^2(M) - 1}{\kappa_+^2(M) + 1} = 1 - \frac{2}{\kappa_+^2(M) + 1}, \tag{6.7.3}$$

where $\kappa_+(M) = \sigma_+/\sigma_-$ is an upper bound on the condition number $\kappa(M) = \sigma_1/\sigma_n$ of the matrix M. (In practice it is usually desirable to avoid approximating σ_1 and particularly σ_n. An alternative choice of the easily computable value $a_0 = 1/(\|M\|_1 \|M\|_\infty)$ may only slightly increase the range $[1 - a_0\sigma_1^2, 1 - a_0\sigma_n^2]$ for the eigenvalues of the matrices $I - X_i M$.) For a *non-singular matrix* M, (6.7.2) and (6.7.3) together imply that $\|I - X_i M\|_2 \le 1/2$ for

$$i \approx 2\log_2 \kappa_+(M), \tag{6.7.4}$$

and $\|I - X_{i+k} M\|_2 \le \epsilon$ for

$$k \approx \log_2 \log_2(1/\epsilon). \tag{6.7.5}$$

The same analysis also immediately applies in the case where M is a *singular matrix*, that is, $\sigma_+ \ge \sigma_1 \ge \ldots \ge \sigma_r \ge \sigma_- > 0, r = \text{rank}(M)$. In this case *iteration defined by (6.2.4) and (6.7.1) converges to the Moore–Penrose generalized inverse M^+ with the same rate* expressed by (6.7.4) and (6.7.5). That is, the eigenvalues $s_j^{(i)}$ of the matrices $X_i M$ equal 0 for all i and for $j > r, r = \text{rank}(M)$ and converge to 1 as $i \to \infty$ for $j \le r$. The convergence rate is defined by relations (6.7.4), (6.7.5), that is, for the eigenvalues $s_j^{(i)}$ we have $|s_j^{(i)} - 1| \le 1/2$ and $|s_j^{(i+k)} - 1| \le \epsilon$ for $j \le r, i$ in (6.7.4) and k in (6.7.5).

In the paper [PS91], various modifications of process (6.2.4) and choices of the initial approximation were proposed. In particular, the paper includes the following results:

a) A scaled version of Newton's iteration,

$$X_{i+1} = a_{i+1} X_i (2I - M X_i), \quad i = 0, 1, \ldots, \tag{6.7.6}$$

where $a_{i+1} = 2/(1 + (2 - a_i^-)a_i^-)$, $a_0^- = 2\sigma_-/(\sigma_+ + \sigma_-)$, $a_{i+1}^- = a_{i+1}(2 - a_i^-)a_i^-$, $a_i^- \le \sigma_j(X_i M) \le a_i$ for $j = 1, \ldots, r$; $r = \text{rank}(M), i = 0, 1, \ldots$, and (6.7.1) holds. The residual $R(X_i, M) = I - X_i M = I - X_0 M p_i(X_0 M) = t_{2^i}(X_0 M)$ becomes the scaled Chebyshev polynomial in $X_0 M$ on the interval (σ_-, σ_+), which has the 2^i-th degree and minimizes the norm $\|p(x)\|_{\infty, \sigma_-, \sigma_+} = \max_{\sigma_- \le x \le \sigma_+} |p(x)|$ among all polynomials $p(x)$ of degree of at most 2^i such that $p(0) = 1$. Since the singular values of $X_0 M$ lie in this interval, the eigenvalues of $X_i M$ converge to 1 most rapidly, provided that the matrices $X_i M$ are polynomials of degree 2^i in $X_0 M$ vanishing where $X_0 M = 0$. It has been estimated in [PS91] that bound (6.7.4) on the number of Newton steps decreases roughly by twice under this scaling policy.

b) Convergence acceleration based on the third order processes of residual correction,

$$X_{i+1} = (a_i(X_i M)^2 + b_i X_i M + c_i I) X_i, \quad i = 0, 1, \ldots \tag{6.7.7}$$

(with specified scalars a_i, b_i and c_i), which shows potential benefits of using higher order iteration, $X_{i+1} = X_i p_k(X_i), i = 0, 1, \ldots$, where $p_k(x)$ is a degree k polynomial, $k > 1$ (see (6.2.9) and (6.2.10));

c) Suppression of the smallest singular values of matrices M and M^+ in a modification of Newton's process. The modification forces convergence to a matrix

$M_\epsilon^+ = V \Sigma_\epsilon^+ U^*$ where

$$M^+ = V\Sigma^+U^*, \quad M_\epsilon = U\Sigma_\epsilon V^*, \quad M = U\Sigma V^*,$$
$$U^*U = UU^* = V^*V = VV^* = I,$$
$$\Sigma = \mathrm{diag}(\sigma_i)_{i=1}^n, \quad \Sigma_\epsilon = \mathrm{diag}(\sigma_i(\epsilon))_{i=1}^n,$$
$$\Sigma^+ = \mathrm{diag}(\Sigma_i^+)_{i=1}^n, \quad \Sigma_\epsilon^+ = \mathrm{diag}(\sigma_i^+(\epsilon))_{i=1}^n,$$
$$\sigma_1 \geq \sigma_2 \geq \ldots \geq \sigma_n \geq 0, \quad \sigma_i(\epsilon) = \begin{cases} \sigma_i & \text{where } \sigma_i \geq \epsilon \\ 0 & \text{otherwise,} \end{cases}$$
$$\sigma_i^+ = \begin{cases} 1/\sigma_i & \text{where } \sigma_i \geq 0 \\ 0 & \text{otherwise,} \end{cases}$$
$$\sigma_i^+(\epsilon) = \begin{cases} 1/\sigma_i & \text{where } \sigma_i \geq \epsilon \\ 0 & \text{otherwise.} \end{cases}$$

Let us show the algorithm referring the reader to [PS91] for motivation and analysis. Iteration (6.7.6) is applied with the scaling of part a) but with

$$a_0 = \sigma_+^2 a, \quad a_0^- = a\epsilon^2, \quad a = \min(2/(\sigma_+^2 + \epsilon^2), \rho/\epsilon^2), \tag{6.7.8}$$

$\rho = (3-\sqrt{5})/2 = 0.3819\ldots$ (Under the scaling in (6.7.8), the value ρ partitions the range for the spectrum of the matrix $X_0 M$ reflecting the respective partition of the singular values of the matrix M by ϵ. σ_- is not needed in this variation of the iteration.) The iteration is performed until $a_i^- \geq \rho$ for some integer i. Then the matrix X_i is scaled, that is, replaced by the matrix $(\rho/a_i^-)X_i$, and the iteration is continued based on the expressions

$$X_{i+1} = \tilde{X}_i M \tilde{X}_i, \quad \tilde{X}_i = (2I - X_i M)X_i, \quad i = 0, 1, \ldots. \tag{6.7.9}$$

The singular values $\sigma_j(M)$ are partitioned by ϵ into two groups: those exceeding ϵ correspond to the eigenvalues $\lambda^{(i)}$ of $X_i M$ lying in the interval $\rho < \lambda^{(i)} \leq 2 - \rho$; iteration (6.7.9) sends them towards 1. The other eigenvalues of $X_i M$ lie in the interval $[0, \rho)$; they correspond to the singular values $\sigma_j(M) < \epsilon$. Iteration (6.7.9) sends them towards 0. This means the desired convergence to the matrix M_ϵ^+. In other words, under iteration (6.7.9), $X_i M = q_i(X_0 M)$ where $q_i(x)$ is a polynomial of degree 4^i,

$$\lim_{i \to \infty} q_i(x) = \begin{cases} 0, & 0 \leq x < \rho \\ 1, & \rho < x \leq 2 - \rho. \end{cases}$$

The convergence is ultimately quadratic but is slow near ρ and $2 - \rho$. The iteration can be immediately extended to the computation of the matrices $M_\epsilon = MM_\epsilon^+ M$ and $\tilde{M}_\epsilon = M - M_\epsilon$.

d) The proof of strong numerical stability of both original and modified iterations.

e) Improved initial scaling for the Hermitian positive definite matrix M. In particular, even having no estimates for the singular values $\sigma_i(M)$, one may choose

$$a_0 = 1/\|M\|_F, \tag{6.7.10}$$

letting $1/a_0$ be the Frobenius norm of the matrix M. Under this choice, we have $\|I - X_0 M\|_2 \leq 1 - n^{-1/2}/\kappa(M)$, and then the number i of Newton's steps (6.2.4) required to achieve the bound $\|I - X_i M\|_2 \leq 1/2$ immediately decreases to

$$i = 0.5 \log_2 \kappa(M) + O(1). \tag{6.7.11}$$

The same process converges to M^+ with the same rate under (6.7.10) where M is a semi-definite Hermitian matrix.

The above results are not valid in the case of Newton-Structured Iteration. Indeed, the compression of the displacement generators caused by application of the Compression Subroutine **R** may perturb the computed approximations X_i, so the basic relations between the singular vectors and/or eigenvectors of X_i and the input matrix M are gradually more and more destroyed by the compression.

6.8 How much should we compress the displacements of the computed approximations?

Suppose that Algorithm 6.3.1 is applied to a structured input matrix M. Then at every iteration step the compression of the displacement perturbs the computed approximation X_i. Of the two Subroutines **R1** and **R2**, the former one allows compression of various levels, measured by the number $l = l(i)$ of the truncated singular values of the displacement $L(X_i)$. The more singular values of the displacements $L(X_i)$ are truncated, the stronger compression we have in each Newton step. This means more structure and faster computations in each step. On the other hand, the stronger perturbation is caused by the transition from X_i to Y_i, the more Newton steps may be required, and the less valid is the convergence analysis of the previous section.

How many smallest singular values of the displacement $L(X_i)$ should we truncate? Surely it is numerically safe to truncate the singular values that are smaller than the *machine epsilon* (also called the *unit roundoff*), that is, the level of a fixed single precision, but is it safe to apply the maximal truncation, that is, to truncate all singular values but the r largest of them, $r = \text{rank}(L(M^{-1}))$? In particular, the initial approximation by X_i to M^{-1} for small i tends to be weak. Therefore, initially the convergence is fragile, and the compression may destroy it. Can we proceed initially with limited compression or with no compression? The next example shows that we cannot generally ensure convergence if we choose the initial approximation as in Section 6.7 and apply compression. That is, one should not expect that both convergence of the iteration and superfast performance of all Newton-Structured steps can be proved simultaneously.

Example 6.8.1. Newton's processes with the maximal compression and with no compression. Let $A = Z_1$, $B = Z_{-1}$, and $L = \nabla_{Z_{-1}, Z_1}$ and consider Algorithm 6.3.1 combined with Subroutine **R1** where the compression is maximal, that is, all singular values of the displacements $L(X_i)$ but the r largest ones are truncated, and where $r = \text{rank}(\nabla_{Z_1, Z_{-1}}(M)) = \text{rank}(\nabla_{Z_{-1}, Z_1}(M^{-1}))$. Then we have $\|L^{-1}\|_F \leq 1/(2 \sin \frac{\pi}{2n})$ according to Table 6.3 and, therefore, $\|L^{-1}\|_2 \leq \sqrt{r}/(2 \sin \frac{\pi}{2n})$ according to Corollary 6.6.3. It follows that

$$\mu \leq (1 + \sqrt{r}/\sin \frac{\pi}{2n})\kappa(M) \tag{6.8.1}$$

for μ in (6.4.5),

$$\mu\hat{\rho}_{2,i+j} \leq (\mu\hat{\rho}_{2,i})^{2^j}, \quad i, j = 0, 1, 2, \ldots \tag{6.8.2}$$

(see Corollary 6.4.3). Bounds (6.8.1) and (6.8.2) ensure global quadratic convergence of Algorithm 6.3.1 (starting with $i = 0$) if $\mu\hat{\rho}_{2,0} \leq \theta < 1$. To satisfy the latter inequality in our case, we should require that

$$\hat{\rho}_{2,0} \leq 1/((1 + \sqrt{r}/\sin\frac{\pi}{2n})\kappa(M)). \tag{6.8.3}$$

It can be seen immediately that all choices from the previous section for X_0 fall short of supporting bound (6.8.3). Suppose that in an attempt to fix the problem we perform the first k Newton steps with no compression of the displacements $L(X_i)$ for $i = 1, 2, \ldots, k$. Assume the favorable case of a positive definite Hermitian input matrix M and the choice of $X_0 = I/\|M\|_F$. Even then, however, we would have needed at least $j \approx \log_2\kappa(M)$ steps to yield the bound $\mu\hat{\rho}_{2,j} \leq 1/2$. That is, we would have ensured quadratic convergence only starting with the j-th Newton step for j of the above order. At this step, however, we cannot ensure the desired structure of the computed approximations X_i if we proceed with no compression. Indeed, Theorem 6.3.2 only implies that $\mathrm{rank}(L(X_i)) \leq r3^i, i = 1, 2, \ldots$, whereas we may already have $r3^i \geq n$ for $i = \log_3(n/r)$, that is, for $i < j \approx \log_2\kappa(M)$ even where $\kappa(M) < n/r$. As soon as the displacement rank of the $n \times n$ matrix X_i reaches the level n, the original matrix structure is entirely lost, and even a single Newton step becomes far too expensive to support the overall superfast performance of Algorithm 6.3.1.   □

Can we still somehow achieve levels (6.7.4) and (6.7.5) of the convergence rate and simultaneously superfast performance of each Newton-Structured step? Yes, by using a *homotopic process with Newton's iteration as a basic subroutine* applied at each homotopic step. Superfast performance of the resulting algorithms for well conditioned Toeplitz and Toeplitz-like input matrices has been proved in [P92b]. We cover this approach in the next section.

Another direction is the *heuristic application of the non-homotopic Algorithm 6.3.1*. Due to Example 6.8.1, proving superfast performance is hardly expected in this direction, but the experimental evidence shows its high efficiency in the Toeplitz case with variable compression based on Subroutine **R1**. Then again how much should the compression be limited? Extensive experiments with Toeplitz input matrices and with various policies of compression reported in the papers [PKRCa] and [BMa] show that usually the iteration converges even with quite a strong initial compression and even for rather ill conditioned input matrices. Table 6.21 in Section 6.11 reflects a very characteristic phenomenon observed in many test runs: the 2-norms of the residuals of the computed approximations to the inverses of random Toeplitz matrices first decreased a little, then grew above 1 (and sometimes well above 1), and then finally converged to 0. It follows that the upper estimates (6.8.1)–(6.8.3) are overly pessimistic, which should be no surprise if one recalls our argument in the beginning of Section 6.6 and extends it to the estimated factor $\kappa(M)$ of μ (see (6.8.1)). The tables in Section 6.11 show the *overall work* of Newton-Structured Iteration measured by the sum of the lengths of the displacement generators for the matrices Y_i over all i required to decrease the residual norm below $1/2$. We observe that the overall work grows very slowly as the condition number $\kappa(M)$ grows; the order of $\log_2 \log_2(1/\epsilon)$ additional steps are sufficient to de-crease the norm further, below a fixed positive ϵ (cf. (6.7.4) and (6.7.5)). Furthermore, the overall work of Newton-Structured Iteration also grows very slowly as the dimen-sion n of the input matrices increases. Similar behavior of the convergence process was experimentally observed for its homotopic variations described in the next section.

To optimize the truncation/compression levels, particularly at the initial Newton steps, one may apply experimental tests under a fixed policy of compression. Here is a high level description of this policy (see more specifics in Section 6.11 and [PKRCa]):

> **A policy of adaptive compression.** *First proceed with the maximal com-pression, that is, truncate all singular values of the displacement $L(X_i)$ but the r largest ones, for $r = \text{rank}(L(M^{-1}))$. As long as the residual norm of the resulting approximation Y_i is not small enough to support (sufficiently rapid) convergence of the iteration, repeat the same step re-cursively, each time truncating fewer and fewer singular values.*

6.9 Homotopic Newton's iteration

Even where no initial approximation is available, rapid convergence of Newton-Structured Iteration can be ensured based on homotopy techniques. These techniques enable faster inversion of general Hermitian indefinite input matrices and, unlike all non-homotopic approaches, enable formal proof of rapid convergence of Newton–Toeplitz-like Iteration. It is still not clear, however, which of all homotopic and non-homotopic processes (including various options of limited compression of the displace-ments in both processes) would optimize computational time for structured input ma-trices in practical implementation.

6.9.1 An outline of the algorithms

Let us next outline homotopic processes for matrix inversion. The processes start with a readily invertible matrix M_0 (say with $M_0 = I$) and then proceeds with a **Basic Subroutine** for recursive iterative inversion of the matrices

$$M_k = M_0 + (M - M_0)\tau_k, \quad k = 1, 2, \ldots, K, \tag{6.9.1}$$

where $0 < \tau_1 < \tau_2 < \ldots < \tau_k = 1$ (see a generalization in Exercise 6.21). For the Basic Subroutine, we choose Algorithm 6.3.1 where Subroutine **R** may perform maximal compression, limited compression, or no compression of the displacements $L(X_i)$. The approximation to the matrix M_{k-1}^{-1} computed at the $(k-1)$-th step is used as an initial approximation X_0 to the inverse M_k^{-1} when the Basic Subroutine is invoked at the next k-th homotopic step. The overall computational cost grows as the overall number K of homotopic steps grows. This suggests increasing the step sizes $\tau_k - \tau_{k-1}$ as much as possible. For every $k < K$, however, we have an upper bound on the increase because the initial approximation X_0 to the inverse M_k^{-1} should be close enough to enable rapid convergence of the Basic Subroutine. The latter requirement also defines a stopping criterion for the Basic Subroutine applied at the $(k-1)$-th homotopic step for $k = 2, \ldots, K$. At the K-th step, the stopping criterion is given by the value of the tolerance to the approximation error for M^{-1}. Let us elaborate upon the presented outline.

6.9.2 Symmetrization of the input matrix

The standard symmetrization techniques [GL96] reduce the inversion of a non-singular matrix M to the Hermitian (real symmetric) case. We have

$$M^{-1} = (MM^*)^{-1}M = M^*(MM^*)^{-1}, \tag{6.9.2}$$

$$M_*^{-1} = \begin{pmatrix} 0 & M^{-1} \\ (M^*)^{-1} & 0 \end{pmatrix}. \tag{6.9.3}$$

M^*M and MM^* are Hermitian (or real symmetric) positive definite matrices, $M_* = \begin{pmatrix} 0 & M^* \\ M & 0 \end{pmatrix}^{-1}$ is an Hermitian (or real symmetric) indefinite matrix (see Definition 4.6.3). The condition number of a matrix M, $\kappa(M)$ is squared in the transition to the matrices M^*M and MM^* (see (4.6.8)) but remains invariant in the transition to the matrix M_*. On the contrary, the size of the matrix M is doubled in the transition to the matrix M_*. The displacement rank of a matrix M is roughly doubled by each of the symmetrizations (6.9.2) and (6.9.3).

Hereafter, to the end of Subsection 6.9.5, we assume that the input matrix $M = M^*$ is Hermitian or real symmetric, with real eigenvalues $\lambda_j = \lambda_j(M)$, $j = 1, \ldots, n$, enumerated so that

$$|\lambda_n| \le |\lambda_{n-1}| \le \ldots \le |\lambda_1|. \tag{6.9.4}$$

6.9.3 Initialization of a homotopic process for the inversion of general and structured matrices

Let us choose the matrix M_0 to keep the matrices M_k in (6.9.1) away from singularity for all k by ensuring that

$$\kappa(M(t)) \le \kappa(M) \text{ for all } t, \ 0 \le t \le 1, \tag{6.9.5}$$

where $M(t) = M_0 + t(M - M_0)$. For a Hermitian positive definite matrix M, bounds (6.9.5) are satisfied under the choice of

$$M_0 = I. \tag{6.9.6}$$

Indeed, in this case, we have

$$\lambda_j(M(t)) = t\lambda_j(M) + 1 - t, \quad j = 1, \ldots, n,$$

$$\kappa(M(t)) = \frac{\lambda_1(M(t))}{\lambda_n(M(t))} = \frac{1 + s\lambda_1(M)}{1 + s\lambda_n(M)}, \quad s = t/(1-t),$$

and the condition number is maximized for $t = 1$. For any Hermitian indefinite matrix M, we satisfy (6.9.5) by choosing

$$M_0 = \sqrt{-1}\, I. \tag{6.9.7}$$

Indeed, in this case, we have

$$\lambda_j(M(t)) = t\lambda_j(M) + (1-t)\sqrt{-1},$$

and therefore,

$$\sigma_j^2(M(t)) = \lambda_j(M(t)M^*(t)) = (t\lambda_j(M))^2 + (1-t)^2 \text{ for } 0 \le t \le 1,$$

$$\kappa(M(t)) = \left|\frac{\lambda_1(M(t))}{\lambda_n(M(t))}\right|^2 = \frac{1 + s^2\lambda_1^2(M)}{1 + s^2\lambda_n^2(M)}, \quad s = t/(1-t).$$

Then again the condition number is maximized for $t = 1$. If the input matrix M is real, assignment (6.9.7) has the drawback of involving imaginary values (see Exercises 2.8(e), (f) and 3.22(a)).

Choices (6.9.6) and (6.9.7) for the matrix M_0 are effective where M is a general or Toeplitz-like input matrix (see Exercise 6.16 (a)) but do not match some other important matrix structures. We may, however, replace expression (6.9.6) by the following more general assignment

$$M_0 = \widehat{M}, \tag{6.9.8}$$

where M and $\widehat{M}$ are a pair of Hermitian positive definite matrices. Likewise, in the more general case where M and $\tilde{M}$ are a pair of Hermitian matrices at least one of which is positive definite, we replace expression (6.9.7) by

$$M_0 = \sqrt{-1}\,\tilde{M}. \tag{6.9.9}$$

The reader is referred to [PKRCa] on the upper bounds on the norms $\|M^{-1}(t)\|_2$ under the two latter assignments.

The more general choices (6.9.8) and (6.9.9) enable a better match between the structures of the matrices M and M_0.

Example 6.9.1. Let M be a positive definite Hankel-like matrix. The matrices $M(t) = M_0 + t(M - M_0)$ are Toeplitz+Hankel-like under (6.9.6) and (6.9.7), but both (6.9.8) and (6.9.9) with $\widehat{M} = aJ$, $\tilde{M} = aJ$ for a scalar a enable preserving the simpler Hankel-like structure of the matrix M (see Exercise 6.16(b)).

Example 6.9.2. Suppose $M = R$ is a Pick matrix in (3.8.1), that is,

$$R = \left(\frac{x_i x_j^* - y_i y_j^*}{z_i + z_j^*}\right)_{i,j=0}^{n-1}, \quad \text{Re } z_i > 0,$$

and x_i and y_j are row vectors for all i and j. Then again one may rely on (6.9.6) and (6.9.7) for the initialization of the homotopic process, but choices (6.9.8) and (6.9.9) for the matrices $\widehat{M}$ and $\tilde{M}$ of the form $\left(\frac{u_i u_j^*}{z_i + z_j^*}\right)_{i,j=0}^{n-1}$ (for row vectors $\mathbf{u}_i$, $i = 0, 1, \ldots, n - 1$) better preserve the Cauchy-like structure of the Pick matrix $M = R$ in the transition to the matrices M_k, provided the dimension of the vectors $\mathbf{u}_i$ is small (see Exercise 6.17).

6.9.4   The choice of step sizes for a homotopic process. Reduction to the choice of tolerance values

To complete the definition of the homotopic process (6.9.1), it remains to choose the step sizes $\tau_k - \tau_{k-1}$ for all k. This can be done based on the following "trial-and-error" approach: first recursively increase the step size until this destroys rapid convergence, then go back a little, that is, reduce the step size. Alternatively, a large initial step size can be recursively decreased until sufficiently rapid convergence is achieved. We prefer, however, to start by fixing a tolerance bound θ_k on the initial residual norm at the k-th homotopic step for every k, and then to calculate the step size $t_{k-1} - t_k$ as a function in θ_k.

To analyze this process, first rewrite (6.9.1) in a more convenient way. That is, divide the equation by the scalar τ_k, then replace the matrices M_k/τ_k by M_k for positive k, write $t_k = (1/\tau_k) - 1$, and obtain that

$$M_k = t_k M_0 + M, \quad k = 1, 2, \ldots, K;$$
$$M_k = M_{k-1} - \Delta_{k-1} M_0, \quad \Delta_{k-1} = t_{k-1} - t_k > 0, \quad k = 2, 3, \ldots, K;$$

$t_K = 0$, $M_K = M$. We assume that the Basic Subroutine is applied to invert the matrix M_k by using the initial approximations $X_0 = (t_1 M_0)^{-1}$ for $k = 1$ and $X_0 = M_{k-1}^{-1}$ for $k > 1$. It is immediately verified that

$$r_1 = \|I - M_1(t_1 M_0)^{-1}\|_2 = \|M M_0^{-1}\|_2/t_1,$$
$$r_k = \|I - M_k M_{k-1}^{-1}\|_2 = \Delta_k \|M_0 M_{k-1}^{-1}\|_2, \quad k = 2, 3, \ldots.$$

Suppose our goal is to ensure some fixed upper bounds θ_k on the initial residual norms r_k in Newton's processes applied at the k-th homotopic steps for $k = 1, 2, \ldots$. According to the above equations, it is sufficient to obtain some upper estimates n_1^+ and n_k^+ for the norms $\|M M_0^{-1}\|_2 \le \|M\|_2 \|M_0^{-1}\|_2$ and $\|M_0 M_{k-1}^{-1}\|_2 \le \|M_0\|_2 \|M_{k-1}^{-1}\|_2$, respectively, for $k = 2, 3, \ldots$, and to choose $t_1 = n_1^+/\theta_1$, $\Delta_k = \theta_k/n_k^+$, $k = 2, 3, \ldots$.

Estimating the norms is a rather simple task. Indeed, the Hermitian matrices M, M_0, and M_0^{-1} are given to us before we begin the computations. The Hermitian matrix M_{k-1}^{-1} is available when we begin the k-th homotopic step (at which we estimate the norm $\|M_{k-1}^{-1}\|$ to choose the step size $t_{k-1} - t_k$). It remains to apply Theorem 6.1.2 (a), (b), the power iteration, or the Lanczos algorithm [GL96].

For the particular matrices M_0 in (6.9.6) and (6.9.7), we have

$$\|M_0\|_2 = \|M_0^{-1}\|_2 = 1.$$

Furthermore, usually we do not have to estimate the norms $\|M_{k-1}^{-1}\|_2$ for all k. Indeed, for a positive definite matrix M and the matrix $M_0 = I$ in (6.9.6), we deduce that

$$\|M_{k-1}^{-1}\|_2 = \lambda_n^{-1}(M_{k-1}) = (t_{k-1} + \lambda_n(M))^{-1} \text{ for all } k \ge 1. \qquad (6.9.10)$$

Thus, we may first closely approximate the 2-norm $\|M_1^{-1}\|_2$ and consequently $\lambda_n(M)$, the smallest eigenvalue of the matrix M, and then the norms $\|M_{k-1}^{-1}\|_2$ for all $k > 1$, due to (6.9.10). (For ill conditioned matrices M, only a crude approximation to the eigenvalue $\lambda_n(M)$ and the norm $\|M_1^{-1}\|_2$ are realistic; they may be close enough for $k = 2$ but a refinement based on approximations to the norms $\|M_{k-1}^{-1}\|_2$ for larger k may be required.)

Similarly, for a Hermitian matrix M, we deduce from equation (6.9.7) that

$$\|M_{k-1}^{-1}\|_2^2 = \lambda_n^{-2}(M_{k-1}) = (t_{k-1}^2 + \lambda_n^2(M))^{-1} \text{ for all } k \geq 1.$$

Then again the latter expression enables us to express first the squared eigenvalue $\lambda_n^2(M)$ via the squared norm $\|M_1^{-1}\|_2^2$ and then the squared norms $\|M_{k-1}^{-1}\|_2^2$ for $k = 2, 3, \ldots, K$ via the squared eigenvalue $\lambda_n^2(M)$.

6.9.5   The choice of tolerance values

Now, to complete the description of the homotopic process, it remains to choose the tolerance values θ_k, $k = 1, 2, \ldots$. In the case of general input matrices M, this choice has been optimized in the paper [PKRCa]. It was shown there that the resulting algorithms invert a general Hermitian positive definite matrix M by using about as many flops as in the best non-homotopic versions of Newton's iteration in [PS91], whereas for Hermitian indefinite input matrices M, the homotopic versions in [PKRCa] use substantially fewer flops than the non-homotopic algorithms in [PS91], that is, roughly by twice fewer flops to decrease the residual norm below $1/2$ and about as many flops as in [PS91] at the subsequent stages.

For Newton-Structured Iteration, the choice of the values θ_k is more involved due to the impact of the compression of the displacements. Nevertheless, in the case of $n \times n$ non-singular Toeplitz-like matrices M, the paper [P92b] proposed a policy under which superfast inversion was proved provided that $\log \kappa(M) = O(\log n)$. As we can see in Example 6.8.1, proving such results should be hard for the known non-homotopic versions of Newton's iteration. On the other hand, as mentioned in Section 6.8, experimental computations with random Toeplitz input matrices M show that under the adaptive policies of the choice of the tolerance values θ_k and/or the compression level, both homotopic and non-homotopic iterations perform substantially better than the worst case analysis of [P92b] predicts.

Extension from the Toeplitz-like case to the case of various other structured input matrices is immediate via the transformation approach (as was pointed out already in [P92b]), but the direct extension of the algorithms in [P92b] is also possible by using the techniques in Subsection 6.9.3 (see (6.9.8), (6.9.9)).

6.9.6   Variations and parallel implementation of the Newton-Structured Iteration

Optimizing Newton-Structured Iteration, one should first choose between its homotopic and non-homotopic versions. In each of them further variations are possible.

The *non-homotopic version* may vary depending on the choice of the level of compression $l = l(i)$ in the i-th Newton step for each i where $l(i)$ equals $\text{rank}(L(Y_i))$, that is, the number of untruncated singular values of the displacement $L(X_i)$. The adaptive choice of $l(i)$ is recommended for smaller i, whereas the choice of $l(i) =$ rank($L(M^{-1})$) is most effective for larger i. The parameter $l = l(i)$ can be used in the Basic Subroutine of the *homotopic version* of Newton-Structured Iteration, where in addition one may choose a tolerance value θ_k as a parameter defining the size of the k-th homotopic step for $k = 0, 1, \ldots$. Then again, an adaptive process is possible where the choice of both parameters θ_k and $l(i)$ is optimized simultaneously. (Applying a homotopic process to a non-Hermitian input matrix M, one has also a choice between the symmetrization policies (6.9.2) and (6.9.3).)

Optimization of all these choices and parameters is an important open problem of ongoing theoretical and experimental research. A convenient optimization criterion is the minimization of the overall work of the iteration, that is, the sum of the compression levels $l(i)$ taken over all Newton's steps of a non-homotopic process or over all Newton's steps in all K applications of the Basic Subroutine of a homotopic process.

The non-homotopic version of the Newton-Structured Iteration is a special case of the homotopic version where $\theta = 1$. We may expect some improvement for $\theta < 1$, but it is still unclear what the best policy is for choosing θ and for the cut-off level of truncation of the singular values. In our experiments with Toeplitz input matrices, we used a very crude policy of adaptively decreasing θ from the initial value $\theta = 0.7$. The truncation level was fixed throughout the computation. We call this policy the *stiff* 0.7-*down policy*. It turned out to be a little inferior to the non-homotopic approach, although the convergence was never destroyed (see Section 6.11). Of course, the choice of this crude policy substantially weakened the homotopic processes, and our experiments only compared the power of compression versus homotopy, but not versus joint application of compression and homotopy. The experimental study based on compression and homotopy combined is the subject of our future research.

At a high level of further study, we may ask which algorithms may compete with Newton-Structured Iteration. Natural candidates are generalizations that use various policies of scaling (see (6.7.6)) and/or higher order iteration (see (6.2.9) and (6.7.7)). Then again a purely theoretical analysis of these generalizations is hard because generally the compressed approximations to the inverse M^{-1} are not polynomials in M or M^*M anymore, but practically the compression may distort the computed approximations much less than the worst case analysis suggests, and then the cited generalizations (including those in Section 6.7) could lead to significant acceleration.

An alternative to Newton's iteration is the *divide-and-conquer algorithm* of Chapter 5. This is a natural counterpart to iterative approaches. Indeed, the divide-and-conquer algorithm is less sensitive to the growth of the condition number $\kappa(M)$ and has an important correlation to the State-Space approach to rational computations (see Section 5.8). On the other hand, Newton's iteration is simpler to program on computers and allows for more effective parallel implementation.

Let us briefly comment on *parallel implementation*. The entire iteration requires $O(\log \kappa(M) + \log\log(1/\epsilon))$ Newton steps. For realistic bounds ϵ on the output residual norm, the latter bound on the number of Newton steps can be viewed as at worst logarithmic in n provided that the dimension n is large or moderately large and the input matrix M is well conditioned or at least not very ill conditioned. Each Newton step essentially amounts to computing two matrix products. Therefore, each step runs in parallel arithmetic time $O(\log n)$ on n processors for $n \times n$ Toeplitz/Hankel and

Toeplitz/Hankel-like matrices, whose multiplication by a vector is reduced to performing only a small number of FFT's. (Recall that FFT on $n = 2^k$ points uses parallel arithmetic time $O(\log n)$ on n processors.) Multiplication of $n \times n$ Vandermonde-like or Cauchy-like matrices by a vector and consequently a Newton step applied to these matrices require parallel arithmetic time $O(\log^2 n)$ by using n processors. Alternatively, we may apply a multiplicative transformation in Section 4.8 to reduce the inversion problem for a matrix of Vandermonde or Cauchy types to the Toeplitz-like case (see Section 4.8) by using parallel arithmetic time $O(\log^2 n)$ on n processors. For comparison, all known algorithms that invert an $n \times n$ Toeplitz matrix on n processors and do not use Newton's iteration (including the algorithms in Chapter 5), require parallel arithmetic time of at least the order of n.

6.10   Newton's iteration for a singular input matrix

The variation (6.7.6), (6.7.8), (6.7.9) of Newton's iteration for general and Hermitian semi-definite inputs matrices converges to the numerical generalized inverse M_ϵ^+ for a singular or non-singular input matrix M and can be combined with homotopic processes. Indeed, the term $(1 - \tau_k)M_0$ for $M_0 = I$ in (6.9.6) or $M_0 = \sqrt{-1}I$ in (6.9.7) contributes the eigenvalues $1 - \tau_k$ or $(1 - \tau_k)\sqrt{-1}$, respectively, to the matrices $M_k = M_0 + (M - M_0)\tau_k$ in (6.9.1) provided the matrix M has eigenvalues 0. The eigenvalues $1 - \tau_k$ have magnitudes below ϵ and are ignored for larger k for which $\tau_k > 1 - \epsilon$. This implies convergence to M_ϵ^+. Furthermore, we observe that $M_\epsilon^+ = M^+$ for smaller ϵ.

Using Newton-Structured Iteration to compute the matrices M^+ and M_ϵ^+ may lead to an additional problem. That is, the compression techniques in Sections 6.4 and 6.5 apply only where the output matrices are structured. If M is a non-singular structured matrix, then M^{-1} is structured as well, but M_ϵ^+ is generally unstructured for $\epsilon \geq 0$ (or at least it is hard to exploit the structure of M_ϵ^+) if $\text{rank}(M_\epsilon^+ M - I)$ is large. If the rank is small, we may exploit the structure of M_ϵ^+ and extend Newton's iteration respectively, based on the following results.

Theorem 6.10.1. *For any non-negative ϵ and any triple of $n \times n$ matrices A, B, and M, we have*

$$\nabla_{B,A}(M_\epsilon^+) = M_\epsilon^+ A(MM_\epsilon^+ - I) - (M_\epsilon^+ M - I)BM_\epsilon^+ - M_\epsilon^+ \nabla_{A,B}(M)M_\epsilon^+.$$

Corollary 6.10.2. *Under the assumptions of Theorem 6.10.1, let $r_\epsilon = \text{rank}(M_\epsilon^+) = \text{rank}(M_\epsilon)$. Then $\text{rank}(\nabla_{B,A}(M_\epsilon^+)) \leq \text{rank}(\nabla_{A,B}(M)) + 2n - 2r_\epsilon$.*

The level of the truncation of the singular values in the approach of Section 6.4 can be defined by Corollary 6.10.2.

Finally, for the special cases of Toeplitz and Hankel singular matrices M, some structure of the matrices M^+ is shown in the papers [HH93], [HH94].

6.11   Numerical experiments with Toeplitz matrices

The presented Newton's iteration algorithms were tested numerically for $n \times n$ Toeplitz input matrices M. As a black box Basic Subroutine, the Newton iteration Algorithm 6.3.1 was applied with the compression Subroutine **R1** implemented as Algorithm 7.5.1 from [PBRZ99]. The tests were performed by M. Kunin at the Graduate Center of CUNY, in cooperation with R. Rosholt of the Lehman College of CUNY.

The tests used the following computational facilities:

OS – Red Hat Linux 7.0

compiler – GCC 2.96 (also using bundled random number generator)

library – CLAPACK 3.0 (routines for computing SVD of real and complex matrices and eigenvalues of symmetric matrices)

Both non-homotopic and stiff homotopic versions of Newton-Structured Iteration were applied to the same input matrices M. In non-homotopic processes the compression level, that is, the number l of untruncated singular values of the displacements, was chosen adaptively to minimize l as long as convergence was achieved. More precisely, for a fixed threshold value ϵ, the candidate compression level l was calculated as follows. Let $\sigma_1^{(i)}, \ldots, \sigma_n^{(i)}$ denote the singular values of X_i written in the non-increasing order. Then we chose $l = l(i)$ satisfying $\sigma_{l+1}^{(i)} < \epsilon \sigma_1^{(i)}$, $\sigma_l^{(i)} \geq \epsilon \sigma_1^{(i)}$. If the Newton process diverged for these ϵ and l, then all results of the computation (obtained by this moment) were discarded, except that the contribution to the overall work of the iteration was counted. Then the iteration process was repeated with the compression level $l/2$. In the case of divergence, this value was recursively halved further until convergence but at most 10 times. The experiments for homotopic processes were limited to recursive optimization of the tolerance values θ_h and consequently the homotopic step sizes, under the stiff 0.7-down policy. This was our simplified preliminary policy, subject to improvement in our future experiments.

In the stiff 0.7-down homotopic processes, the initial value of θ was always set to 0.7 and never increased. In the case of divergence, θ was recursively halved, and the process was repeated until convergence. For a fixed θ, the step sizes were calculated using LAPACK for computing the eigenvalues of the matrix M. The value l of the compression level was fixed and remained invariant in all Newton steps throughout the entire homotopic process. This value was chosen experimentally when convergence with this value was observed for one or two test runs.

The computations stopped at the final homotopic and non-homotopic steps where the residual norm decreased to the single precision 0; at all other homotopic steps, the computations stopped where the residual norm decreased below 10^{-6}. (The latter bound was a little smaller than was necessary for convergence.)

The algorithm was tested for $n \times n$ Toeplitz matrices M of the following classes (see details below).

1. Real symmetric tridiagonal Toeplitz matrices $(t_{i,j})_{i,j=0}^{n-1}$, $t_{i,j} = 0$ where $|i - j| > 1$, $t_{i+1,i} = t_{i,i+1} = 1$, $t_{i,i}$ equals 4 or -2.

2. The matrices $\left(\frac{1}{1+|i-j|}\right)_{i,j=0}^{n-1}$.

3. Randomly generated Toeplitz matrices.

4. Randomly generated real symmetric positive definite matrices with a specified condition number.

5. Randomly generated real symmetric indefinite Toeplitz matrices.

The tests results were differentiated further according to the condition number of the matrix M and the size $n \times n$, for n ranging from 50 to 350.

An $n \times n$ random real symmetric Toeplitz matrix of class 5 was defined by generating the n random entries of its first row; for an unsymmetric Toeplitz matrix of class 3, also the $n - 1$ remaining entries of its first column were generated. The random entries were generated as the random normally distributed variables with the mean 0 and the standard deviation 1. At this stage, a random number generator was applied with the **rand()** function from the standard C library that comes with the GCC compiler for Cygwin on Windows 2000. The condition number was most frequently quite small for random matrices, and then the algorithms converged very rapidly. To make the results more meaningful, part of the experiments was restricted to the matrices with larger condition numbers. To form a matrix of class 4, that is, to achieve positive definiteness and a desired condition number, we computed the two extremal eigenvalues of a random real symmetric Toeplitz matrix and then added the matrix aI for an appropriate positive a.

For unsymmetric and symmetric indefinite Toeplitz matrices M, the non-homotopic Newton process was initialized with the matrices $X_0 = M^T/(\|M\|_1\|M\|_\infty)$. For a symmetric positive definite matrix M, the same process was applied with $X_0 = I/\|M\|_F$. For the homotopic processes with the same symmetric input matrices M, the same Newton processes were applied with the invariant truncation level $l = 2$, except that the initial matrices X_0 were determined by the homotopic rules and the choice of the matrix M_0. For unsymmetric input matrices M, both homotopic and non-homotopic processes also were applied to two symmetrized matrices $M^T M$ and $M_* = \begin{pmatrix} 0 & M \\ M^T & 0 \end{pmatrix}$ (cf. (6.1.2) and (6.8.3)). For homotopic processes, in these two cases, the invariant truncation levels $l = 12$ and $l = 6$ were selected, respectively. In the symmetric indefinite case, the initial choice (6.9.7) was used for the matrix M_0. In the positive definite case, the matrix $M_0 = I$ was selected according to (6.9.6). The initial threshold bound ϵ in non-homotopic Newton processes was selected at the levels 0.025 for the unsymmetric and symmetric indefinite matrices, 0.00005 for the symmetrized matrices $\begin{pmatrix} 0 & M \\ M^T & 0 \end{pmatrix}$.

In the column K we show the number of test runs where the random input matrices M have the condition numbers bounded from below by the value in the respective line of the column κ. The columns N and L show the number of Newton's steps and the overall work (that is, the sum of the truncation levels in all Newton's steps) in non-homotopic processes; the columns N_H and L_H show the same data for homotopic processes. H is the number of homotopic steps, s_θ is the number of times the tolerance value θ was halved (split) during the homotopic process. All these data are averaged over the K test runs for the selected range of the condition number, except that for symmetric positive definite matrices M the column κ shows the fixed condition numbers $\kappa(M)$. For non-random matrices M of classes 1 and 2, we have $K = 1$, that is, the tests run for a single fixed matrix M. For each class of input matrices, the truncation level for the singular values was decreased recursively as long as this did not destroy the convergence in experimental tests.

In addition to the data shown in Tables 6.4–6.20, we include a table showing an interesting phenomenon observed in more than 25% test runs for non-homotopic processes with the random Toeplitz input matrices of class 3. That is, in these test runs the residual 2-norm was growing above 1 (and sometimes well beyond 1) and then decreased down to 0 in the next Newton steps. This phenomenon must be attributed to compression because otherwise the residual 2-norm would have been squared in each Newton's step, which would have automatically implied the divergence of Newton's iteration. Table 6.21 demonstrates this phenomenon in some details. That is, for each test run it shows the condition number of the input matrix M and the maximum residual norm ρ_i reached during the iteration process, which finally converged. The columns marked by ρ_1, ρ_2, and ρ_3 show the values of the residual norm for non-homotopic process with no symmetrization of M and with symmetrization based on (6.9.3) and (6.9.2), respectively.

6.12 Conclusion

Newton's iteration is a fundamental tool for numerical solution of equations and systems of equations. We first applied this tool to integer inversion (for demonstration) and then studied its application to approximating the inverse M^{-1} or the Moore–Penrose generalized inverse M^+ of a structured matrix M. Besides having strong numerical stability, the iteration is very simple to implement and has the advantage of allowing efficient parallelization, unlike other known superfast algorithms for the inversion of structured matrices. $i = O(\log \kappa(M) + \log \log(1/\epsilon))$ Newton's iteration steps are sufficient to force all eigenvalues of the matrices $X_i M$ to be within a positive $\epsilon < 1$ from the respective eigenvalues 1 and 0 of the matrix $M^+ M$, where M is a general unstructured matrix, X_i is the computed approximation, and $\kappa(M)$ denotes the condition number of the input matrix. For a non-singular M, the latter estimates mean that $\|I - X_i M\|_2 \le \epsilon$.

Every iteration step amounts essentially to performing matrix multiplication twice. To yield superfast algorithms in the case where M is a well conditioned structured matrix, we compress the computed approximations, which otherwise tend to lose their structure rapidly. Strong compression may, however, destroy the approximation. The balance between compression and approximation is a delicate subject requiring further theoretical and experimental study. In this chapter, we described two compression techniques, one of which allowed compression of various levels. We analyzed both approaches and estimated the overall numbers of Newton steps required for convergence. This involved estimates for the norms of the inverse displacement operators, and we showed five techniques for obtaining these estimates. We also recalled some results on convergence acceleration and on the initial approximation to the inverse. The avail-

Table 6.4: Results for the tridiagonal Toeplitz matrices M of class 1 with the entries on the main diagonal equal to 4 and on the first super and subdiagonals equal to 1.

n	κ	H	N_H	L_H	N	L
50	2.992	3	21	42	8	16
100	2.998	3	20	40	9	18
150	2.999	3	21	42	9	18
200	3.000	3	20	40	9	18
250	3.000	3	20	40	9	18
300	3.000	3	21	42	9	18
350	3.000	3	20	40	10	20

Table 6.5: Results for the tridiagonal Toeplitz matrices M of class 1 with the entries on the main diagonal equal to 2 and on the first super and subdiagonals equal to -1.

n	κ	H	N_H	L_H	N	L
50	1053.480	8	46	92	16	32
100	4133.640	9	56	112	19	38
150	9240.230	9	56	112	20	40
200	16373.200	10	59	118	21	42
250	25532.700	10	60	120	22	44
300	36718.500	11	63	126	23	46
350	49930.800	11	67	134	23	46

Table 6.6: The matrices $M = \left(\frac{1}{|i-j|}\right)_{i,j}$ **of class 2**

n	κ	H	N_H	L_H	N	L
50	16.221	4	25	50	9	18
100	19.642	4	25	50	9	18
150	21.680	4	26	52	10	20
200	23.138	4	27	54	10	20
250	24.274	4	26	52	10	20
300	25.205	4	27	54	10	20

Table 6.7: Random Toeplitz matrices M **of class 3,** $n = 100$ **(non-homotopic iteration)**

κ	K	ϵ	N	L
10000	1	0.025	32.00	124.00
5000	2	0.013	61.00	261.00
1000	7	0.022	41.71	185.86
500	14	0.018	42.36	187.36
100	69	0.025	23.83	101.96
50	56	0.025	20.91	88.86
10	45	0.025	18.82	76.11

Table 6.8: Symmetric indefinite Toeplitz matrices $\begin{pmatrix} 0 & M \\ M^T & 0 \end{pmatrix}$ **for a random Toeplitz matrix** M **of class 3,** $n = 100$ **(the column** K **is the same as in Table 6.4)**

κ	H	N_H	L_H	s_θ	ϵ	N	L
10000	10.00	67.00	402.00	0.00	6.3e-04	102.00	1115.00
5000	19.50	112.50	675.00	1.00	1.3e-03	63.00	695.00
1000	22.00	119.86	719.14	1.43	2.1e-03	44.00	497.57
500	16.64	97.36	584.14	1.36	2.4e-03	29.64	329.14
100	13.57	82.46	494.78	1.52	2.5e-03	23.64	276.00
50	11.39	70.34	422.04	1.61	2.5e-03	20.93	249.79
10	9.40	56.96	341.73	1.49	2.5e-03	18.76	220.82

Table 6.9: Symmetric positive definite Toeplitz matrices $M^T M$ for a random Toeplitz matrix M of class 3, $n = 100$ (the column K is the same as in Table 6.4) (Note: in this case no decrease of the initial tolerance value $\theta = 0.7$ was required in the experiments, so the resulting table does not contain the s_θ column)

κ	H	N_H	L_H	ϵ	N	L
10000	18.00	106.00	1272.00	1.3e-05	84.00	690.00
5000	16.50	99.50	1194.00	2.8e-05	65.00	584.00
1000	14.71	88.00	1056.00	1.4e-05	79.71	737.29
500	12.50	76.14	913.71	3.6e-05	42.64	379.50
100	10.42	63.57	762.78	4.8e-05	23.62	214.30
50	8.86	54.70	656.36	5.0e-05	19.12	180.95
10	7.69	47.13	565.60	5.0e-05	16.73	162.96

Table 6.10: Results for random symmetric positive definite Toeplitz matrices M of class 4, $n=50$

κ	K	H	N_H	L_H	ϵ	N	L
250000	19	12	73.16	146.32	0.050	24.00	49.58
10000	19	9	56.42	112.84	0.050	19.74	40.95
5000	19	9	54.63	109.26	0.048	19.89	42.74
1000	20	8	47.40	94.80	0.050	16.00	33.85
500	20	7	43.25	86.50	0.050	15.10	31.40
100	18	6	36.28	72.56	0.050	12.94	26.56
50	20	5	31.70	63.40	0.050	12.00	24.15
10	19	4	25.89	51.79	0.050	9.63	19.26

Table 6.11: Results for random symmetric positive definite Toeplitz matrices M **of class 4,** $n=100$

κ	K	H	N_H	L_H	ϵ	N	L
250000	19	12	73.68	147.37	0.048	26.11	54.53
10000	17	9	56.24	112.47	0.044	24.06	53.24
5000	19	9	54.21	108.42	0.038	28.00	66.16
1000	14	8	47.64	95.29	0.048	18.36	38.93
500	19	7	43.53	87.05	0.049	16.74	35.47
100	16	6	36.50	73.00	0.050	13.19	27.88
50	19	5	31.84	63.68	0.050	12.11	24.68
10	18	4	25.72	51.44	0.050	10.00	20.00

Table 6.12: Results for random symmetric positive definite Toeplitz matrices M **of class 4,** $n=150$

κ	K	H	N_H	L_H	ϵ	N	L
250000	15	12	73.53	147.07	0.036	40.00	97.53
10000	11	9	56.09	112.18	0.044	24.18	53.82
5000	14	9	54.64	109.29	0.035	30.36	72.29
1000	17	8	47.65	95.29	0.049	18.06	38.53
500	18	7	43.39	86.78	0.049	17.56	37.72
100	13	6	36.54	73.08	0.050	13.85	29.46
50	18	5	32.11	64.22	0.050	12.94	26.39
10	18	4	25.39	50.78	0.050	10.00	20.00

Table 6.13: Results for random symmetric positive definite Toeplitz matrices M of class 4, n=200

κ	K	H	N_H	L_H	ϵ	N	L
250000	12	12	73.50	147.00	0.035	39.17	93.33
10000	11	9	56.82	113.64	0.048	21.64	45.45
5000	12	9	54.17	108.33	0.038	29.75	70.33
1000	13	8	47.46	94.92	0.035	30.23	72.23
500	13	7	43.54	87.08	0.044	20.38	45.77
100	15	6	36.53	73.07	0.050	14.40	29.67
50	12	5	32.00	64.00	0.050	13.17	26.75
10	12	4	25.58	51.17	0.050	10.17	20.33

Table 6.14: Results for random symmetric indefinite Toeplitz matrices M of class 5, n=50

κ	K	H	N_H	L_H	s_θ	ϵ	N	L
1000	6	8.00	50.00	100.00	0.00	0.015	52.00	235.17
500	9	7.00	45.00	90.00	0.00	0.018	39.89	179.00
100	46	6.33	40.41	80.83	0.00	0.022	28.67	126.48
50	25	5.28	35.36	70.72	0.04	0.025	20.00	83.76
10	14	4.93	32.21	64.43	0.00	0.025	17.50	69.64

Table 6.15: Results for random symmetric indefinite Toeplitz matrices M of class 5, n=100

κ	K	H	N_H	L_H	s_θ	ϵ	N	L
10000	2	10.00	60.50	121.00	0.00	0.019	45.50	204.00
5000	3	9.00	57.00	114.00	0.00	0.003	118.00	605.67
1000	7	8.14	50.71	101.43	0.00	0.013	64.71	336.29
500	9	7.00	46.11	92.22	0.00	0.018	42.89	209.22
100	64	6.67	43.05	86.09	0.09	0.022	29.22	136.97
50	12	6.58	43.83	87.67	0.42	0.025	21.00	93.50
10	2	5.00	33.50	67.00	0.00	0.025	19.50	78.50

Table 6.16: Results for random symmetric indefinite Toeplitz matrices M of class 5, $n=150$

κ	K	H	N_H	L_H	s_θ	ϵ	N	L
10000	8	10.25	64.12	128.25	0.00	0.015	83.00	437.00
5000	3	9.00	59.33	118.67	0.00	0.010	88.00	479.00
1000	22	8.50	54.68	109.36	0.09	0.012	75.41	406.23
500	15	7.20	49.40	98.80	0.07	0.019	42.00	212.47
100	45	7.11	45.67	91.33	0.24	0.022	30.22	147.13
50	5	5.80	38.80	77.60	0.20	0.022	26.00	118.40
10	1	5.00	35.00	70.00	0.00	0.025	20.00	84.00

Table 6.17: Results for random symmetric indefinite Toeplitz matrices M of class 5, $n=200$

κ	K	H	N_H	L_H	s_θ	ϵ	N	L
10000	2	11.00	71.00	142.00	0.00	0.001	194.00	1150.50
5000	2	9.00	60.00	120.00	0.00	0.016	63.50	330.50
1000	19	8.47	54.58	109.16	0.11	0.012	67.32	356.74
500	25	8.68	56.68	113.36	0.48	0.018	44.48	228.84
100	48	7.50	48.35	96.71	0.33	0.023	30.44	149.83
50	1	6.00	38.00	76.00	0.00	0.025	22.00	100.00
10	1	5.00	33.00	66.00	0.00	0.025	20.00	89.00

Table 6.18: Results for random symmetric indefinite Toeplitz matrices M of class 5, $n=250$

κ	K	H	N_H	L_H	s_θ	ϵ	N	L
10000	6	13.00	77.83	155.67	0.33	0.005	111.33	634.17
5000	4	9.00	61.00	122.00	0.00	0.003	150.00	921.50
1000	30	9.70	61.33	122.67	0.50	0.016	56.97	293.50
500	17	8.59	57.29	114.59	0.53	0.015	52.29	281.18
100	40	8.28	52.67	105.35	0.57	0.020	35.98	185.47

Table 6.19: Results for random symmetric indefinite Toeplitz matrices M of class 5, n=300

κ	K	H	N_H	L_H	s_θ	ϵ	N	L
10000	5	13.20	83.00	166.00	0.60	0.004	139.60	871.80
5000	4	12.50	75.75	151.50	0.50	0.011	82.50	457.00
1000	26	9.96	62.85	125.69	0.50	0.013	68.85	382.31
500	32	9.12	59.44	118.88	0.62	0.017	49.25	258.75
100	28	8.54	53.86	107.71	0.64	0.021	35.64	182.25
50	1	6.00	40.00	80.00	0.00	0.025	22.00	97.00

Table 6.20: Results for homotopic and nonhomotopic processes applied to symmetric positive definite matrix obtained from the original by premultiplying with its transpose. (Note: in this case no decrease of the initial tolerance value $\theta = 0.7$ was required in the experiments, so the resulting table does not contain the s_θ column)

κ	H	N_H	L_H	ϵ	N	L
10000	18.00	106.00	1272.00	1.3e-05	84.00	690.00
5000	16.50	99.50	1194.00	2.8e-05	65.00	584.00
1000	14.71	88.00	1056.00	1.4e-05	79.71	737.29
500	12.50	76.14	913.71	3.6e-05	42.64	379.50
100	10.42	63.57	762.78	4.8e-05	23.62	214.30
50	8.86	54.70	656.36	5.0e-05	19.12	180.95
10	7.69	47.13	565.60	5.0e-05	16.73	162.96

Table 6.21: Maximal values of the residual norms for non-homotopic processes with the original, symmetrized double-sized, and symmetrized positive definite matrices, respectively. Whenever the value did not exceed 1, it is denoted by **** in the table.**

κ	ρ_1	ρ_2	ρ_3
13117.400	1.01437	1.24521	2.29178
9997.910	1.00164	5.04473	2.87706
6038.480	1.00109	******	1.00278
3859.550	2.65327	3.61647	6.60274
2610.770	1.04845	2.71020	1.79991
2464.170	1.01631	1.00020	******
2324.950	2.43046	2.42135	1.49457
1848.360	1.00955	1.00033	1.13965
1597.240	1.00009	******	4.52156
1188.030	1.19201	******	1.57259
907.966	1.00097	1.01116	******
856.188	1.00250	******	1.10969
841.877	1.00008	******	1.88333
788.226	1.00240	******	******
740.068	1.00006	4.63459	1.49978
695.662	1.00002	******	******
689.868	1.00671	2.39270	3.50735
627.683	1.00044	******	2.46652
616.774	1.00332	******	1.18659
575.609	******	******	5.43561
554.912	1.17903	1.46120	3.31692
514.378	1.08366	******	******
512.452	1.06951	******	******
508.758	1.00109	******	4.44255
487.164	1.00206	******	1.49436
472.963	1.01446	******	1.61860
465.666	1.00001	******	******
441.397	1.00312	******	1.63125
428.375	1.00073	******	******
341.934	1.25551	******	1.01262
340.355	1.00045	******	******
302.378	1.05618	******	1.10584
282.635	1.00102	******	******
278.193	1.16907	******	******

κ	ρ_1	ρ_2	ρ_3
264.911	1.02277	******	******
261.404	1.00168	******	******
259.630	1.00628	******	2.75387
252.902	1.00218	******	1.22844
248.518	1.00040	******	1.69287
243.424	1.00110	******	******
242.588	2.09438	******	******
228.158	1.10735	******	******
223.702	1.01272	******	1.21397
202.115	1.00013	******	******
200.249	1.00212	******	******
197.326	1.00032	******	******
196.148	2.36707	******	******
190.886	1.00005	******	******
190.090	1.00010	******	2.96128
188.992	1.00323	******	******
174.938	1.00159	******	******
172.100	1.00260	******	******
171.442	1.05699	******	******
171.416	1.00030	******	******
166.156	1.00021	******	******
165.404	1.00381	******	******
162.923	1.00054	******	******
160.472	1.00127	******	******
158.845	1.00001	******	******
158.335	1.00005	******	******
147.662	1.00834	******	******
143.380	1.00044	******	******
141.694	1.00001	******	******
136.384	1.00002	******	******
132.914	1.00011	******	******
128.667	1.00001	******	******
125.703	1.00001	******	******
114.952	1.00022	******	******
102.699	1.00023	******	******
69.764	1.00001	******	******
61.406	1.19951	******	******
58.648	1.87989	******	******

able techniques of the analysis, however, rely on estimating how much compression perturbs the computed approximation, and here theoretical estimates tend to be overly pessimistic according to experimental computations with Toeplitz matrices. Thus the actual power of the iteration is stronger than the theoretical analysis certifies, and this power can be further enhanced by scaling and other modifications. Furthermore, we also showed a homotopic variation of Newton's iteration, which has proved to be superfast for a well conditioned Toeplitz-like input. Then again some adaptive policies for directing the homotopic process may improve its performance further. Some results of experimental tests have been included in this chapter.

6.13   Notes

Section 6.1   On the definitions and properties of matrix norm, see [GL96], [Hi96]. Theorem 6.1.3 follows from [GL96, Theorems 2.5.3, 8.6.2, and 8.6.4].

Section 6.2   On the origin of Newton's iteration, see [B40], [Bo68]. The integer division algorithm is due to S. A. Cook (see [AHU74], [BM75], [K98] for details). Newton's iteration for matrix inversion is due to Schultz [S33]. The residual correction algorithm in (6.2.9) is covered in [IK66, pages 86-88].

Section 6.3–6.5   The idea of using Newton's method for the solution of Toeplitz linear systems has arisen in a discussion between the present author and F. G. Gustavson [BGY80, page 293]. The idea was developed for Toeplitz and Toeplitz-like matrices in [P89a], [P92], [P92b], [P93], [P93a], [PBRZ99], and extended to other structured matrices in [BP93] (Toeplitz+Hankel-like case) [PZHD97] (Cauchy-like case), and [PR01], [PRW00], [PRWa] (unified approach). The current presentation follows [PRWa], in Section 6.4 also [PR01]. Theorem 6.3.2 is from [PZHD97]. On the extension to Stein type operators, see [PRWa]. The results of Section 6.4 extend similar results of [P92b], [P93a], [PBRZ99] proved in the Toeplitz-like case. On the SVD based compression, see also [P93]. Subroutine **R2** was used in [PZHD97] in the Cauchy-like case and in [PBRZ99] in the Toeplitz-like case. In the Toeplitz+Hankel-like case and in a distinct algebraic context, it was applied in [BP93], [BP94]. The two variants (6.5.7) and (6.5.8) of Newton–Toeplitz iteration are from [PRWa]; they are variations of similar results from [PBRZ99] proved in the Toeplitz-like case.

Section 6.6   Subsections 6.6.2 and 6.6.3 are from [PRWa]. Subsections 6.6.4 and 6.6.5 are from [PWa].

Section 6.7   The first part of the section is from [B-I66], [B-IC66], [SS74]. The second part is from [PS91].

Section 6.8   Variation of the compression level for improving performance was proposed in [P92b]. Experimental tests of some adaptive policies of compression were reported in [BMa] and [PKRCa].

Section 6.9 Homotopic Newton's iteration was proposed and studied in [P92b], [P01], and [PKRCa] for general matrices, in [P92b] for Toeplitz-like matrices, and in [PKRCa] in a unified way for various classes of structured matrices. Detailed estimates for the overall number of Newton's steps in homotopic processes for general matrices and the results of experimental tests with Toeplitz matrices can be found in [PKRCa]. On parallel arithmetic complexity of FFT, see [L92], [BP94, Chapter 4], [Q94]; on parallel multiplication of structured matrices by vectors, see [BP94, Chapter 4].

Sections 6.10 and 6.11 follow [PKRCa].

6.14 Exercises

6.1 Complete all omitted proofs of the results claimed in this chapter.

6.2 Specify the constants hidden in the "O" notation in the arithmetic computational complexity estimates of this chapter.

6.3 Write programs to implement the algorithms of this chapter. Then apply the algorithms to some specified structured input matrices.

6.4 Estimate the work space required for the algorithms of this chapter.

6.5 Show that the real value x_{i+1} defined by (6.2.1) is the intersection of the tangent line for the function $f(x)$ at $x = x_i$ and the x-coordinate line.

6.6 [PRWa] Extend Newton's iteration and its analysis in Sections 6.4–6.9 to the case of singular but internally regular operators in Section 4.5 and/or Stein type displacement operators. For internally regular operators L use their norms defined on the space of matrices orthogonal to the null space of L.

6.7 Extend Subroutine **R1** where all but $3r$ largest singular values of the displacement $L(X_i)$ are truncated. Estimate the convergence rate and the computational cost per iteration step in flops.

6.8 Assume the above truncation policy in the first Newton steps and then shift to the truncation of all but the r largest singular values. Specify a criterion for shifting from the $3r$ to the r-truncation policy that minimizes the overall work required for convergence within a fixed tolerance bound.

6.9 (a) Extend the Newton–Toeplitz Iteration in Section 6.5 by applying Toeplitz inversion formulae in [VHKa], [BMa] and other papers cited in the Notes to Section 1.5.

(b) Extend Toeplitz inversion formulae to the inversion of Toeplitz+Hankel matrices M (see Examples 4.2.1 and 4.4.8). Then extend the Newton–Toeplitz Iteration of Section 6.5 respectively.

6.10 Estimate the overall work required for approximating a given Toeplitz-like matrix M within a fixed bound ϵ on the error norm $\| Y_i - M^{-1} \|$. Assume Newton-Structured Iteration with each of the Subroutines **R1** and **R2**. In the case where $M = T$ is a Toeplitz matrix, also compare the numbers of flops used where the modifications of Subroutine **R2** in Section 6.5 and Exercise 6.9(a) are applied.

6.11 Compare bounds (6.4.6) and (6.4.7) with (6.5.5) and (6.5.6). Which of them is stronger for Toeplitz matrices M? For Toeplitz-like matrices M? How will bounds (6.4.6) and (6.4.7) change in the shift to the $3r$-policy of Exercises 6.7 and 6.8?

6.12 [PRWa], [PWa]. Let $L = \Delta_{A,B}$.

(a) Specify the estimates of Corollary 6.6.1 where $A = D(\mathbf{t})$, $B = \hat{Z} = 2 \sum_{i=1}^{\lfloor n/2 \rfloor} (-1) Z^{2i-1}$.

Specify the estimates in Theorem 6.6.2 where either

(b) $A = D(\mathbf{s})$, $B = D(\mathbf{t})$

or

(c) $A = Y_{00}^{-1}$, $B = Y_{11}$, $Y_0 = Z + Z^T$, $Y_{11} = Y_{00} + \mathbf{e}_0 \mathbf{e}_0^T + \mathbf{e}_{n-1} \mathbf{e}_{n-1}^T$.

(d) Compare the estimates of parts (b)–(c) with the estimates based on the results in Sections 6.6.4 and 6.6.5.

(e) Relate to each other the bounds on the norms $\nu_{r,l}(\nabla_{Z_f, Z_f}^{-1})$ and $\nu_{r,l}(\nabla_{Z_e, Z_f}^{-1})$ for $e \neq f$.

6.13 (a) Substitute the estimates for the norm ν^- in Section 6.6 into Corollaries 6.4.4 and 6.5.3. Compare the resulting pairs of estimates for the errors and convergence rate for Toeplitz-like matrices.

(b) Can you improve the latter estimates in the case of Toeplitz matrices based on the three variants of Newton's iteration defined by (6.5.7), (6.5.8), and (6.5.9)?

(c) Extend parts (a) and (b) to internally regular input operators L.

6.14 Consider the scaled residual correction (RC) processes:

$$X_0 = M^* / (\|M\|_1 \|M\|_\infty), \quad X_{i+1} = a_{i+1} X_i \sum_{k=0}^{p-1} R_i^k, \quad i = 0, 1, \ldots$$

where $R_i = I - M X_i$ (see (6.2.6), (6.2.9)).

(a) [IK66]. Assume that $a_i = 1$ for all i and observe that in this case the choice of $p = 3$ optimizes the number of flops required to yield a desired upper bound ϵ on the output residual norm for a general matrix M.

(b) Compare the latter flop count to the case of the scaled Newton iteration in [PS91].

(c) Extend the proof of numerical stability of this processes for $p = 2$ [PS91] to larger values of p.

(d) Assume a fixed single precision and estimate the best choice of the parameters p and a_i, $i = 0, 1, \ldots$ in the RC process above to minimize the number of flops required to yield convergence.

(e) Estimate the convergence rate of iteration (6.7.9) in terms of the distance of the eigenvalues of the matrix $X_0 M$ from $\rho = (3 - \sqrt{5})/2 = 0.3819 \ldots$ and $2 - \rho$. Consider a modified iteration from [PS91] where (6.7.9) is replaced by the equation

$$X_{i+1} = (-2X_i M + 3I)X_i M X_i.$$

The latter iteration performs matrix multiplication only three times per step (versus four in (6.7.9)) and sends the eigenvalues of the matrix $X_0 M$ to 0 and 1 if they lie in the open intervals (0,1/2) and (1/2, $\tilde{\rho}$), respectively, where $\tilde{\rho} = (1 + \sqrt{3})/2 = 1.366\ldots$. Can you utilize this process to decrease the overall work required for achieving convergence?

6.15 How should the scaling policies, the choice of the parameter p, and the answers to Exercise 6.14 change in the case of a non-singular Toeplitz input matrix M? Assume that in each Newton iteration step the displacement $L(X_i)$ is compressed based on:

(a) Subroutine **R1** with truncation to the r or the $3r$ largest singular values of the displacement $L(X_i)$, $r = \text{rank}(L(M^{-1}))$,

(b) Subroutine **R1** where all singular values below $\epsilon/\|L(X_i)\|_2$ are truncated and where ϵ is the "machine epsilon",

(c) Subroutine **R2**,

(d) (6.5.7), (6.5.8) or (6.5.9) (see Exercise 6.13(b)).

6.16 How much will the L-ranks of the two non-singular matrices M_1 and M_2 differ from each other provided that

(a) $L = \nabla_{Z_e, Z_f}$, $M_1 - M_2 = aI$ for any triple of scalars a, e, and f,

(b) $L = \nabla_{Z_e^T, Z_f}$, $M_1 - M_2 = aJ$ for any triple of scalars a, e, and f,

(c) $L = \nabla_{Y_{00}, Y_{11}}$, $M_1 + M_2 = aJ$ for any scalar a?

6.17 Estimate from above the displacement rank of the matrices M, $\widetilde{M}$, and $M_1 = M_0 + \tau_1(M - M_0)$ in Example 6.9.2 in terms of the dimensions of vectors $\mathbf{x}_i$, $\mathbf{y}_i$, and $\mathbf{u}_i$.

6.18 (a) Elaborate upon the homotopic Newton processes for Vandermonde-like and Cauchy-like input matrices.

(b) Choose an initial matrix M_0 for the homotopic process (with Algorithm 6.3.1 as the Basic Subroutine) applied to invert the Hilbert matrix $H = \left(\frac{1}{i+j+1}\right)_{i,j=0}^{n-1}$. **Hint:** Shift from the matrix H to the Toeplitz matrices JH or HJ.

6.19 [PKRCa].

    (a) Assume an $n \times n$ general Hermitian positive definite input matrix M. Write $\theta_k = \theta$ for all k. Choose the positive $\theta < 1$ which minimizes the overall number a of times we perform matrix multiplication in a homotopic process based on a scaled Newton iteration. Assume most effective scaling policies of [PS91]. Estimate a.

    (b) Do the same exercise for a general $n \times n$ Hermitian matrix M.

6.20   (a) Study experimentally how convergence of Newton-Structured (non-homotopic) Iteration is affected by the compression of the displacement. Use the maximal compression to the length equal to the $\nabla_{A,B}$-rank of the Toeplitz input matrix M (and $\nabla_{B,A}$-rank of its inverse M^{-1}) and limited compressions to larger values of the length.

    (b) In the case of homotopic processes, study experimentally the efficiency of various policies for the choice of tolerance values (and consequently step sizes). Start with general matrices M. Then examine the case where the matrix M is structured and estimate the effect of the maximal and limited compression on the choice of step sizes. Compare the results of the policy that relies on bounds (6.4.6) and (6.4.7) or (6.5.5) and (6.5.6) and the dynamic adaptive policy. (According to the latter policy, these bounds are either ignored or used only as the starting points for an adaptive policy of increasing, say doubling the step size.)

6.21 Consider the following generalization of the homotopic process (6.9.1):

$$M_{h+1} = M_0 F_{h+1} + M = M_h + M_0(F_{h+1} - F_h)$$

where $F_0 = I, F_1, F_2, \ldots$ is a sequence of matrices converging to 0 and satisfying $\|M_h^{-1} M_0(F_{h+1} - F_h)\| \leq \theta_h$ for fixed scalars $\theta_h < 1, h = 0, 1, \ldots,$ and chosen to keep the trajectory $\{M_{h+1}, h = 0, 1, \ldots\}$ away from singularities. Try to choose the matrices $F_0, F_1, \ldots$ to decrease the overall work required to approximate M^{-1}.

6.22 The matrices M and M_0 in Section 6.9.3 can be simultaneously diagonalized by an invertible matrix U under the choices (6.9.6)–(6.9.9), that is, $D = UMU^*$ and $D_0 = UM_0U^*$ are diagonal matrices for some invertible matrix U (cf. [Par80, Theorem 15-3-3]).

    (a) Emulate the homotopic processes in Section 6.9 assuming no compression and applying them for the matrices D and D_0 instead of M and M_0, and estimate the number of Newton's steps involved in terms of the eigenvalues of D and D_0.

    (b) Analyze the homotopic processes in Section 6.9 applied to a Hermitian positive definite Toeplitz matrix M and the tridiagonal matrix M_0 sharing its three non-zero diagonals with M. If M_0 is not positive definite, replace it by $M_0 + aI$ for a positive a.

Chapter 7

Newton Algebraic Iteration and Newton-Structured Algebraic Iteration

Summary   Newton iteration is a fundamental tool of computer algebra. We specify this iteration for the general problem of algebraic rootfinding and apply it first to the solution of some important problems of computations with polynomials and integers. Then it is applied to computations with general and structured matrices M: we compute the characteristic polynomial $\det(xI - M)$, the minimum polynomial, the inverse M^{-1}, and the Krylov matrix $K(M, \mathbf{v}, q)$ (for a vector $\mathbf{v}$ and a positive integer q). The matrix computation algorithms are fast (although not always superfast) and allow most effective parallelization.

For the inversion of matrices filled with integers or rational numbers, Newton's algebraic iteration enables dramatic decrease of the precision of the computations. If in addition, the input matrix is structured, then the algorithm reaches the superfast level in terms of its bit-operation complexity, which is nearly linear in the bit-complexity of the representation of the output. That is, Newton's algebraic iteration nearly reaches optimality. The approach requires computations modulo a random prime and its powers, to avoid singularities. For some problems the algorithm is nearly optimal also in terms of arithmetic complexity. At the end of the chapter, in particular in Exercise 7.15, we briefly recall some advanced techniques that yield effective parallelization of the superfast sequential computations with integer Toeplitz-like matrices.

7.1   Some introductory remarks

Newton's algebraic iteration is a basic tool of computer algebra and is our next subject. In the next two sections, we outline the method and apply it to some fundamental problems of computer algebra; thereafter, we focus on its application to matrix computations.

Graph for selective reading

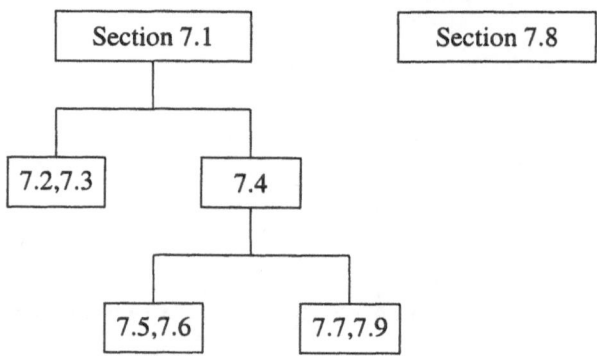

In its both algebraic and numerical versions, the iteration,

$$w_{h+1} = w_h - f(w_h)/f'(w_h), \quad h = 0, 1, \dots,$$

rapidly improves an initial approximation w_0 to the solution w of an equation $f(w) = 0$. Let w_h denote the output of the h-th step of the iteration. Then in the algebraic version, we have

$$w_h - w = 0 \bmod z^{j2^h} \text{ provided that } w_0 - w = 0 \bmod z^j, \quad h = 1, 2, \dots,$$

where $w = w(z)$, $w_0 = w_0(z)$, and $w_h = w_h(z)$ are polynomials in z, whereas in the numerical version we have

$$w_h - w = O(w_0 - w)^{2^h}, \quad h = 1, 2, \dots,$$

where w, w_0, and w_h are numbers. By viewing these numbers as the values of the above polynomials at a point z sufficiently close to 0, we can see the correlation between the two versions. The main technical difference between them is due to numerical truncation of the binary values in the numerical version.

We study the algebraic version assuming error-free computations with infinite precision. In Sections 7.6 and 7.8, we estimate the bit-cost of the computations by counting the bit-operations involved. To make the estimates uniform throughout the chapter, we rely on the bound $\mu(q)$ on the bit-operation cost of performing an arithmetic operation modulo $2^q + 1$, where

$$\mu(q) = O((q \log q) \log \log q), \tag{7.1.1}$$

although for small and moderately large values q, it is more realistic to assume $\mu(q)$ of the order of q^2 and $q^{\log_2 3}$, respectively. We extrapolate these estimates to the bit-cost of an arithmetic operation with q-bit operands. (See Exercise 2.9 and the references therein for proofs and details.)

7.2   Newton's algebraic iteration: generic algorithm

Hereafter, $h'(x)$ denotes a formal derivative of a formal power series or a polynomial $h(x)$.

Problem 7.2.1. Polynomial equation modulo a power.

INPUT: two non-negative integers j and K, $K > j$, a rational function $f(w) = n(w)/d(w)$ with the coefficients of the polynomials $n(w)$ and $d(w)$ in the ring of formal power series in a variable z, and a polynomial $w_0(z)$ such that

$$f'(w_0(0)) \neq 0. \qquad (7.2.1)$$

((7.2.1) implies that the polynomial $f'(w_0(z))$ mod z^j has the reciprocal modulo z^j.)

OUTPUT: a polynomial $w(z)$ mod z^K such that $(f(w_k(z)))$ mod $z^K = 0$ for $k = \lceil \log_2(K/j) \rceil$, or equivalently, the first K coefficients of a formal power series $w(z)$ such that

$$f(w(z)) = 0. \qquad (7.2.2)$$

Solution: Newton's algebraic iteration solves Problem 7.2.1 as follows.

Algorithm 7.2.2. Newton's algebraic iteration. Write $w_h(z) = w(z)$ mod $z^{j 2^h}$, $h = 1, 2, \ldots, k$, $k = \lceil \log_2(K/j) \rceil$. Successively compute the polynomials

$$w_{h+1}(z) = w_h(z) - f(w_h(z))/f'(w_h(z)) \text{ mod } z^{j 2^{h+1}}, \quad h = 0, 1, \ldots, k - 1. \quad (7.2.3)$$

Output the polynomial $w_k(z)$ mod z^K.

The proof of correctness I. To prove correctness of the algorithm, introduce auxiliary rational function $g(x) = f(x)/(x - w)$ where $w(z)$ is an unknown formal power series satisfying (7.2.2). We have $g'(x) = f'(x)/(x - w) - f(x)/(x - w)^2$; therefore, $f(x) = f'(x)(x - w) - g'(x)(x - w)^2$. Substitute the latter expression for $f(x)$ and then $x = w_h(z)$ into (7.2.3), and after some obvious cancellations obtain that

$$w_{h+1} = w + (w_h - w)^2 g'(w_h)/f'(w_h) \text{ mod } z^{j 2^{h+1}}. \qquad (7.2.4)$$

We have $w(0) = w_h(0)$ for all h, and $f'(w(0)) \neq 0$ due to (7.2.1). Therefore, $f'(w_h(0)) \neq 0$ and the function $f'(w_h(z))$ is invertible modulo z^K for all h. Now it follows from (7.2.4) that *each iteration step (7.2.3) doubles the number of the computed coefficients of the formal power series* $w(z)$. This proves the correctness of Newton's algebraic iteration (7.2.3) and also shows that the solution of Problem 7.2.1 amounts to $\lceil \log_2(K/j) \rceil$ Newton steps (7.2.3). $\square$

The proof of correctness II. Here is an alternative proof of correctness, which better shows correlation to the numerical version of Newton's iteration. Expand a function $f(x)$ as a formal power series in $x - y$,

$$f(x) = f(y) + \sum_{h=1}^{+\infty} \frac{(x - y)^h}{h!} f^{(h)}(y). \qquad (7.2.5)$$

Approximate $f(x)$ by the two-term partial sum

$$f_1(x) = f(y) + (x - y)f'(y) = f(x) \bmod (x - y)^2$$

(see (6.2.2)). Let $f(w(z)) = 0$, $w_h(z) = w(z) \bmod z^{j2^h}$ for all $h \geq 0$. Now substitute $x = w_{h+1}(z)$ and $y = w_h(z)$ and obtain successively that

$$(x - y)^2 = 0 \bmod z^{j2^{h+1}},$$

$$f_1(w_{h+1}) = f(w_{h+1}) \bmod z^{j2^{h+1}} = f(w) \bmod z^{j2^{h+1}} = 0 \bmod z^{j2^{h+1}},$$

$$f(w_{h+1}) + (w_{h+1} - w_h)f'(w_h) = 0 \bmod z^{j2^{h+1}}.$$

Recall that the function $f'(w)$ has the reciprocal modulo $z^{j2^{h+1}}$, divide the latter equation by $f'(w_h)$, and arrive at (7.2.3). □

Newton's iteration (7.2.3) is generalized in several iterative processes having similar properties; some processes approximate the formal power series $f(x)$ in (7.2.5) by its higher order partial sums.

Newton's algebraic iteration has a counterpart called *Hensel's lifting* where one new coefficient of the power series $w(z)$ is computed in each step but this step uses fewer ops (see Algorithm 7.7.3 in Section 7.7). The Newton and Hensel lifting processes can also be applied to computations with integers where the solutions of equations and systems of equations are initially given modulo a prime p. Then the processes yield the solutions modulo a higher power of p. Here is an example.

Algorithm 7.2.3. Evaluation of all zeros of a polynomial having only integer zeros.

INPUT: natural n and N and the coefficients of a monic polynomial $q(x) = \sum_{i=1}^{n} q_i x^i$, $q_n = 1$, that has only integer zeros, which lie between $-N$ and N.

OUTPUT: the set of all zeros of $q(x)$.

COMPUTATION:

1. Compute the discriminant d of $q(x)$ (that is, the resultant of $q(x)$ and of its derivative $q'(x)$ times $(-1)^{n(n-1)/2}$).

2. Compute the smallest prime p that does not divide d.

3. By means of the exhaustive search, compute modulo p the set S_1 of all zeros of $q(x)$.

4. Recursively compute the sets $S_{j+1} = \{(x_j + p^J r_j) \bmod p^{2J} \mid x_j \in S_j, r_j = -u_j v_j \bmod p^J, u_j = q(x_j)/p^J, v_j = 1/(q'(x_j)) \bmod p^J, J = 2^j\}$, for $j = 1, 2, \ldots, k - 1$, choosing k to be the smallest integer such that p^K for $K = 2^k$ exceeds $2N$, $k = \lceil (\log(2N))/\log p \rceil$. (This process is actually Newton's iteration, and S_j is the set of zeros of $q(x)$ defined up to their reduction modulo p^J.)

5. Output the set of zeros of $q(x)$ recovered from the set S_k by means of *modular rounding-off*, that is, output $h = h \bmod m$ if $h \bmod m < m/2$ and output $(h \bmod m) - m$ otherwise (in both cases provided that $m = p^J > 2|h|$).

7.3 Specific examples of Newton's algebraic iteration

Example 7.3.1. Polynomial reciprocal by Newton's iteration (cf. Cook's algorithm for Problem 6.2.1 of integer division). The computation of the first K coefficients of the formal power series $t(z) = 1/v(z)$ for a given polynomial or formal power series $v(z)$, $v(0) \neq 0$ (see Problem 2.5.2) amounts to the solution of the equation $f(t) = 0$ where $f(t) = v - 1/t$ over the ring of formal power series in z.

Newton's iteration takes the following form: $j = 1$, $t_0 = t(z) \bmod z = 1/v(0)$, $t_{h+1}(z) = t_h(z)(2 - v(z)t_h(z))$, $h = 0, 1, \ldots, k$, $t_k(z) = t(z) \bmod z^K$, $k = \lceil \log_2 K \rceil$. For $K = 2^k$, the resulting algorithm and the algorithm of Section 2.5 for Problems 2.5.1 and 2.5.2 are computationally equivalent (that is, both algorithms compute exactly the same intermediate and output values) but have distinct derivations and interpretations.

Problem 7.3.2. Generalized reversion of power series. Given a natural K and two formal power series,

$$s(z) = \sum_{i=1}^{+\infty} s_i z^i, \quad t(w) = w + \sum_{i=2}^{+\infty} t_i w^i,$$

find the polynomial $w(z)$ of degree less than K satisfying

$$s(z) = t(w(z)) \bmod z^K.$$

Problem 7.3.2 is a generalization of the customary reversion problem where $s(z) = z$ and where for a given formal power series $t(w)$, we seek a formal power series $w(z)$ such that $z = t(w(z))$.

Solution: Since $s(0) = t(0) = 0$ and since we seek solution modulo z^K, the terms $t_i w^i$ of $t(w)$ and $s_i z^i$ of $s(z)$ for $i \geq K$ do not influence the solution and can be ignored. Thus, the problem can be stated for polynomials rather than for the formal power series $s(z)$ and $t(w)$. Apply Newton's method for $j = 2$, $w_1(z) = s_1 z = w(z) \bmod z^2$ to the equation $f(w) = s(z) - t(w) = 0$ and arrive at the following iteration:

$$w_{h+1}(z) = w_h(z) + \frac{s(z) - t(w_h(z))}{t'(w_h(z))} \bmod z^{2^h}, \quad h = 1, 2, \ldots.$$

Here, $f'(w) = t'(w) = 1 + \sum_{i=1}^{2^h-1} i t_i w^{i-1}$. Since $w(0) = w_1(0) = 0$, we have $f'(w_1(0)) = f'(w(0)) = f'(0) = t'(0) = 1$. Therefore, the formal power series $f'(w(z))$ and $f'(w_j(z))$ have reciprocals, and we deduce from (7.2.3) that $w_{h+1}(z) = w(z) \bmod z^H$, $H = 2^h$, $h = 1, \ldots, \lceil \log K \rceil$. Given the polynomial $w_h(z) \bmod z^H$, compute the coefficients of the polynomial $w_{h+1}(z) \bmod z^{2H}$ by applying (7.2.3) for $j = 1$ and $f(w) = s - t(w)$. First compute the polynomial $r_h(z) = s(z) - t(w_h(z)) \bmod z^{2H}$, by applying the algorithm for Problem

3.4.3 of computing the composition of power series. Then compute the polynomial $q_h(z) = t'(w_h(z)) \bmod z^{2H}$ and its reciprocal modulo z^{2H} (see Problem 2.5.2 and Example 7.3.1). Finally, compute the polynomial $w_{h+1}(z) = w(z) \bmod z^{2H} = w_h(z) + r_h(z)/q_h(z) \bmod z^{2H}$ for $h = k$, $H = K$, which is Newton's step (7.2.3). The solution uses $O((K \log K)^{3/2})$ ops, with a substantial overhead constant hidden in the "O" notation. The cost bound is dominated by the cost of the composition of power series. □

Problem 7.3.3. Evaluation of a root of a polynomial. Given natural d and n and a polynomial $p(z)$ of degree of at most n such that $p_0 = p(0) = 1$, compute the unique polynomial $w(z)$ of degree of at most n such that $w_0 = w(0) = 1$ and $(w(z))^d = p(z) \bmod z^{n+1}$.

Solution: Apply Newton's iteration (7.2.3) with $j = w_0(z) = 1$ to the equation $f(w) = w^d - p = 0$. The solution algorithm uses $O(M(n))$ ops (for $M(n)$ in (2.4.1), (2.4.2)) with a small overhead and is used in the solution of the next problem. □

Problem 7.3.4. Polynomial decomposition. Given two natural numbers m and n and the coefficients of a monic polynomial $u(z)$ of degree $N = mn$, either find two monic polynomials, $s(t)$ of degree m and $t(z)$ of degree n, such that $u(z) = s(t(z))$, or prove that there exist no such pairs of polynomials.

Solution: First assume that $u(z) = s(t(z))$ for some monic polynomials, $s(t)$ of degree m and $t(z)$ of degree n. Then the polynomials $u(z)$ and $(t(z))^m$ must agree on their n highest degree terms. That is, the polynomial $u(z) - (t(z))^m$ has degree of at most $N - n$. Replace $u(z)$ and $t(z)$ by the reverse polynomials, $U(z) = z^N u(1/z)$ and $T(z) = z^n t(1/z)$. Then $T(0) = 1$, $U(z) - (T(z))^m = 0 \bmod z^n$. Based on the latter relations, evaluate the coefficients of the polynomial $T(z) \bmod z^n$ at the cost of performing $O(M(n))$ ops (see Problem 7.3.3) and then immediately recover the coefficients of the candidate polynomial $t(z)$. Then solve Problem 2.5.5 of the computation of the generalized Taylor expansion of $u(z)$ as a polynomial in $t(z)$. The arithmetic computational cost of performing this stage, $O(M(N) \log m)$ ops, with a small overhead, dominates the asymptotic cost of the computation of the polynomial $t(z)$. This defines the unique set of polynomials $s_i(z)$, $i = 0, 1, \ldots, m$, of degrees less than m satisfying the polynomial identity of Problem 2.5.5. If the polynomials $s_i(z) = s_i$ are constants for all i, then we arrive at the desired polynomial $s(t) = \sum_{i=0}^m s_i t^i$, such that $u(z) = s(t(z))$; otherwise, we conclude that there exists no desired decomposition $u(z) = s(t(z))$ for the given input triple of $u(z)$, m and n. □

The solution shows that the coefficients of the polynomials $s(t)$ and $t(z)$ of decomposition $u(z) = s(t(z))$ are rational functions in the coefficients of the polynomial $u(z)$.

On an alternative probabilistic method for the evaluation of the coefficients of the polynomial $s(t)$ for given polynomials $u(z)$ and $t(z)$, which is numerically less stable but asymptotically faster if $\log m = o(\log N)$, see [BP94, pages 52–53].

Our next problem is an extension of Problem 7.3.4.

Problem 7.3.5. Complete decomposition of a polynomial. Given the coefficients of a polynomial $u(z)$ of degree N and a prime factorization of the integer N, compute a complete decomposition of the polynomial $u(z)$ into a composition of indecomposable polynomials.

Solution: Solve Problem 7.3.4 of polynomial decomposition for the input polynomial $u(z)$ and for all natural m dividing N until a decomposition $u(z) = s(t(z))$ is found. (This must occur unless for all divisors m of the integer N there is no decomposition of the polynomial $u(z)$, in which case we arrive at a trivial complete decomposition $u(z) = u(z)$.) Then recursively apply the same process to the polynomials $s(t)$ and $t(z)$ until the complete decomposition of the input polynomial $u(z)$ is obtained. It can be estimated that the overall cost of the solution is $O(N^{1+\epsilon})$ ops, for any fixed positive ϵ.   □

Theorem 7.3.6. *Complete decomposition of any polynomial is unique up to within the trivial ambiguities of the three following groups:*

   i) $z = s(t(z))$ *for linear polynomials* $s(t) = t - a$, $t(z) = z + a$ *and for any constant* a;

   ii) $u(z) = s(t(z)) = s_1(t_1(z))$ *provided that* $s(t) = t^m$, $t(z) = z^h q(z^m)$, $s_1(t) = t^h q^m(t)$, $t_1(z) = z^m$, h *is a natural number, and* $q(z)$ *is a polynomial;*

   iii) $s(t(z)) = t(s(z))$ *for* $s(t) = T_m(t)$, $t(z) = T_n(z)$ *where* $T_h(x)$ *denotes the* h-*th degree Chebyshev polynomial of the first kind,* $T_{h+1}(x) = 2x T_h(x) - T_{h-1}(x)$, $h = 1, 2, \ldots, T_0(x) = 1, T_1(x) = x$.

Problem 7.3.7. Evaluation of algebraic functions. Given the coefficients of $n + 1$ polynomials $f_i(z)$, $i = 0, \ldots, n$, compute the first K coefficients of a fractional power series $w = w(z) = \sum_{i=0}^{+\infty} w_i z^{i/d}$, $d \le n$, satisfying the equation

$$f(w) = f_n(z)w^n + \cdots + f_0(z) = 0. \tag{7.3.1}$$

($w(z)$ may have several expansions into the fractional power series; one of them is to be computed.)

Solution: (7.3.1) is a special case in (7.2.2), so the general solution scheme of Section 7.2 is applied.

STAGE A). Without loss of generality, assume that $f_n(0) \ne 0$. Write $w_0 = w(0)$, reduce (7.3.1) modulo z, and obtain that $f_n(0)w_0^n + f_{n-1}(0)w_0^{n-1} + \cdots + f_0(0) = 0$. Then the value w_0 can be computed as a zero of the above n-th degree polynomial. In the singular case, where the functional power series $f'(w)$ is not invertible (that is, where $f'(w) = 0 \bmod z$), apply *Newton's polygon process* [KT78] to reduce the original problem to a regular one, where $f'(w) \ne 0 \bmod z$. The subsequent application of the algebraic function theory yields that $d = 1$ and $w = w(z)$ is a formal power series. Newton's polygon process defines one of the expansions of $w(z)$ by computing the j

coefficients of its lowest degree terms for a fixed j. The arithmetic cost of computing each coefficient is dominated by the arithmetic cost of computing a zero of a polynomial of degree of at most n; the integer $j - 1$ does not exceed the lowest degree in z of the terms of the resultant of the polynomials $f(w)$ and $f'(w)$ (see the definition of the resultant in Section 2.7); the degree does not exceed $(2n - 1)m$ where m is the maximum degree of $f_i(z)$ for all i.

STAGE B).   Given $w(z) \bmod z^j$, the polynomial $w(z) \bmod z^K$ is computed in at most $\log(K/j) \le \log K$ iteration steps (7.2.3). The h-th iteration is reduced to multiplication of $O(n)$ pairs of polynomials modulo z^{2jH}, $H = 2^h$, due to (7.2.3), (7.2.4), and (7.3.1), so the overall cost of performing Stage B is $O(nM(K))$ ops, for $M(n)$ in (2.4.1)–(2.4.2). The latter bound dominates the estimated cost of performing Stage A for large K.

A trade-off is possible: we may avoid solving a polynomial equation at Stage A (to decrease its cost) and carry out the polynomial zeros symbolically, by operating in an algebraic extension $\mathbb{E}$ of the original field $\mathbb{F}$. Here $\mathbb{E}$ is the field of polynomials modulo the (irreducible) minimum polynomial of these zeros, that is, the monic polynomial with coefficients in $\mathbb{F}$ having these zeros and having the minimum degree d, $d \le n$. Then the cost of performing Stage B increases by the factor of $\tilde{M}(n)$, representing the cost of multiplication and addition of polynomials modulo $m(z)$, rather than the cost of similar operations with their zeros.   □

Here are some sample problems that can be reduced to the solution in (7.3.1) (compare Problems 7.3.2 and 7.3.3):

1. Computation of an n-th root $w(x)$ of a polynomial $p(x)$ satisfying (7.3.1), such that $f_0(z) = p(x)$, $f_n(z) = 1$, $f_i(z) = 0$ for $i = 1, 2, \ldots, n - 1$ in (7.3.1).

2. Reversion of a polynomial modulo z^K: $f_0(z) = -s(z)$, $f_i(z) = t_i$, where t_i are constants for $i = 1, 2, \ldots, n$; $t_1 = 1$.

3. Computation of the Legendre polynomials: $n = 2$, $f_2(z) = 1 - 2tz + z^2$, $f_1(z) = 0$, $f_0(z) = -1$. The i-th coefficient of $w(z)$ is the i-th degree Legendre polynomial in t; $(1 - 2tz + z^2)^{-1/2}$ is the generating function of Legendre polynomials.

7.4   Newton's algebraic iteration for the inversion of general matrices

Let $M(x)$ be a polynomial matrix, which can be viewed as a polynomial with matrix coefficients,

$$M(x) = M_0 + \sum_{i=1}^{N} M_i x^i, \quad \det M_0 \ne 0, \tag{7.4.1}$$

or as a matrix with polynomial entries. Then its inversion is reduced to the computation of the matrix $X_0 = M_0^{-1}$, followed by performing Newton's algebraic iteration:

$$X_0 = M_0^{-1}, \quad X_{i+1}(x) = 2X_i(x) - X_i(x)M(x)X_i(x) \bmod x^{2^{i+1}}, \tag{7.4.2}$$

$i = 0, 1, \ldots$ The iteration performs matrix multiplication twice per step and produces a sequence of iterates

$$X_{i+1}(x) = M^{-1}(x) \bmod x^{2^{i+1}}, \quad i = 0, 1, \ldots. \tag{7.4.3}$$

7.5 Newton and Newton-Structured Algebraic Iterations for characteristic polynomials and the Krylov spaces

Let us apply iteration (7.4.2) to the computation of the characteristic polynomial and the Krylov spaces of a matrix (see Definition 2.1.1).

Problem 7.5.1. Krylov space (matrix) computation. Given an $n \times n$ matrix M, a vector $\mathbf{v}$ of dimension n, and a positive integer q, compute the Krylov matrix $K(M, \mathbf{v}, q)$.

Problem 7.5.2. Computing the characteristic polynomial of a matrix. Given an $n \times n$ matrix M, compute its characteristic polynomial $c_M(x) = \det(xI - M)$.

Solution of Problems 7.5.1 and 7.5.2 by Newton's algebraic iteration.   Substitute $N = 1, M_0 = I, M_1 = -M$ into (7.4.1) and apply Newton's algebraic iteration (7.4.2) to the resulting input matrix $M(x) = I - xM$. Then $X_i(x) = (I - xM)^{-1} \bmod x^{2^i} = \sum_{j=0}^{2^i-1}(xM)^j$, and we obtain the values of trace(M^j) for $j = 0, 1, \ldots, 2^i - 1$ as well as the Krylov matrices $K(M, \mathbf{v}, q) = \{M^j\mathbf{v}, \ j = 0, 1, \ldots, q - 1\}$ for a fixed pair of a vector $\mathbf{v}$ and an integer i and for $q = 2^i$. Having the values of trace(M^j) available for $j = 0, 1, \ldots, 2n$, we may recover the coefficients of $\det(xI - M)$, the characteristic polynomial of M over any field of constants by using $O(M(n) \log n)$ ops for $M(n)$ in (2.4.1), (2.4.2) (this is the inverse of the power sums Problem 2.5.4 because trace$(M^j) = \sum_i \lambda_i^j$, $\det(\lambda_i I - M) = 0$ for all i).

The overall cost of the computation of the characteristic polynomial of an $n \times n$ matrix by this algorithm is dominated by the cost of performing steps (7.4.2) for $i = 0, 1, \ldots, k$ (that is, essentially performing matrix multiplication twice per step and $2k$ times in k steps), where $k = O(\log n)$. $\square$

In this approach, the computation of the Krylov matrix $K(M, \mathbf{v}, q)$ is reduced to performing matrix multiplication $2\lceil \log_2 q \rceil$ times and is similar to the well-known recursive algorithm that successively computes the matrices

$$M\mathbf{v}, \quad M^2, \quad M^2(\mathbf{v}, M\mathbf{v}) = (M^2\mathbf{v}, M^3\mathbf{v}), \quad M^4,$$
$$M^4(\mathbf{v}, M\mathbf{v}, M^2\mathbf{v}, M^3\mathbf{v}) = (M^4\mathbf{v}, M^5\mathbf{v}, M^6\mathbf{v}, M^7\mathbf{v}), \quad \ldots \tag{7.5.1}$$

Is this any better than the successive computation of the vectors $M^{i+1}\mathbf{v} = M(M^i\mathbf{v})$, $i = 0, 1, \ldots$? Yes, its parallel implementation immediately achieves arithmetic time of $O(\log^2 n)$ on n^3 or even fewer processors, whereas the straightforward way requires more than linear parallel time.

In the case of a structured matrix M, the straightforward computation of the Krylov sequence $M^i\mathbf{v}, i = 1, \ldots, q$, requires $qv_{r,n}(L)$ ops for $v_{r,n}(L)$ of Definition 4.7.1 but uses parallel arithmetic time greater than q. By operating with short L-generators of the matrices involved, Newton's algebraic iteration uses about as many ops but much smaller parallel time (see Remark 7.5.5). Furthermore, it also enables computing the characteristic polynomial of a structured matrix M using smaller sequential arithmetic time than all other known algorithms. Let us specify the iteration and the arithmetic complexity estimates assuming a Sylvester type regular operator $L = \nabla_{A,B}$.

Algorithm 7.5.3. Newton-Structured Algebraic Iteration. Write

$$L(M(x)) = G(x)H^T(x), \quad L(X_j(x)) = G_j(x)H_j^T(x), \quad j = 0, 1, \ldots,$$

recall (7.4.3), and Theorem 1.5.3, and compress the computation of iteration (7.4.2) as follows (see (6.5.2) and (6.5.4)):

$$G_{i+1}(x) = -(2X_i(x) - X_i(x)M(x)X_i(x))G(x) \bmod x^{2^{i+1}}, \qquad (7.5.2)$$

$$H_{i+1}^T(x) = H^T(x)(2X_i(x) - X_i(x)M(x)X_i(x)) \bmod x^{2^{i+1}}, \qquad (7.5.3)$$

$i = 0, 1, \ldots$.

The i-th iteration step essentially amounts to performing pre- and post-multiplication of the matrices $X_i(x)$ and $M(x)$ by $3r$ vectors provided that the matrices $H^T(x)$ and $G(x)$ consist of r rows and r columns, respectively.

By exploiting the matrix structure in the above way and applying some additional techniques, we decrease the estimated computational cost as follows.

Theorem 7.5.4. *Let a structured $n \times n$ matrix M associated with a regular Sylvester type operator $L = \nabla_{A,B}$ be given with its L-generator of length l. Write $L_- = \nabla_{B,A}$ and $r = \mathrm{rank}(L(M))$. Then one may compute the characteristic polynomial of the matrix M using $O(l^2n + (v_{r,n}(L_-) + v_{r,n}(L))rM(n))$ ops and compute the matrix $K(M, \mathbf{v}, q) = \{M^j\mathbf{v}, j = 0, 1, \ldots, q - 1\}$ for a fixed vector $\mathbf{v}$ of dimension n and a fixed positive integer q using $O(l^2n + (v_{r,n}(L_-) + v_{r,n}(L))rM(q))$ ops, for $r = \mathrm{rank}(L(M))$ and $v_{r,n}(L)$ of Definition 4.7.1. Furthermore, for these two computational tasks, it is sufficient to use $O(l^2n + r^2M(n^2))$ and $O(l^2n + r^2M(nq))$ ops, respectively, for $M(n)$ in (2.4.1), (2.4.2) provided that the input matrix M is a Toeplitz-like, Hankel-like, or a Toeplitz-like+Hankel-like matrix.*

Even for a sructured matrix M, the $n \times q$ Krylov matrix is generally unstructured and has nq distinct entries, which implies the lower bound of nq ops on the complexity of Problem 7.5.1. That is, the bound of Theorem 7.5.4 is nearly optimal and Algorithm 7.5.3 is superfast concerning Problem 7.5.1. For Problem 7.5.2 the lower bound has the order of n and is not matched by the known upper bounds, which have the order of at best $M(n^2)$ ops.

Remark 7.5.5. Algorithm 7.5.3 allows effective parallel implementation, yielding polylogarithmic parallel arithmetic time on n^2 or nq processors for the computation of the characteristic polynomial and the Krylov space $K(M, \mathbf{v}, q)$ of an $n \times n$ structured input matrix M, respectively. Indeed, the algorithm multiply matrices $2\lceil \log_2(1 + q) \rceil - 1$ times to compute the Krylov matrix $K(M, \mathbf{v}, q)$. For $q = 2n - 1$, this is also the hardest stage of computing the characteristic polynomial of M. On the other hand, $n \times n$ matrix multiplication uses arithmetic time $O(\log n)$ on n^3 or fewer processors. All other known methods for the characteristic polynomial as well as the Krylov sequence, including the straightforward computation and the algorithm in (7.5.1), use about as many or more ops but have much inferior parallel performance, requiring at least linear parallel arithmetic time or blowing up the processor bounds. In other words, our solution of Problem 7.5.2 uses $O(M(n^2))$ ops but can be viewed as *superfast parallel algorithms*. Similar property holds for the algorithms of the next section.

7.6   Extensions of Krylov space computation

Krylov spaces and matrices play a prominent role in the study of iterative methods for numerical solution of linear system of equations; we study the algebraic counterparts of these methods.

Definition 7.6.1. The *minimum polynomial* of an $n \times n$ matrix M is the minimum degree monic polynomial $m_M(x) = \sum_{i=0}^{k} m_i x^i$, $m_k = 1$, such that $m_M(M) = 0$.

Given the minimum polynomial $m_M(x)$ of a non-singular matrix M, we have $m_0 \neq 0$, $M^{-1} = -\sum_{i=1}^{k} (m_i/m_0) M^{i-1}$, and consequently, $M^{-1} \mathbf{v} = -\sum_{i=1}^{k} (m_i/m_0) M^{i-1} \mathbf{v}$ for any vector $\mathbf{v}$. That is, knowing the minimum polynomial $m_M(x)$ and the Krylov matrix $K(M, \mathbf{b}, k-1)$, one may immediately compute the solution $\mathbf{x} = M^{-1} \mathbf{b}$ to a non-singular linear system $M\mathbf{x} = \mathbf{b}$ over any field of constants.

Furthermore, if the Krylov matrix $K(M, \mathbf{v}, 2n-1)$ is available for a random vector $\mathbf{v}$, one may similarly sample another random vector $\mathbf{u} = (u_i)_{i=0}^{n-1}$ and compute the coefficients of the lowest degree monic polynomial $m_{\mathbf{u}, \mathbf{v}, M}(x)$ satisfying the vector equation $\mathbf{u}^T m_{\mathbf{u}, \mathbf{v}, M}(M) \mathbf{v} = 0$. It is immediately recognized that these coefficients represent the minimum span for a linear recurrence sequence and therefore can be computed as the solution of Problem 2.9.7 (the Berlekamp-Massey problem). Algorithm 7.5.3 and some non-trivial effective parallel algorithms can be applied to solve this problem efficiently, but how are the polynomials $m_{\mathbf{u}, \mathbf{v}, M}(x)$ and $m_M(x)$ related to each other? We easily prove that the polynomial $m_{\mathbf{u}, \mathbf{v}, M}(x)$ always divides the minimum polynomial $m_M(x)$. Furthermore, we have $m_{\mathbf{u}, \mathbf{v}, M} = m_M(x)$ with a probability of at least $1 - 2 \deg m_M(x)/|S|$ (under our assumption that the entries of the vectors $\mathbf{u}$ and $\mathbf{v}$ are randomly sampled from a finite set S of cardinality $|S|$).

For a structured linear system of equations, the above algorithm is not superfast because slow computation of the Krylov matrix $K(M, \mathbf{v}, 2n-1)$ is involved, but according to Remark 7.5.5, this stage can be effectively parallelized. The remaining stages of the solution of a linear system $M\mathbf{x} = \mathbf{b}$ amount to pre-multiplication of the Krylov matrix $K(M, \mathbf{b}, k-1)$ by the row vector $-(m_i/m_0)_{i=1}^{k}$ and the solution of Problem 2.9.7 and thus can be also effectively parallelized. Furthermore, this approach has been extended to randomized parallel computation of the determinant, the rank, and the null space of a general $n \times n$ matrix also by using polylogarithmic arithmetic time and n^3 or fewer processors. For Toeplitz-like (and consequently, due to the transformation techniques, for Hankel-like, Vandermonde-like, and Cauchy-like) input matrices

M, the number of arithmetic processors used in all these polylogarithmic parallel time computations can be decreased to n^2.

The Krylov space computation is the most expensive stage of the reduction of a matrix to normal forms. Acceleration at this stage is translated into improved computation of the matrix normal forms and, further, of the solution of the matrix eigenproblem.

7.7 Inversion of general integer matrices and solution of general integer linear systems

Let us describe Newton's algebraic iteration (*Newton's lifting*) for the inversion of general (unstructured) integer matrices. In Section 7.9, this study is extended to yield superfast inversion of structured integer input matrices in nearly optimal time.

Algorithm 7.7.1. Inversion of an integer matrix modulo a prime power.

INPUT: A positive integer N, a prime p, the matrices M_i in (7.4.1) filled with integers, for $i = 0, 1, \ldots, N$, such that the matrix M_0 is non-singular modulo p, that is,

$$(\det M_0) \bmod p \neq 0, \tag{7.7.1}$$

and the matrix $M_0^{-1} = M^{-1} \bmod p$.

OUTPUT: The matrices $M^{-1} \bmod p^{2^i}$, for $M = M_0, i = 1, \ldots, h$.

COMPUTATION: Apply iteration (7.4.2) for $x = p$ and obtain that

$$X_i(p) = M^{-1}(p) \bmod p^{2^i} = M^{-1} \bmod p^{2^i}, \quad i = 0, 1, \ldots, h \tag{7.7.2}$$

(see (7.4.3)).

Now recall that the algorithm in Section 2.9 supporting the solution of Problem 2.9.8 uses $O(\mu(\log K) \log K)$ bit-operations for $\mu(q)$ in (7.1.1) to recover the ratio n/d of two positive integers (not exceeding a fixed positive integer K) from the value $(n/d) \bmod (2K^2)$. That is, the algorithm recovers the entries of the matrix M^{-1} represented as the ratios of bounded integers provided that we are given the matrix $X_i(p)$ in (7.7.2) for a sufficiently large integer i. To bound the integers n and d, we recall that

$$M^{-1} = (\operatorname{adj} M)/\det M \tag{7.7.3}$$

where adj M is the adjoint (adjugate) matrix of $M = (\mathbf{m}_1, \ldots, \mathbf{m}_n)$, and apply Hadamard's bound

$$|\det M| \leq \prod_{j=1}^{n} \|\mathbf{m}_j\|_2. \tag{7.7.4}$$

Summarizing, we obtain the following result.

Theorem 7.7.2. *Given an $n \times n$ matrix M filled with integers in the range from $-k$ to k and a prime p such that (7.7.1) holds, it is sufficient to apply $O(i(n)\mu(\log p) + \tilde{m}(n)\log K + n^2\mu(\log K)\log\log K)$ bit-operations for $\mu(q)$ in (7.1.1) to compute the matrix M^{-1} provided that*

$$K = (k\sqrt{n})^n \qquad (7.7.5)$$

and $i(n)$ and $\tilde{m}(n)$ ops are sufficient for $n \times n$ matrix inversion and multiplication, respectively.

Proof. First compute the matrix $M^{-1} \bmod p$ using $i(n)$ ops modulo p and $O(\mu(\log p))$ bit-operations per op modulo p, that is, a total of $O(i(n)\mu(\log p))$ bit-operations. Then compute the integer

$$h = \lceil \log_2((\log(2K^2))/\log p)\rceil, \qquad (7.7.6)$$

perform h Newton's steps (7.4.2) for x replaced by p, and obtain the matrix $X_h = M^{-1} \bmod p^{2^h}$. At the i-th step, we perform $O(\tilde{m}(n))$ ops modulo p^{2^i}, that is, we use $O(\tilde{m}(n)\mu(2^i\log p))$ bit-operations, for $i = 0, 1, \ldots, K$. This means a total of $O(\tilde{m}(n)\mu(2^h\log p)) = O(\tilde{m}(n)\mu(\log K))$ at all steps. By Hadamard's bound (7.7.4), we have $K \geq |\det M|$ for K in (7.7.5). By (7.7.6), $p^{2^h} \geq 2K^2$. Therefore, we recover every entry of the matrix M^{-1} from the respective entry of the matrix $X_h = M^{-1} \bmod p^{2^h}$ by computing a unique solution to Problem 2.9.8 where the explicit reconstruction conditions (2.9.6) is satisfied since $(\det M) \bmod p \neq 0$. The recovery requires $O(\mu(\log K)(\log\log K))$ bit-operations per entry (see (2.9.7)), and this completes the proof of Theorem 7.7.2. $\qquad\square$

It should be instructive to compare Newton's iteration of equations (7.4.2) with the next algorithm, which like Newton's, can be followed by the recovery of the rational output by solving Problem 2.9.8 for every component of the output vector.

Algorithm 7.7.3. Hensel's lifting for a linear system.

INPUT: three natural numbers h, n, and s, an integer vector $\mathbf{b}$ of dimension n, and an integer $n \times n$ matrix M such that $(\det M) \bmod s \neq 0$.

OUTPUT: the vector $\mathbf{x}_h = M^{-1}\mathbf{b} \bmod s^h$.

COMPUTATION: First compute the matrices $M_1 = M \bmod s$, $M_1^{-1} \bmod s$, and $\Delta = (M_1 - M)/s$ and the vector $\mathbf{x}_1 = (M_1^{-1}\mathbf{b}) \bmod s$. Then recursively compute the vectors $\mathbf{x}_{j+1} = M_1^{-1}(s\Delta\mathbf{x}_j + \mathbf{b}) \bmod s^{j+1}$, $j = 1, \ldots, h - 1$.

The correctness of the algorithm is proved in [MC79]. To decrease its computational cost, we compute the vector $\mathbf{x}_{j+1}$ as follows:

$$\mathbf{u}_j = (\mathbf{b} - M\mathbf{x}_j) \bmod s^{j+1},$$

$$\mathbf{v}_j = (M_1^{-1}\mathbf{u}_j/s^j) \bmod s^2,$$

$$\mathbf{x}_{j+1} = \mathbf{x}_j + s^j\mathbf{v}_j \bmod s^{j+1}, \quad j = 0, 1, \ldots, h - 1.$$

Here the vectors $\mathbf{x}_j$ are updated implicitly, that is, the vectors $(\mathbf{x}_{j+1} - \mathbf{x}_j)/s^j$ are computed modulo s^2, and similarly, the vectors $(\mathbf{u}_j - \mathbf{u}_{j-1})/s^{j-1}$ are computed modulo s^2. All ops are performed modulo s or s^2, that is, with the precision of at most $\lceil 4 \log_2 s \rceil$ bits. The computation of the vector $\mathbf{x}_1$ requires $O(i(n))$ ops, whereas for each $j \geq 1$, the j-th step involves $O(v(n))$ ops. Here $i(n)$ and $v(n)$ are the numbers of ops required for the inversion of M_1 and multiplication of the inverse matrix M_1^{-1} by a vector, respectively, and all ops are performed modulo s or s^2, that is, by using $O(\mu(\log(s))$ bit-operations per an op. Even though the number h of Hensel's steps is of the order of $(\log(2K^2))/\log s$ versus $\log((\log(2K^2))/\log p)$ in (7.7.6), the saving in the precision of computation in Algorithm 7.7.3 more than compensates us for this deficiency provided that $\mu(q)$ is of the order of q^2 or $q^{1.58}$ (see Exercise 2.9) and M is an unstructured matrix.

7.8 Sequences of primes, random primes, and non-singularity of a matrix

If assumption (7.7.1) does not hold for a selected prime $p = p_1$, then the computation of the matrix M^{-1} mod p detects this, and we repeat the computation recursively for distinct primes $p_2, p_3, \ldots$ replacing p until we obtain the matrix M^{-1} mod p_i. Unless $\det M = 0$, we stop for some $i \leq q$ where

$$p_1 \cdots p_q > |\det M|. \qquad (7.8.1)$$

Clearly, bound (7.8.1) is satisfied for

$$q = 1 + \lfloor \log |\det M| / \log(\min_i p_i) \rfloor \leq 1 + \lfloor \log_2 |\det M| \rfloor. \qquad (7.8.2)$$

Now, suppose that a single prime is chosen at random in a fixed range and estimate the probability that (7.7.1) holds for a fixed non-zero value of $\det M$. We rely on the next result.

Theorem 7.8.1. *Let $f(n)$ be a positive increasing function defined on the set of positive integers, such that $f(n) > 0$ and $\lim_{n \to \infty} f(n) = \infty$. Then there exist two positive constants C and $n_0 > 1$ such that the interval $J = \{p : f(n)/n < p < f(n)\}$ contains at least $f(n)/(C \log_2 f(n))$ distinct primes for any $n > n_0$.*

Proof. Theorem 7.8.1 is an immediate corollary of the following well-known bounds: $c_- < (\pi(x) \ln x)/x < c_+$ where $x \geq 2$; c_- and c_+ are two positive constants, and $\pi(x)$ is the number of primes not exceeding x. $\qquad \square$

Corollary 7.8.2. *Let $f(n)$, $h(n)$, and $k(n)$ be three functions in n such that the function $f(n)$ is positive and increasing, $\lim_{n \to \infty} f(n) = \infty$, $h(n) \neq 0$, $k(n) > 0$, and*

$$0 < (h(n))^{1/k(n)} \leq f(n)/n. \qquad (7.8.3)$$

Let p be a random prime in the interval J of Theorem 7.8.1. Then for the positive constants C and n_0 in Theorem 7.8.1 and for any fixed $n > n_0$, we have $h(n)$ mod $p \neq 0$ with a probability exceeding $1 - Ck(n)(\log f(n))/f(n)$.

Proof. Suppose that exactly $l(n)$ primes from the interval J divide $h(n)$. Then their product also divides $h(n)$ and therefore cannot exceed $h(n)$. Since all these primes lie in the interval J, each of them exceeds $f(n)/n$. Therefore, their product exceeds $(f(n)/n)^{l(n)}$. Hence, $(f(n)/n)^{l(n)} < h(n)$. Compare this inequality with the assumed bound $h(n) \leq (f(n)/n)^{k(n)}$ and obtain that $l(n) < k(n)$. Therefore, the number of primes lying in the interval J and dividing the integer $h(n)$ is less than $k(n)$. Compare this number with the overall number of primes in the interval J estimated in Theorem 7.8.1 and obtain the claimed probability estimate. $\qquad\square$

Corollary 7.8.3. *Let C and n_0 denote the two constants of Theorem 7.8.1. Fix a real $b \geq 2$, an integer $n > n_0$, a matrix norm, and a non-singular $n \times n$ integer matrix M, with $\|M\| \geq 2$. Let a prime p be chosen at random (under the uniform probability distribution) in the range from $n^{b-1} \log \|M\|$ to $n^b \log \|M\|$. Then (7.7.1) holds with a probability exceeding $1 - 2Cn^{1-b}$.*

Proof. Apply Corollary 7.8.2 for $f(n) = n^b \log_2 \|M\|$, $h(n) = |\det M|$, and $k(n) = (n \log \|M\|)/\log(n^{b-1} \log_2 \|M\|)$. Recall that $|\det M| \leq \|M\|^n$, substitute the above expressions for $f(n)$, $h(n)$, and $k(n)$ into the second inequality in (7.8.3), and verify it by comparing the logarithms of the values on its both sides. Therefore, by Corollary 7.8.2, (7.7.1) holds for the assumed choice of a random prime p with a probability exceeding $1 - P$ where

$$P \leq C \frac{n \log \|M\|}{\log(n^{b-1} \log_2 \|M\|)} \frac{\log(n^b \log_2 \|M\|)}{n^b \log \|M\|)} = Cn^{1-b}\left(1 + \frac{\log n}{\log(n^{b-1} \log_2 \|M\|)}\right).$$

It follows that $P \leq 2Cn^{1-b}$ because by assumption, $b \geq 2$, $\|M\| \geq 2$, and therefore, $\log n \leq \log(n^{b-1} \log_2 \|M\|)$. $\qquad\square$

7.9 Inversion of structured integer matrices

Surely, any structured rational matrix M associated with an operator $L = \nabla_{A,B}$ defined by two rational matrices A and B can be scaled to turn the displacement $L(M)$ into an integer matrix, but the scaling may blow up the magnitude of the entries of the matrices M and $L(M)$, which would make the subsequent computations too costly. This is the case for scaling Cauchy and Cauchy-like matrices, although the situation is much more favorable for Toeplitz-like and Hankel-like matrices.

Example 7.9.1. $L = \nabla_{A,B}$, $A \in \{Z_e, Z_e^T\}$, $B \in \{Z_f, Z_f^T\}$ for some fixed integers e and f. (These operators L are associated with Toeplitz-like and Hankel-like matrices.) Let an $n \times n$ matrix M be filled with integers in the range from $-k$ to k, for a positive k. Then the matrix $L(M)$ is filled with integers in the range from $-R$ to R, for $R = (|e| + |f|)k$.

Let us next extend Theorem 7.7.2 to the case where a structured $n \times n$ integer matrix M is given by its shortest L-generator (G, H) and the entries of the matrices G and H are bounded integers. In the next two simple theorems, we estimate the bit-operation complexity of the Newton lifting and the recovery stages, respectively.

Theorem 7.9.2. *Suppose that p is a prime, j is a positive integer, and M is an $n \times n$ integer matrix such that $(\det M) \bmod p \neq 0$. Let $L = \nabla_{A,B}$ and $L_- = \nabla_{B,A}$ be regular linear operators. Furthermore, let $L(M) = GH^T$ and $L_-(M^{-1}) = G_- H_-^T$ where G, H, G_-, and H_- are $n \times r$ rational matrices and $r = \text{rank}(L(M))$, and let us be given the matrices $G_- \bmod p$, $H_- \bmod p$, $G \bmod p^{2^j}$, $H \bmod p^{2^j}$. Then in j steps, the Newton-Structured Algebraic Iteration (7.5.1), (7.5.2) for $x = p$ and $i = 0, 1, \ldots, j$ outputs the matrices G_j and H_j by performing*

$$\beta_{r,n,j,p} = O((v_{r,n}(L) + v_{r,n}(L_-))r\mu(2^j \log p)) \tag{7.9.1}$$

bit-operations for $\mu(q)$ in (7.1.1).

Based on Theorem 7.9.2 and the solution algorithm for Problem 2.9.8, we now estimate the bit-operation cost of computing the inverse of an integer matrix given with its shortest rational L-generator. The estimates are in terms of the magnitudes of the input entries.

Theorem 7.9.3. *Under the assumptions of Theorem 7.9.2, let the matrices M, G, and H have integer entries lying in the range from $-k$ to k, from $-g$ to g, and from $-h$ to h, respectively, for some positive k, g, and h. Let $K = n(k\sqrt{n})^{n-1} \max\{g, h, k/\sqrt{n}\}$, $j = \lceil \log_2(\log(2K^2+1)/\log p) \rceil$. Then $\gamma_{r,n,j,p} = \beta_{r,n,j,p} + O(nr \log(2^j \log p)\mu(2^j \log p))$ bit-operations for $\beta_{r,n,j,p}$ in (7.9.1) and $\mu(q)$ in (7.1.1) are sufficient to compute all rational entries of a $\nabla_{B,A}$-generator of length r for the matrix $(d_g d_h M)^{-1}$.*

Proof. Recall Hadamard's bound (7.7.4) and deduce that all entries of the generator matrices $G_- = -M^{-1}G$ and $H_- = M^{-1}H$ for the displacement of M^{-1} are the ratios of the pairs of integers not exceeding K. The L_--generator matrices $G_j(p)$ and $H_j^T(p)$ for the matrix $X_j(p)$ are computed by using $\beta_{r,n,j,p}$ bit-operations and define the entries modulo N where $N = p^{2^j} > 2K^2$. Then the solution algorithm for Problem 2.9.8 enables us to recover the exact values of these entries by using $O(nr \log(2^j \log p)\mu(2^j \log p))$ bit-operations. $\square$

For the linear operators L in Examples 1.4.1, 4.4.1–4.4.9, and Section 4.5, we have $v_{r,n}(L) = O(rn \log^d n)$ for $d \leq 2$. Therefore, for the associated structured matrices, *the algorithm supporting Theorem 7.9.3 is superfast at the bit-operation complexity level.* Indeed, even the representation of the output requires the order of $N = nr2^j \log_2 p$ bits, which gives us an information lower bound of the order of N on the number of bit-operations required to compute this output, and we have $\gamma_{r,n,j,p} = O((rN \log^2 N) \log \log N)$.

Remark 7.9.4. The Newton-Structured Algebraic Iteration (7.5.1), (7.5.2) as well as Theorems 7.9.2 and 7.9.3 can be easily extended to the case of Stein type operators and the partly regular operators in Sections 4.4 and 4.5.

For the sake of completeness, let us estimate the *complexity of the reduction* to the above case where a shortest L-generator is given to us from the case where only a short or even long L-generator is available. Given a short (or long) integer L-generator (G, H) of the input matrix M, we apply our algorithm of the proof of Theorem 4.6.4

to compute a shortest L-generator $(\widetilde{G}, \widetilde{H})$ of M. Expressions (4.6.11) enable us to estimate the magnitudes of the entries of the integer matrices $d_g\widetilde{G}$ and $d_h\widetilde{H}$ and the integer multipliers d_g and d_h. Furthermore, the algorithm for the transition from the original generator (G, H) to the new generator $(\widetilde{G}, \widetilde{H})$ in (4.6.11) essentially amounts to Gaussian elimination with complete pivoting, so application of the Bareiss modification [B68] enables us to bound also the bit-operation cost of the transition to $\widetilde{G}, \widetilde{H}$. The next theorem summarizes the resulting estimates and shows that the compression is also superfast at the bit-operation level.

Theorem 7.9.5. *Given an L-generator (G, H) of length l for an $n \times n$ matrix M and two positive integers, g and h such that the entries of the matrices G and H are integers in the ranges from $-g$ to g and from $-h$ to h, respectively, it is sufficient to perform $O(nl^2)$ ops by using $O(nl^2\mu(\log q))$ bit-operations for $\mu(q)$ in (7.1.1) to compute two positive integers d_g and d_h and an L-generator $(\widetilde{G}, \widetilde{H})$ of the minimum length $r = \mathrm{rank}(L(M))$ for the matrix M where the entries of the matrices $d_g\widetilde{G}$ and $d_h\widetilde{H}$ are integers in the ranges from $-\tilde{g}$ to $\tilde{g}$ and from $-\tilde{h}$ to $\tilde{h}$, respectively, and where $q = \max\{\tilde{g}, \tilde{h}, d_g, d_h\}$, $\tilde{g} \le (g\sqrt{l})^{l+1}hl^2\sqrt{l}$, $\tilde{h} \le (ghl\sqrt{r})^r\sqrt{r}$, $d_g \le (g\sqrt{l})^l$, $d_h \le (ghl\sqrt{r})^r$.*

Proof. The Bareiss algorithm of [B68] exploits all cancellations of the numerators and denominators of the operands involved in Gaussian elimination. It remains to estimate the magnitudes of the output ratios. Let $|W|$ denote $\max_{i,j}|w_{ij}|$ for a matrix $W = (w_{ij})$. Observe that $|T| \le |\widehat{H}| \le l|H|\cdot|(S, E)| \le ghl, \tilde{h} \le r|\widehat{H}|\cdot|\operatorname{adj}T|, d_h \le |\det T|$ for the matrices E, H, $\widehat{H}$, S, and T in equations (4.6.11) and Theorem 4.6.6. Combine the above bounds with Hadamard's bound (7.7.4) and obtain that $\tilde{h} \le (ghl\sqrt{r})^r\sqrt{r}$, $d_h \le (ghl\sqrt{r})^r$. Similarly, we obtain $\tilde{g} \le l^2|(T, F)|\cdot|\widehat{G}|\cdot|\operatorname{adj}S| \le g^2hl^3(g\sqrt{l})^{l-1} = hl^2(g\sqrt{l})^{l+1}$, $d_g \le |\det S| \le (g\sqrt{l})^l$. $\square$

The algorithms supporting Theorems 7.9.2 and 7.9.5 immediately allow for their superfast parallelization, but not so for the solution of Problem 2.9.8. Moreover, no effective parallel rational number reconstruction using less than linear parallel time is known so far.

Superfast parallel solution of structured integer linear systems and several other fundamental computational tasks for integer structured matrices is possible, however. We end this section, the chapter, and essentially the book by citing the algorithm of the paper [P00b], which solves the latter tasks based on *a non-trivial combination of both numerical and algebraic versions of Newton's iteration, extending our variable diagonal technique of [P85], [P87], with the divide-and-conquer construction in Chapter 5.* The algorithm enables effective exact parallel inversion of any non-singular (but not necessarily well conditioned) integer Toeplitz and Toeplitz-like matrices (with extensions to several fundamental problems of polynomial computation in Chapter 2). The algorithm yields *randomized polylogarithmic arithmetic parallel time bound and at the same time involves a nearly linear number of ops, performed with the bit-precision that does not exceed the bit-precision of the output.* The algorithm supports the same cost bounds for the computation of the rank and a compressed basis for the null space of a singular integer Toeplitz or Toeplitz-like matrix, except that the upper bound on the bit-precision decreases substantially for the computation of the rank. This level of parallel complexity has not been achieved by any other known means so far. We outline the algorithm in Exercise 7.15.

7.10 Conclusion

Newton's algebraic iteration is an effective basic tool of computer algebra. We analyzed its general scheme and several specific applications to fundamental computations with integers, polynomials, and matrices (both general and structured). Newton's iteration and its counterpart of Hensel's lifting enable variation of the precision to perform the most intensive part of the computation with a lower precision. For matrix computations, the iteration is superfast in terms of the bit-operation cost (in some cases also in terms of the arithmetic cost) and allows most efficient parallelization. We demonstrated these properties in the solution of specific fundamental computational problems. At some point, we applied special randomization techniques to avoid singularities in algebraic matrix computations.

7.11 Notes

Section 7.2   Newton's iteration is studied in many computer algebra texts and papers (see, e.g., [GG99] and [Y76]). On higher order Newton's algebraic iteration, see [BK78], [KT78]. On Hensel's lifting, see [Y76], [Z93], [GG99]. Algorihm 7.2.3 is from [L83]; otherwise the presentation follows [BP94].

Section 7.3   The algorithm for Problem 7.3.2 is from [BK78], for Problem 7.3.3 from [Y76], [KT78], for Problem 7.3.4 from [GKL87], for Problem 7.3.5 from [GKL87] and [G90], for Problem 7.3.7 from [KT78]. The composition can be reduced to the reversion as well [BK78], [K98], so both problems have the same asymptotic complexity. (See [K98] for the solution of the reversion problem by using about $K^3/6$ ops, which is a better bound for smaller K.) On the extension of Problem 7.3.4 to the computation of approximate polynomial decomposition (with application of Newton's numerical iteration), see [CGJW99]. Theorem 7.3.6 is from [E41], [R22].

Section 7.4   The iteration can already be found in [MC79].

Section 7.5   Theorem 7.5.4 and the results of Remark 7.5.5 were obtained in [P92] in the Toeplitz-like case and in [BP93], [BP94] in the Toeplitz+Hankel-like case.

Section 7.6   On the application of Krylov spaces and matrices to an iterative solution of linear systems of equations and the matrix eigenproblem, see [GL96], [S96], [G98]. On parallel algorithms for the Berlekamp–Massey Problem 2.9.7, see [P96] (cf. also [KP94], [P97a], [P00a]). The probability that $m_{\mathbf{u},\mathbf{v},M}(x) = m_M(x)$ was estimated in [KP91]. On $n \times n$ matrix multiplication using parallel arithmetic time $O(\log n)$ and n^3 processors, see [Q94]; on using fewer processors, see [K99], [LP01]. The Krylov space approach to the solution of linear systems of equations in finite fields was proposed in [W86]; it is related to the Lanczos algorithm (cf. [CEKSTV00]). On the extension to parallel computation of the determinant, the rank, and the null space of a general matrix, see [KP91], [KP92]; of a Toeplitz-like matrix, see [P96] (also see [KP94], [P97a], [P00a]). For structured input matrices, the resulting algorithms are accelerated by the factor of n versus general matrices. Consequently, computation of matrix normal forms can be accelerated. Furthermore, the entire solution of the matrix eigenproblem is accelerated from the order of n^3 to $O(n^2)$ ops, up to polylogarithmic factors [PC99], [PCZ98].

Section 7.7   Newton's and Hensel's lifting algorithms for linear systems and matrix inversion are from [MC79]. An interesting recent result on Hadamard's bound can be found in [ABM99].

Section 7.8   The presentation follows [P00a] (cf. [P87], [GP88]). The well-known bounds on $\pi(x)$ used in the proof of Theorem 7.8.1 can be found in [IR82], [N00].

Section 7.9   The results are straightforward except for those of [P00b] sketched at the end. (Also see [P96b], the proceedings version of [P00b].)

7.12   Exercises

7.1 Complete all omitted proofs of the results claimed in this chapter.

7.2 Specify the constants hidden in the "O" notation in the arithmetic computational complexity estimates of this chapter.

7.3 Write programs to implement the algorithms of this chapter. Then apply the algorithms to some specified polynomials and matrices.

7.4 Estimate the work space required for the algorithms of this chapter.

7.5 (a) In order to compute modulo x^m the square root $w(x)$ of a polynomial $a(x)$ of degree n in x, apply Newton's iteration to the equations $f(w) = w^2 - a = 0$ and $g(w) = a/w^2 - 1 = 0$, obtaining the two recurrences $w_{i+1} = \frac{1}{2}(w_i + \frac{a}{w_i})$ and $w_{i+1} = \frac{1}{2}(3w_i - \frac{w_i^3}{a})$, respectively. Estimate the arithmetic computational costs of the two algorithms.

    (b) Extend the algorithms of part (a) to computing the k-th roots of a positive integer for integers $k > 1$.

7.6 (a) Show that the power sums of the roots of a polynomial may non-uniquely define its coefficients in a finite field. Consider

      (i) the polynomials of degree 3 in the field $\mathbb{Z}_2$, and

      (ii) the polynomials x^m and $(x + 1)^m$ in the field $\mathbb{Z}_k$ where k divides m.

(b) [P97a], [P00a] Show a superfast algorithm for the inverse power sums problem over any field.

7.7 Reduce polynomial composition and polynomial reversion problems to each other.

7.8 Recall the Cayley–Hamilton theorem, which states that $c_M(M) = 0$ for the characteristic polynomial $c_M(x) = \det(xI - M)$, and deduce that $m_M(x)$ divides $c_M(x)$.

7.9 Estimate parallel arithmetic time for the computation (over any field) of the characteristic polynomial, minimum polynomial, and the Krylov matrix $K(M, \mathbf{v})$:

(a) for an $n \times n$ general matrix M using n^3 processors,

(b) for an $n \times n$ Toeplitz matrix M using n^2 processors.

7.10 Define $m_{\mathbf{v},M}(x)$ as the monic polynomial of the minimum degree such that $m_{\mathbf{v},M}(M)\mathbf{v} = \mathbf{0}$. Show that $m_{\mathbf{u},\mathbf{v},M}(x)$ divides $m_{\mathbf{x},M}(x)$, whereas $m_{\mathbf{v},M}(x)$ divides $m_M(x)$.

7.11 Compare the bit-operation complexity of matrix inversion based on Newton's and Hensel's algorithms in the cases of a general and structured integer input matrix. Make this comparison based on each of the three estimates for the bit-complexity $\mu(d)$ of an arithmetic operation of Exercise 2.9.

7.12 [W86], [KP91], [KP92], [KP94], [P96]. Extend the computation of the Krylov space of a matrix M to randomized computation of its rank, null space, and minimum polynomial $m(x)$ as well as to the solution of a linear system $M\mathbf{x} = \mathbf{b}$. Furthermore achieve polylogarithmic parallel arithmetic time bound in the implementation of the resulting extension algorithms on n^2 processors (not including the computational cost of the generation of random parameters).

7.13 (a) [W86] If M is a strongly non-singular $n \times n$ matrix, $y_0, \ldots, y_{n-1}$ are indeterminates, and $\mathbf{y} = (y_i)_{i=0}^{n-1}$, then we have $m_{D(\mathbf{y})M}(x) = c_{D(\mathbf{y})M}(x)$ and $m_{MD(\mathbf{y})}(x) = c_{MD(\mathbf{y})}(x)$.

(b) [KP92]. Suppose that $\mathbf{s} = (s_i)_{i=0}^{n-1}$, $t = (t_i)_{i=0}^{n-1}$, $s_0 = 1$, $t_0, s_1, t_1, \ldots,$ s_{n-1}, t_{n-1} are randomly sampled elements of a finite set $\mathbf{S}$ of a cardinality $|\mathbf{S}|$. Prove that $m_{\widetilde{M}}(x) = c_{\widetilde{M}}(x)$ with a probability of at least $1 - 4n^2/|\mathbf{S}|$ for any fixed matrix M and the matrix $\widetilde{M} = Z^T(\mathbf{s})MZ(\mathbf{t})$.

7.14 Extend Corollary 7.8.3 to estimate the probability that a strongly non-singular integer matrix remains strongly non-singular modulo p for a prime p sampled randomly in the range from $n^{b-2} \log \|M\|$ to $n^b \log \|M\|$ for $b \geq 3$.

7.15 [P00b]. Devise a randomized superfast parallel algorithm for the inversion of an $n \times n$ non-singular Toeplitz-like matrix M filled with integers. The algorithm should run in polylogarithmic arithmetic time $O(\log^d n)$, $d = O(1)$, on n processors and use precision of computation at the level of the output precision. Then extend the algorithm further to the computation of the rank and a short displacement generator of M where M is singular. Proceed as follows.

(a) First observe that $((M + apI)^{-1} - M^{-1}(p)) \bmod p = 0$ for any integer a provided that $(\det M) \bmod p \neq 0$. Further observe that the matrix $I/(ap)$ is a close approximation to the matrix $(M + apI)^{-1}$ for a sufficiently large real value of ap. Then choose a sufficiently large integer a and prime p and estimate the probability that $(\det M) \bmod p \neq 0$.

(b) Assume that the latter inequality holds and apply a numerical version of Newton's iteration (6.3.1) with $X_0 = I/(ap)$ to approximate the matrix $(M + apI)^{-1}$ closely enough to enable the recovery of its entries by application of the solution algorithm for Problem 2.9.8. Estimate the number of ops and the bit-precisions required in this computation except for its latter step of recovering the entries. (This *technique of variable diagonal* originated in [P85] (see also [P87]).) Show that the algorithm is superfast, requires the precision of computation of the order of the estimated output precision, and, except for its stage of the recovery of the rational output, can be implemented in polylogarithmic parallel arithmetic time on n processors.

(c) No known algorithm for the solution of Problem 2.9.8 runs faster than in linear parallel time. To yield a polylogarithmic time bound for the entire construction of part (b), modify it to avoid the solution of Problem 2.9.8. Achieve this by combining the BCRTF algorithms of Chapter 5 with the techniques of part (b). Based on this combination, recursively compute, modulo high powers of the prime p, the determinants and inverses of the northwestern blocks and their Schur complements involved in the process of the computation of the BCRTF. Use the determinants for scaling to yield integer adjoint matrices; the algorithm computes them modulo high powers of p. Recover the entries of these matrices and the determinants from their values modulo high powers of p.

(d) Extend the resulting algorithm to the computation of the rank of a Toeplitz-like integer matrix and short displacement generator for its null space. Estimate the parallel bit-complexity.

(e) Try to extend the algorithms to other classes of structured matrices based on the transformation approach of Section 4.8. Estimate the parallel bit-complexity.

Conclusion

The modern study of structured matrices has been greatly influenced by the introduction in 1979 (by Kailath, Kung, and Morf) of the fundamental concept of displacement rank, which specified the matrices with the structure of Toeplitz and Hankel types and enabled more efficient computations with these matrices. Subsequently, various effective algorithms were developed along this line for several other classes of structured matrices as well. This included superfast algorithms, running in nearly linear time, up to (poly)logarithmic factors and using small (linear) memory space.

We covered the fundamentals of computations with structured matrices, the displacement rank approach, and two groups of superfast algorithms based on this approach, that is, the divide-and-conquer algorithms and Newton's iteration. We described both numerical and algebraic versions of these algorithms, observing the correlation and establishing links between these two computational fields, and studied applications to some fundamental computations with general and structured matrices. Unlike the previous books and surveys, we presented superfast algorithms both in some detail and, based on the displacement rank approach in a unified way over various classes of structured matrices.

We also included detailed study of the correlation between computations with structured matrices and some of the most fundamental problems of polynomial and rational computations. These problems covered a substantial part of computer algebra as well as the tangential Nevanlinna–Pick and matrix Nehari problems of numerical rational interpolation and approximation. Consequently, superfast structured matrix computations have been extended to a superfast solution of the latter problems. For demonstration, we also applied structured matrices to loss-resilient encoding/decoding.

Bibliography

[A63]   S. J. Aarseth, Dynamical Evaluation of Clusters of Galaxies-I, *Monthly Notices of the Royal Astronomical Society*, **126**, 223–255, 1963.

[A78]   K. E. Atkinson, *An Introduction to Numerical Analysis*, Wiley, New York, 1978.

[A85]   A. W. Appel, An Efficient Program for Many-Body Simulation, *SIAM Journal on Scientific and Statistical Computing*, **6**, 85–103, 1985.

[A86]   C. R. Anderson, A Method of Local Corrections for Computing the Velocity Field due to a Distribution of Vortex Blobs, *Journal of Computational Physics*, **62**, 111–123, 1986.

[A89]   J. Abbott, Recovery of Algebraic Numbers from their p-adic Approximations, *Proceedings of International Symposium on Symbolic and Algebraic Computation (ISSAC'89)*, 112–120, ACM Press, New York, 1989.

[AB91]   I. Adler, P. Beiling, Polynomial Algorithms for LP over a Subring of the Algebraic Integers with Applications to LP with Circulant Matrices, *Proceedings of 32nd Annual IEEE Symposium on Foundations of Computer Science (FOCS'91)*, 480–487, IEEE Computer Society Press, Washington, DC, 1991.

[ABM99]   J. Abbott, M. Bronstein, T. Mulders, Fast Deterministic Computations of the Determinants of Dense Matrices, *Proceedings of International Symposium on Symbolic and Algebraic Computation (ISSAC'99)*, 197–204, ACM Press, New York, 1999.

[ABKW90] A. C. Antoulas, J. A. Ball, J. Kang, J. C. Willems, On the Solution of the Minimal Rational Interpolation Problem, *Linear Algebra and Its Applications*, **137–138**, 479–509, 1990.

[AG89]   G. S. Ammar, P. Gader, New Decompositions of the Inverse of a Toeplitz Matrix, *Proceedings of 1989 International Symposium on Mathematical Theory of Networks and Systems (MTNS'89)*, **III**, 421–428, Birkhäuser, Boston, 1989.

[AG91]   G. S. Ammar, P. Gader, A Variant of the Gohberg–Semencul Formula Involving Circulant Matrices, *SIAM Journal on Matrix Analysis and Applications*, **12, 3**, 534–541, 1991.

[AGr88]   G. S. Ammar, W. G. Gragg, Superfast Solution of Real Positive Definite Toeplitz Systems, *SIAM Journal on Matrix Analysis and Applications*, **9, 1**, 61–76, 1988.

[AHU74]   A. V. Aho, J. E. Hopcroft, J. D. Ullman, *The Design and Analysis of Computer Algorithms*, Addison–Wesley, Reading, Massachusetts, 1974.

[ANR74]   N. Ahmed, T. Natarajan, K. Rao, Discrete Cosine Transforms, *IEEE Transaction on Computing*, **23**, 90–93, 1974.

[Al85]   H. Alt, Multiplication Is the Easiest Nontrivial Arithmetic Function, *Theoretical Computer Science*, **36**, 333–339, 1985.

[ASU75]   A. V. Aho, V. Steiglitz, J. D. Ullman, Evaluation of Polynomials at Fixed Sets of Points, *SIAM Journal on Computing*, **4**, 533–539, 1975.

[B40]   E. T. Bell, *The Development of Mathematics*, McGraw-Hill, New York, 1940.

[B68]   E. H. Bareiss, Sylvester's Identity and Multistep Integer-Preserving Gaussian Elimination, *Mathematics of Computation*, **22**, 565–578, 1968.

[B76]   R. P. Brent, Fast Multiple-Precision Evaluation of Elementary Functions, *Journal of ACM*, **23**, 242–251, 1976.

[B81]   S. Barnett, Generalized Polynomials and Linear Systems Theory, *Proceedings of 3rd IMA Conference on Control Theory*, 3–30, 1981.

[B83]   D. A. Bini, On a Class of Matrices Related to Toeplitz Matrices, TR-83–5, *Computer Science Department, SUNYA*, Albany, New York, 1983.

[B84]   D. A. Bini, Parallel Solution of Certain Toeplitz Linear Systems, *SIAM Journal on Computing*, **13, 2**, 268–279, 1984.

[B85]   J. R. Bunch, Stability of Methods for Solving Toeplitz Systems of Equations, *SIAM Journal on Scientific and Statistical Computing*, **6, 2**, 349–364, 1985.

[B88]   J. Barnes, Encounters of Disk/Halo Galaxies, *The Astrophisical Journal*, **331**, 699, 1988.

[Ba]   D. J. Bernstein, Multidigit Multiplication for Mathematicians, *Advances in Applied Mathematics*, to appear.

[BA80]   R. R. Bitmead, B. D. O. Anderson, Asymptotically Fast Solution of Toeplitz and Related Systems of Linear Equations, *Linear Algebra and Its Applications*, **34**, 103–116, 1980.

[BB84]   J. W. Borwein, P. B. Borwein, The Arithmetic-Geometric Mean and Fast Computation of Elementary Functions, *SIAM Review*, **26**, **3**, 351–366, 1984.

[BC83]   D. A. Bini, M. Capovani, Spectral and Computational Properties of Band Symmetric Toeplitz Matrices, *Linear Algebra and Its Applications*, **52/53**, 99–126, 1983.

[Be68]   E. R. Berlekamp, *Algebraic Coding Theory*, McGraw-Hill, New York, 1968.

[BDB90]   D. A. Bini, F. Di Benedetto, A New Preconditioner for the Parallel Solution of Positive Definite Toeplitz Systems, *Proceedings of 2nd ACM Symposium on Parallel Algorithms and Architectures*, 220–223, ACM Press, New York, 1990.

[BE73]   A. Björck, T. Elfving, Algorithms for Confluent Vandermonde Systems, *Numerische Mathematik*, **21**, 130–137, 1973.

[BEPP99]   H. Brönnimann, I. Z. Emiris, S. Pion, V. Y. Pan, Sign Determination in Residue Number Systems, *Theoretical Computer Science*, **210**, **1**, 173–197, 1999.

[BF93]   D. A. Bini, P. Favati, On a Matrix Algebra Related to the Discrete Hartley Transform, *SIAM Journal on Matrix Analysis and Applications*, **14**, 500–507, 1993.

[BG95]   D. A. Bini, L. Gemignani, Fast Parallel Computation of the Polynomial Remainder Sequence via Bezout and Hankel Matrices, *SIAM Journal on Computing*, **24**, 63–77, 1995.

[BGH82]   A. Borodin, J. von zur Gathen, J. Hopcroft, Fast Parallel Matrix and GCD Computation, *Information and Control*, **52**, **3**, 241–256, 1982.

[BG-M96]   G. A. Backer Jr., P. Graves-Morris, *Padé Approximants*, in *Encyclopedia of Mathematics and Its Applications*, **59**, Cambridge University Press, Cambridge, UK, 1996 (2nd edition; 1st edition by Addison-Wesley, Reading, Massachusetts, 1982, 2 volumes).

[BGR90]   J. A. Ball, I. Gohberg, L. Rodman, Interpolation of Rational Matrix Functions, *Operator Theory: Advances and Applications*, **45**, Birkhäuser, Basel, 1990.

[BGR90a]   J. A. Ball, I. Gohberg, L. Rodman, Nehari Interpolation Problem for Rational Matrix Functions: the Generic Case, H_∞-*control Theory* (E. Mosca and L. Pandolfi, editors), 277–308, Springer, Berlin, 1990.

[BGY80]   R. P. Brent, F. G. Gustavson, D. Y. Y. Yun, Fast Solution of Toeplitz Systems of Equations and Computation of Padé Approximations, *Journal of Algorithms*, 1, 259–295, 1980.

[B-I66]   A. Ben-Israel, A Note on Iterative Method for Generalized Inversion of Matrices, *Mathematics of Computation*, **20**, 439–440, 1966.

[B-IC66]   A. Ben-Israel, D. Cohen, On Iterative Computation of Generalized Inverses and Associated Projections, *SIAM Journal on Numerical Analysis*, **3**, 410–419, 1966.

[BK78]   R. P. Brent, H. T. Kung, Fast Algorithms for Manipulating Formal Power Series, *Journal of ACM*, **25**, **4**, 581–595, 1978.

[BK87]   A. Bruckstein, T. Kailath, An Inverse Scattering Framework for Several Problems in Signal Processing, *IEEE ASSP Magazine*, 6–20, January 1987.

[BM71]   A. Borodin, I. Munro, Evaluation of Polynomials at Many Points, *Information Processing Letters*, **1, 2**, 66–68, 1971.

[BM74]   A. Borodin, R. Moenck, Fast Modular Transforms, *Journal of Computer and Systems Sciences*, **8, 3**, 366–386, 1974.

[BM75]   A. Borodin, I. Munro, *The Computational Complexity of Algebraic and Numeric Problems*, American Elsevier, New York, 1975.

[BM99]   D. A. Bini, B. Meini, Fast Algorithms with Applications to Markov Chains and Queuing Models, *Fast Reliable Algorithms for Matrices with Structures* (T. Kailath and A. H. Sayed, editors), 211–243, SIAM Publications, Philadelphia, 1999.

[BM01]   D. A. Bini, B. Meini, Solving Block Banded Block Toeplitz Systems with Structured Blocks: Algorithms and Applications, in *Structured Matrices: Recent Developments in Theory and Computation* (D. A. Bini, E. Tyrtyshnikov, and P. Yalamov, editors), Nova Science, 2001.

[BMa]   D. A. Bini, B. Meini, Approximate Displacement Rank and Applications, preprint.

[Bo68]   C. A. Boyer, *A History of Mathematics*, Wiley, New York, 1968.

[B-O83]   M. Ben-Or, Lower Bounds for Algebraic Computation Trees, *Proceedings of 15th Annual ACM Symposium on Theory of Computing (STOC'83)*, 80–86, ACM Press, New York, 1983.

[B-OT88]   M. Ben-Or, P. Tiwari, A Deterministic Algorithm for Sparse Multivariate Polynomial Interpolation, *Proceedings of 20th Annual ACM Symposium on Theory of Computing (STOC'88)*, 301–309, ACM Press, New York, 1988.

[BP70]   A. Björck, V. Pereyra, Solution of Vandermonde Systems of Equations, *Mathematics of Computation*, **24**, 893–903, 1970.

[BP86]   D. A. Bini, V. Y. Pan, Polynomial Division and Its Computational Complexity, *J. Complexity*, **2**, 179–203, 1986.

[BP88]   D. A. Bini, V. Y. Pan, Efficient Algorithms for the Evaluation of the Eigen-values of (Block) Banded Toeplitz Matrices, *Mathematics of Computation*, **50, 182**, 431–448, 1988.

[BP91]   D. A. Bini, V. Y. Pan, Parallel Complexity of Tridiagonal Symmetric Eigen-value Problem, *Proceedings of 2nd Annual ACM-SIAM Symposium on Discrete Algorithms (SODA'91)*, 384–393, ACM Press, New York, and SIAM Publications, Philadelphia, January 1991.

[BP91a]   D. A. Bini, V. Y. Pan, On the Evaluation of the Eigenvalues of a Banded Toeplitz Block Matrix, *Journal of Complexity*, **7**, 408–424, 1991.

[BP92]   D. A. Bini, V. Y. Pan, Practical Improvement of the Divide-and-Conquer Eigenvalue Algorithms, *Computing*, **48**, 109–123, 1992.

[BP93]   D. A. Bini, V. Y. Pan, Improved Parallel Computations with Toeplitz-like and Hankel-like Matrices, *Linear Algebra and Its Applications*, **188/189**, 3–29, 1993.

[BP94]   D. A. Bini, V. Y. Pan, *Polynomial and Matrix Computations, Vol.1: Fundamental Algorithms*, Birkhäuser, Boston, 1994.

[BP96]   D. A. Bini, V. Y. Pan, Graeffe's Chebyshev-like, and Cardinal's Processes for Spliting a Polynomial into Factors, *Journal of Complexity*, **12**, 492–511, 1996.

[BP98]   D. A. Bini, V. Y. Pan, Computing Matrix Eigenvalues and Polynomial Zeros Where the Output is Real, *SIAM Journal on Computing*, **27, 4**, 1099–1115, 1998.

[BPa]   D. A. Bini, V. Y. Pan, *Polynomial and Matrix Computations, Vol.2: Fundamental and Practical Algorithms*, Birkhäuser, Boston, to appear.

[BT71]   W. S. Brown, J. F. Traub, On Euclid's Algorithm and the Theory of Subresultants, *Journal of ACM*, **18, 4**, 505–514, 1971.

[C47/48] S. Chandrasekhar, On the Radiative Equilibrium of a Stellar Atmosphere, *Astrophysical Journal*, Pt. XXI, **106**, 152–216, 1947; Pt. XXXII, **107**, 48–72, 1948.

[C73]   A. J. Chorin, Numerical Study of Slightly Viscous Flaw, *Fluid Mechanics*, **57**, 785–796, 1973.

[C80]   G. Cybenko, The Numerical Stability of the Levinson-Durbin Algorithm for Toeplitz Systems of Equations, *SIAM Journal on Scientific and Statistical Computing*, **1**, 301–310, 1980.

[C87]   J. W. Cooley, The Re-Discovery of the Fast Fourier Transform Algorithm, *Mikrochimica Acta*, **3**, 33–45, 1987.

[C90]   J. W. Cooley, How the FFT Gained Acceptance, In *A History of Scientific Computing* (S. G. Nash, editor), 133–140, ACM Press, New York, and Addison-Wesley, Reading, Massachusetts, 1990.

[C96]   J. P. Cardinal, On Two Iterative Methods for Approximating the Roots of a Polynomial, *Proceedings of AMS-SIAM Summer Seminar: Mathematics of Numerical Analysis: Real Number Algorithms* (J. Renegar, M. Shub, and S. Smale, editors), Park City, Utah, 1995. *Lectures in Applied Mathematics*, **32**, 165–188, American Mathematical Society, Providence, Rhode Island, 1996.

[CdB80]   C. D. Conte, C. de Boor, *Elementary Numerical Analysis: an Algorithmic Approach*, McGraw-Hill, New York, 1980.

[CEKSTV00] L. Chen, W. Eberly, E. Kaltofen, B. D. Saunders, W. J. Turner, G. Villard, Efficient Matrix Preconditioners for Black Box Linear Algebra, preprint, 2000.

[CGG84]   B. W. Char, K. O. Geddes, G. H. Gonnet, QCDHUE: Heuristic Polynomial GCD Algorithm Based on Integer GCD Computation, *Proceedings of EUROSAM'84, Lecture Notes in Computer Science*, **174**, 285–296, Springer, New York, 1984.

[CGJW99]   R. M. Corless, M. W. Giesbrecht, D. J. Jeffrey, S. M. Watt, Approximate Polynomial Decomposition, *Proceedings of International Symposium on Symbolic and Algebraic Computation (ISSAC'99)*, 213–219, ACM Press, New York, 1999.

[CGTW95]   R. M. Corless, P. M. Gianni, B. M. Trager, S. M. Watt, The Singular Value Decomposition for Polynomial Systems, *Proceedings of International Symposium on Symbolic and Algebraic Computation (ISSAC'95)*, 195–207, ACM Press, New York, 1995.

[CHQZ87]   C. Canuto, M. Y. Hussaini, A. Quateroni, T. A. Zang, *Spectral Methods for Fluid Dynamics*, Springer, New York, 1987.

[CK91]   D. G. Cantor, E. Kaltofen, On Fast Multiplication of Polynomials over Arbitrary Rings, *Acta Informatica*, **28, 7**, 697–701, 1991.

[CKL89]   J. F. Canny, E. Kaltofen, Y. Lakshman, Solving Systems of Non-Linear Polynomial Equations Faster, *Proceedings of International Symposium on Symbolic and Algebraic Computation (ISSAC'89)*, 121–128, ACM Press, New York, 1989.

[CKL-A87] J. Chun, T. Kailath, H. Lev-Ari, Fast Parallel Algorithm for QR-factorization of Structured Matrices, *SIAM Journal on Scientific and Statistical Computing*, **8, 6**, 899–913, 1987.

[CN96]   R. H. Chan, M. K. Ng, Conjugate Gradient Methods for Toeplitz Systems, *SIAM Review*, **38**, 427–482, 1996.

[CN99]   R. H. Chan, M. K. Ng, Iterative Methods for Linear Systems with Matrix Structures, *Fast Reliable Algorithms for Matrices with Structure* (T. Kailath and A. H. Sayed, editors), 117–152, SIAM, Philadelpha, 1999.

[CPW74] R. E. Cline, R. J. Plemmons, G. Worm, Generalized Inverses of Certain Toeplitz Matrices, *Linear Algebra and Its Applications*, **8**, 25–33, 1974.

[D59]   J. Durbin, The Fitting of Time-Series Models, *Review of International Statistical Institute*, **28**, 229–249, 1959.

[D79]   P. Davis, *Circulant Matrices*, John Wiley, New York, 1979.

[D83]   J. M. Dawson, Particle Simulation of Plasmas, *Review of Modern Physics*, **55**, 403–447, 1983.

[D89]   H. Dym, J-Contractive Matrix Functions, Reproducing Kernel Hilbert Space and Interpolation, *CBMS Regional Conference in Mathematics*, **71**, American Mathematical Society, Providence, Rhode Island, 1989.

[D89a]   H. Dym, On Reproducing Kernel Spaces, J-Unitary Matrix Functions, Interpolation and Displacement Rank, *Operator Theory: Advances and Applications*, **41**, 173–239, Birkhäuser, Basel, 1989.

[DGK79] Ph. Delsarte, Y. Genin, Y. Kamp, The Nevanlinna–Pick Problem for Matrix-Valued Functions, *SIAM Journal on Applied Mathematics*, **36**, 47–61, 1979.

[DGK81] Ph. Delsarte, Y. Genin, Y. Kamp, On the Role of the Nevanlinna–Pick Problem in Circuit and System Theory, *Circuit Theory and Applications*, **9**, 177–187, 1981.

[DGKF89] J. C. Doyle, K. Glover, P. Phargonekar, B. A. Francis, State-Space Solutions to Standard H_2 and H_∞ Problems, *IEEE Transactions on Automatic Control*, **AC–34**, 831–847, 1989.

[dH87]   F. R. de Hoog, On the Solution of Toeplitz Systems, *Linear Algebra and Its Applications*, **88/89**, 123–138, 1987.

[DK69]   C. A. Desoer, E. S. Kuh, *Basic Circuit Theory*, McGraw-Hill, New York, 1969.

[DL78]   R. A. Demillo, R. J. Lipton, A Probabilistic Remark on Algebraic Program Testing, *Information Processing Letters*, **7**, **4**, 193–195, 1978.

[DvdV98] P. Dewilde, A.-J. van der Veen, *Time-Varying Systems and Computations*, Kluwer Academic Publishers, Boston, 1998.

[DV87]   P. Duhamel, M. Vetterli, Improved Fourier and Hartley Transform Algorithms, *IEEE Transactions on Acoustic, Speech, and Signal Processing*, **35**, 849–863, 1987.

[E41]   H. T. Engström, Polynomial Substitutions, *American Journal of Mathematics*, **63**, 249–255, 1941.

[EG97]   Y. Eidelman, I. Gohberg, Inversion Formula and Linear Complexity Algorithm for Diagonal Plus Semiseparable Matrices, *Computers and Mathematics (with Applications)*, **35**, **10**, 25–34, 1997.

[EG97a]   Y. Eidelman, I. Gohberg, Fast Inversion Algorithms for Diagonal Plus Semiseparable Matrices, *Integral Equations and Operator Theory*, **27**, 165–183, Birkhäuser, Basel, 1997.

[EG98]   Y. Eidelman, I. Gohberg, A Look-ahead Block Schur Algorithm for Diagonal Plus Semiseparable Matrices, *Computers and Mathematics (with Applications)*, **35**, **10**, 25–34, 1998.

[EG99]   Y. Eidelman, I. Gohberg, On a New Class of Structured Matrices, *Integral Equations and Operator Theory*, **34**, 293–324, Birkhäuser, Basel, 1999.

[EG99a]   Y. Eidelman, I. Gohberg, Linear Complexity Inversion Algorithms for a Class of Structured Matrices, *Integral Equations and Operator Theory*, **35**, 28–52, Birkhäuser, Basel, 1999.

[EGL96]   I. Z. Emiris, A. Galligo, H. Lombardi, Numerical Univariate Polynomial GCD, *Proceedings of AMS-SIAM Summer Seminar: Mathematics of Numerical Analysis: Real Number Algorithms* (J. Renegar, M. Shub, and S. Smale, editors), Park City, Utah, 1995, *Lectures in Applied Mathematics*, **32**, 323–343, American Mathematical Society, Providence, Rhode Island, 1996.

[EGL97]   I. Z. Emiris, A. Galligo, H. Lombardi, Certified Approximate Polynomial GCDs, *Journal of Pure and Applied Algebra*, **117/118**, 229–251, 1997.

[EP97]   I. Z. Emiris, V. Y. Pan, The Structure of Sparse Resultant Matrices, *Proceedings of ACM International Symposium on Symbolic and Algebraic Computation (ISSAC'97)*, 189–196, ACM Press, New York, 1997.

[EPa]   I. Z. Emiris, V. Y. Pan, Symbolic and Numeric Methods for Exploiting Structure in Constructing Resultant Matrices, *J. of Symbolic Computation*, to appear.

[ER82]   I. Z. Elliot, K. R. Rao, *Fast Transform Algorithms, Analysis, and Applications*, Academic Press, New York, 1982.

[F72]   C. M. Fiduccia, Polynomial Evaluation via the Division Algorithm: The Fast Fourier Transform Revisited, *Proc. 4th Annual ACM Symp. on Theory of Computing (STOC'72)*, 88–93, 1972.

[F75]   I. P. Fedchina, Tangential Nevanlinna–Pick Problem with Multiple Points, *Doklady Akademii Nauk (Armenian SSR)*, **61**, 214–218, 1975 (in Russian).

[F85]   M. Fidler, Hankel and Loewner Matrices, *Linear Algebra and Its Applications*, **58**, 75–95, 1985.

[F94]   R. W. Freund, A Look-ahead Bareiss Algorithm for General Toeplitz Matrices, *Numerische Mathematik*, **68**, 35–69, 1994.

[F95]   B. Fornberg, *A Practical Guide to Pseudospectral Methods*, Cambridge University Press, Cambridge, UK, 1995.

[FD89]   P. A. Fuhrmann, N. B. Datta, On Bezoutians, Vandermonde Matrices, and the Lienard–Chipart Stability Criterion, *Linear Algebra and Its Applications*, **120**, 23–27, 1989.

[FF63]   D. K. Faddeev, V. N. Faddeeva, *Computational Methods of Linear Algebra*, W. H. Freeman, San Francisco, 1963.

[FF90]   C. Foias, A. E. Frazho, The Commutant Lifting Approach to Interpolation Problems, *Operator Theory: Advances and Applications*, **44**, Birkhäuser, Basel, 1990.

[FFGK98] C. Foias, A. E. Frazho, I. Gohberg, M. A. Kaashoek, Metric Constrained Interpolation, Commutant Lifting and Systems, *Integral Equations and Operator Theory*, **100**, Birkhäuser, Basel, 1989.

[FHR93]   T. Fink, G. Heinig, K. Rost, An Inversion Formula and Fast Algorithms for Cauchy–Vandermonde Matrices, *Linear Algebra and Its Applications*, **183**, 179–191, 1993.

[FMKL79] B. Friedlander, M. Morf, T. Kailath, L. Ljung, New Inversion Formulas for Matrices Classified in Terms of their Distances from Toeplitz Matrices, *Linear Algebra and Its Applications*, **27**, 31–60, 1979.

[FOa]   D. Fasino, V. Olshevsky, *How Bad Are Symmetric Pick Matrcies*, preprint, 2000.

[FP74]   M. J. Fischer, M. S. Paterson, String Matching and Other Products, *SIAM-AMS Proceedings*, **7**, 113–125, 1974.

[FP88]   M. Fiedler, V. Ptak, Loewner and Bezout Matrices, *Linear Algebra and Its Applications*, **101**, 187–220, 1988.

[FZ93]   R. W. Freund, H. Zha, Formally Biorthogonal Polynomials and a Look-ahead Levinson Algorithm for General Toeplitz Systems, *Linear Algebra and Its Applications*, **188/189**, 255–303, 1993.

[G60]   N. Gastinel, Inversion d'une matrice généralisant la matrice de Hilbert, *Chiffres*, **3**, 149–152, 1960.

[G62]   W. Gautschi, On Inverse of Vandermonde and Confluent Vandermonde Matrices, *Numerische Mathematik*, **4**, 117–123, 1962.

[G72]   W. B. Gragg, The Padé Table and its Relation to Certain Algorithms of Numerical Analysis, *SIAM Review*, **14, 1**, 1–62, 1972.

[G86]   J. von zur Gathen, Representations and Parallel Computations for Rational Functions, *SIAM Journal on Computing*, **15, 2**, 432–452, 1986.

[G84]   J. von zur Gathen, Parallel Algorithms for Algebraic Problems, *SIAM Journal on Computing*, **13, 4**, 802–824, 1984.

[G88]   A. Gerasoulis, A Fast Algorithm for the Multiplication of Generalized Hilbert Matrices with Vectors, *Mathematics of Computation*, **50, 181**, 179–188, 1988.

[G90]   J. von zur Gathen, Functional Decomposition of Polynomials: the Tame Case, *Journal of Symbolic Computation*, **9**, 281–299, 1990.

[G92]   M. H. Gutknecht, A Complete Theory of the Unsymmetric Lanczos Process and Related Algorithms, Part I, *SIAM Journal on Matrix Analysis and Applications*, **13, 2**, 594–639, 1992.

[G97]   A. Greenbaum, *Iterative Methods for Solving Linear Systems*, SIAM, Philadelphia, 1997.

[G98]   M. Gu, Stable and Efficient Algorithms for Structured System of Linear Equations, *SIAM Journal on Matrix Analysis and Applications*, **19, 2**, 279–306, 1998.

[Gad90] P. Gader, Displacement Operator Based Decomposition of Matrices Using Circulant or Other Group Matrices, *Linear Algebra and Its Applications*, **139**, 111–131, 1990.

[Gau90] W. Gautschi, How Unstable are Vandermonde Systems? *International Symposium on Asymptotic and Computational Analysis: Conference in Honor Frank W. J. Olver's 65th Birthday (R. Wong, editor)*, Lecture Notes in Pure and Applied Mathematics, **124**, 193–210, Marcel Dekker, New York, 1990.

[GE95]   M. Gu, S. C. Eisenstat, A Divide-and-Conquer Algorithm for the Symmetric Tridiagonal Eigenproblem, *SIAM Journal on Matrix Analysis and Applications*, **16, 1**, 172–191, 1995.

[GF74]   I. C. Gohberg, I.A. Feldman, Convolution Equations and Projection Methods for Their Solutions, *Translations of Mathematical Monographs*, **41**, American Mathematical Society, Providence, Rhode Island, 1974.

[GG99]   J. von zur Gathen, J. Gerhard, *Modern Computer Algebra*, Cambridge University Press, Cambridge, UK, 1999.

[GGS87] A. Gerasoulis, M. D. Grigoriadis, L. Sun, A Fast Algorithm for Trummer's Problem, *SIAM Journal on Scientific and Statistical Computing*, **8, 1**, 135–138, 1987.

[GI88]   W. Gautschi, G. Inglese, Lower Bounds for the Condition Number of Vandermonde Matrcies, *Numerische Mathematik*, **52**, 241–250, 1988.

[GK89]   I. Gohberg, I. Koltracht, Efficient Algorithm for Toeplitz Plus Hankel Matrices, *Integral Equations and Operator Theory*, **12**, 136–142, 1989.

[GKK86]   I. Gohberg, T. Kailath, I. Koltracht, Efficient Solution of Linear Systems of Equations with Recursive Structure, *Linear Algebra and Its Applications*, **80**, 81–113, 1986.

[GKKL87]   I. Gohberg, T. Kailath, I. Koltracht, P. Lancaster, Linear Complexity Parallel Algorithms for Linear Systems of Equations with Recursive Structure, *Linear Algebra and Its Applications*, **88/89**, 271–315, 1987.

[GKL87]   J. von zur Gathen, D. Kozen, S. Landau, Functional Decomposition of Polynomials, *Proceedings of 28th Annual IEEE Symposium on Foundation of Computer Science (FOCS'87)*, 127–131, IEEE Computer Society Press, Washington, DC, 1987.

[GKO95]   I. Gohberg, T. Kailath, V. Olshevsky, Fast Gaussian Elimination with Partial Pivoting for Matrices with Displacement Structure, *Mathematics of Computation*, **64**, 1557–1576, 1995.

[GKS90]   D. Y. Grigoriev, Y. Karpinski, M. Singer, Fast Parallel Algorithms for Sparse Multivariate Polynomial Interpolation over Finite Fields, *SIAM Journal on Computing*, **19, 6**, 1059–1063, 1990.

[GKZ94]   I. M. Gel'fand, M. M. Kapranov, A. V. Zelevinski, *Discriminants, Resultants, and Multidimensional Determinants*, Birkhäuser, Boston, 1994.

[GL83]   W. B. Gragg, A. Lindquist, On the Partial Realization Problem, *Linear Algebra and Its Applications*, **50**, 277–319, 1983.

[GL96]   G. H. Golub, C. F. Van Loan, *Matrix Computations*, Johns Hopkins University Press, Baltimore, Maryland, 1989 (2nd edition), 1996 (3rd edition).

[GO92]   I. Gohberg, V. Olshevsky, Circulants, Displacements and Decompositions of Matrices, *Integral Equations and Operator Theory*, **15, 5**, 730–743, 1992.

[GO93]   I. Gohberg, V. Olshevsky, Fast Algorithm for Matrix Nehari Problem, *Proceedings of MTNS–93, Systems and Networks: Mathematical Theory and Applications*, **2**, *Invited and Contributed Papers* (U. Helmke, R. Mennicken, and J. Sauers, editors), 687–690, Academy Verlag, 1994.

[GO94]   I. Gohberg, V. Olshevsky, Complexity of Multiplication with Vectors for Structured Matrices, *Linear Algebra and Its Applications*, **202**, 163–192, 1994.

[GO94a]   I. Gohberg, V. Olshevsky, Fast Algorithms with Preprocessing for Matrix-Vector Multiplication Problem, *Journal of Complexity*, **10**, **4**, 411–427, 1994.

[GO94b]   I. Gohberg, V. Olshevsky, Fast State Space Algorithms for Matrix Nehari and Nehari–Takagi Interpolation Problems, *Integral Equations and Operator Theory*, **20**, **1**, 44–83, 1994.

[GO96]   I. Gohberg, V. Olshevsky, Fast Inversion of Vandermonde and Vandermonde-like Matrices, *Communications, Computation, Control and Signal Processing: A Tribute to Thomas Kailath* (A. Paulraj, V. Roychowdhury, and C. Shaper, editors), 205–221, Kluwer Academic Publisher, 1996.

[GP71]   G. Galimbert, V. Pereyra, Solving Confluent Vandermonde Systems of Hermite Type, *Numerische Mathematik*, **18**, 44–60, 1971.

[GP88]   Z. Galil, V. Y. Pan, Improved Processor Bounds for Combinatorial Problems in RNC, *Combinatorica*, **8**, **2**, 189–200, 1988.

[GR87]   L. Greengard, V. Rokhlin, A Fast Algorithm for Particle Simulation, *Journal of Computational Physics*, **73**, 325–348, 1987.

[GS72]   I. Gohberg, A. Semencul, On the Inversion of Finite Toeplitz Matrices and Their Continuous Analogs, *Matematicheskiie Issledovaniia*, **2**, 187–224, 1972.

[GS84]   U. Grenander, G. Szegö, *Toeplitz Forms and Their Applications*, Chelsea Publishing, New York, 1984 (2nd edition).

[GY79]   F. G. Gustavson, D. Y. Y. Yun, Fast Algorithms for Rational Hermite Approximation and Solution of Toeplitz Systems, *IEEE Transactions on Circuits and Systems*, **26**, **9**, 750–755, 1979.

[GY86]   D. H. Green, F. F. Yao, Computational Geometry, *Proceedings of 27th Annual Symposium on Foundation of Computer Science (FOCS'86)*, 143–152, IEEE Computer Society Press, Washington, DC, 1986.

[H52]   D. A. Huffman, The Method for the Construction of Minimum Redundancy Codes, *Proceedings of ORE*, **40**, 1098–1101, 1952.

[H70]   A. S. Householder, *The Numerical Treatment of a Single Nonliear Equation*, McGraw-Hill, New York, 1970.

[H70a]   A. S. Householder, Bezoutians, Elimination and Localization, *SIAM Review*, **12**, 73–78, 1970.

[H72]   E. Horowitz, A Fast Method for Interpolation Using Preconditioning, *Information Processing Letters*, **1**, **4**, 157–163, 1972.

[H79]   G. Heinig, Beitrage zur spektraltheorie von Operatorbuschen und zur algebraischen Theorei von Toeplitzmatrizen, Dissertation **B**, *TH Karl-Marx-Stadt*, 1979.

[H88]   N. J. Higham, Fast Solution of Vandermonde-like Systems Involving Orthogonal Polynomials, *IMA Journal on Numerical Analysis*, **8**, 473–486, 1988.

[H90]   N. J. Higham, Stability Analysis of Algorithm for Solving Confluent Vandermonde-like Systems, *SIAM Journal on Matrix Analysis and Applications*, **11**, 23–41, 1990.

[H95]   G. Heinig, Inversion of Generalized Cauchy Matrices and the Other Classes of Structured Matrices, *Linear Algebra for Signal Processing, IMA Volume in Mathematics and Its Applications*, **69**, 95–114, Springer, 1995.

[H96]   G. Heinig, Solving Toeplitz Systems after Extension and Transformation, *Calcolo*, **33**, 115–129, 1996.

[Ha96]   S. Haykin, *Adaptive Filter Theory*, 3rd edition, Prentice-Hall, Englewood Cliffs, New Jersey, 1996.

[HB97]   G. Heinig, A. Bojanczyk, Transformation Techniques for Toeplitz and Toeplitz-plus-Hankel Matrices: I. Transformations, *Linear Algebra and Its Applications*, **254**, 193–226, 1997.

[HB98]   G. Heinig, A. Bojanczyk, Transformation Techniques for Toeplitz and Toeplitz-plus-Hankel Matrices: II. Algorithms, *Linear Algebra and Its Applications*, **278**, 11–36, 1998.

[HE81]   R. W. Hockney, J. W. Eastwood, *Computer Simulation Using Particles*, McGraw-Hill, New York, 1981.

[HF89]   U. Helmke, P. A. Fuhrmann, Bezoutians, *Linear Algebra and Its Applications*, **124**, 1039–1097, 1989.

[HH93]   G. Heinig, F. Hellinger, On the Bezoutian Structure of the Moore–Penrose Inverses of Hankel Matrices, *SIAM J. on Matrix Analysis and Applications*, **14**, **3**, 629–645, 1993.

[HH94]   G. Heinig, F. Hellinger, Moore–Penrose Generalized Inverse of Square Toeplitz Matrices, *SIAM J. on Matrix Analysis and Applications*, **15**, **2**, 418–450, 1994.

[Hi96]   N. J. Higham, *Accuracy and Stability of Numerical Algorithms*, SIAM, Philadelphia, 1996.

[HJR88] G. Heinig, P. Jankowski, K. Rost, Fast Inversion of Toeplitz-Plus-Hankel Matrices, *Numerische Mathematik*, **52**, 665–682, 1988.

[HR84]   G. Heinig, K. Rost, Algebraic Methods for Toeplitz-like Matrices and Operators, *Operator Theory*, **13**, Birkhäuser, 1984.

[HR98]   G. Heinig, K. Rost, DFT Representation of Toeplitz-plus-Hankel Bezoutians with Application to Fast Matrix-Vector Multiplications, *Linear Algebra and Its Applications*, **284**, 157–175, 1998.

[HW79]   G. H. Hardy, E. M. Wright, *An Introduction to the Theory of Numbers*, Oxford Univeristy Press, Oxford, UK, 1979.

[IK66]   E. Isaacson, H. B. Keller, *Analysis of Numerical Methods*, Wiley, New York, 1966.

[IK93]   D. Ierardi, D. Kozen, Parallel Resultant Computation, *Synthesis of Parallel Algorithms* (J. H. Reif, editor), 679–720, Morgan Kaufman Publishers, San Mateo, California, 1993.

[IR82]   K. Ireland, M. Rosen, *A Classical Introduction to Modern Number Theory*, Springer, Berlin, 1982.

[J89]   A. Jain, *Fundamentals of Digital Signal Processing*, Prentice-Hall, Englewood Cliffs, New Jersey, 1989.

[JO89]   P. Junghanns, D. Oestreich, Numerische Lösung des Staudammproblems mit Drainage, *Z. Angew. Math. Mech.*, 69, 83–92,1989.

[K68]   D. E. Knuth, *The Art of Computer Programming, Vol. 1*, Addison Wesley, Reading, Massachusetts, 1968 (1st edition).

[K70]   D. E. Knuth, The Analysis of Algorithms, *Proceedings of International Congress of Mathematicians*, Nice, France, **3**, 269–274, 1970.

[K74]   T. Kailath, A View of Three Decades of Linear Filtering Theory, *IEEE Transactions on Information Theory*, **20**, 146–181, 1974.

[K87]   T. Kailath, Signal Processing Applications of Some Moment Problems, *Proceedings of AMS Symposium in Applied Mathematics*, **37**, 71–100, 1987.

[K80]   T. Kailath, *Linear Systems*, Prentice-Hall, Englewood Cliffs, New Jersey, 1980.

[K88]   E. Kaltofen, Greatest Common Divisor of Polynomials Given by Straight-line Program, *Journal of ACM*, **35**, **1**, 231–264, 1988.

[K91]   T. Kailath, Remarks on the Origin of the Displacement-Rank Concept, *Applied Mathematics and Computation*, **45**, 193–206, 1991.

[K94]   E. Kaltofen, Asymptotically Fast Solution of Toeplitz-like Singular Linear Systems, *Proceedings of International Symposium on Symbolic and Algebraic Computation (ISSAC'94)*, 297–304, ACM Press, New York, 1994.

[K95]   E. Kaltofen, Analysis of Coppersmith's Block Wiedemann Algorithm for the Parallel Solution of Sparse Linear Systems, *Mathematics of Computation*, **64, 210**, 777–806, 1995.

[K98]   D. E. Knuth, *The Art of Computer Programming, Volume 2: Seminumerical Algorithms*, Addison-Wesley, Reading, Massachusetts, 1998 (3rd Edition).

[K99]   I. Kaporin, A Practical Algorithm for Fast Matrix Multiplication, *Numerical Linear Algebra with Applications*, **6, 8**, 687–700, 1999.

[KAGKA89] R. King, M. Ahmadi, R. Gorgui–Naguib, A. Kwabwe, M. Azimi–Sadjadi, *Digital Filtering in One and Two Dimensions: Design and Applications*, Plenum Press, New York, 1989.

[Kar87]   N. Karcanias, Invariance Properties and Characterisations of the Greatest Common Divisor of a Set of Polynomials, *International Journal of Control*, **40**, 1751–1760, 1987.

[KC89]   T. Kailath, J. Chun, Generalized Gohberg–Semencul Formulas for Matrix Inversion, *Operator Theory: Advances and Applications*, **40**, 231–246, 1989.

[KKM79] T. Kailath, S. Y. Kung, M. Morf, Displacement Ranks of Matrices and Linear Equations, *Journal of Mathematical Analysis and Applications*, **68, 2**, 395–407, 1979.

[KL88]   E. Kaltofen, Y. N. Lakshman, Improved Sparse Multivariate Polynomial Interpolation Algorithms, *Proceedings of International Symposium on Symbolic and Algebraic Computation (ISSAC'89), Lecture Notes in Computer Science*, **358**, 467–474, Springer, Berlin, 1988.

[KL92]   D. Kapur, Y. N. Lakshman, Elimination methods: An Introduction, *Symbolic and Numerical Computation for Artificial Intelligence (B. Donald, D. Kapur, and J. Mundy, editors)*, 45–89. Academic Press, New York, 1992.

[KL96]   N. Karmarkar, Y. N. Lakshman, Approximate Polynomial Greatest Common Divisor and Nearest Singular Polynomials, *Proceedings of International Symposium on Symbolic and Algebraic Computation (ISSAC'98)*, 35–39, ACM Press, New York, 1996.

[KLL00] E. Kaltofen, W.-s. Lee, A. A. Lobo, Early Termination in Ben-Or/Tiwari Sparse Interpolation and a Hybrid of Zippel's Algorithm, *Proceedings of International Symposium on Symbolic and Algebraic Computation (ISSAC'2000)*, 192–201, ACM Press, New York, 2000.

[KLM78] T. Kailath, L. Ljung, M. Morf, A New Approach to the Determination of Fredholm Resolvents of Nondisplacement Kernels, *Topics in Functional Analysis* (I. Gohberg and M. Kac, editors), 169–184, Academic Press, New York, 1978.

[KLW90]   E. Kaltofen, Y. N. Lakshman, J. M. Wiley, Modular Rational Sparse Mul-
          tivariate Polynomial Interpolation, *Proceedings of International Sympo-
          sium on Symbolic and Algebraic Computation (ISSAC'90)*, 135–139, ACM
          Press, New York, 1990.

[KO63]   A. Karatsuba, Yu. Ofman, Multiplication of Multidigit Numbers on
          Automata, *Soviet Physics Doklady*, **7**, 595–596, 1963.

[KO96]   T. Kailath, V. Olshevsky, Displacement Structure Approach to Discrete
          Transform Based Preconditioners of G. Strang Type and of T. Chan Type,
          Calcolo, **33**, 191–208, 1996.

[KO97]   T. Kailath, V. Olshevsky, Displacement Structure Approach to Polynomial
          Vandermonde and Related Matrices, *Linear Algebra and Its Applications*,
          285, 37–67, 1997.

[KO97a]   T. Kailath, V. Olshevsky, Bunch–Kaufman Pivoting for Partially Recon-
          structable Cauchy-like Matrices with Application to Toeplitz-like Matrices
          and to Boundary Rational Matrix Interpolation Problems, *Linear Algebra
          and Its Applications*, **254**, 251–302, 1997.

[KP74]   I. V. Kovalishina, V. P. Potapov, Indefinite Metric in the Nevanlinna–Pick
          Problem, *Translations of American Mathematical Society*, **138, 2**, 15–19,
          1988. (Russian original 1974.)

[KP91] · E. Kaltofen, V. Y. Pan, Processor Efficient Parallel Solution of Linear
          Systems over an Abstract Field, *Proceedings of 3rd Annual ACM Sympo-
          sium on Parallel Algorithms and Architectures (SPAA'91)*, 180–191, ACM
          Press, New York, 1991.

[KP92]   E. Kaltofen, V. Y. Pan, Processor Efficient Parallel Solution of Linear Sys-
          tems II. The Positive Characteristic and Singular Cases, *Proceedings of
          33rd Annual Symposium on Foundations of Computer Science (FOCS'92)*,
          714–723, IEEE Computer Society Press, Los Alamitos, California, 1992.

[KP94]   E. Kaltofen, V. Y. Pan, Parallel Solution of Toeplitz and Toeplitz-like
          Linear Systems over Fields of Small Positive Characteristic, *Proceed-
          ings of First International Symposium on Parallel Symbolic Computation
          (PASCO'94)* (Hoon Hong, editor), *Lecture Notes in Computing*, **5**, 225–
          233, World Scientific Publishing Company, Singapore, 1994.

[KR89]   E. Kaltofen, H. Rolletscheck, Computing Greatest Common Divisors and
          Factorizations in Quadratic Number Fields, *Mathematics of Computation*,
          53, 188, 697–720, 1989.

[KS91]   E. Kaltofen, B. D. Saunders, On Wiedemann's Method for Solving Sparse
          Linear Systems, *Proceedings of AAECC–5, Lecture Notes in Computer Sci-
          ence*, **536**, 29–38, Springer, Berlin, 1991.

[KS95]   T. Kailath, A. H. Sayed, Displacement Structure: Theory and Applications, *SIAM Review*, **37, 3**, 297–386, 1995.

[KS99]   T. Kailath, A. H. Sayed (editors), *Fast Reliable Algorithms for Matrices with Structure*, SIAM Publications, Philadelphia, 1999.

[KT78]   H. T. Kung, J. F. Traub, All Algebraic Functions Can Be Computed Fast, *Journal of ACM*, **25, 2**, 245–260, 1978.

[KT90]   E. Kaltofen, B. M. Trager, Computing with Polynomial Given by Black Boxes for Their Evaluations: Greatest Common Divisors, Factorization, Separation of Numerators and Denominators, *Journal of Symbolic Computation*, **9, 3**, 301–320, 1990.

[Ku74]   H. T. Kung, On Computing Reciprocals of Power Series, *Numerische Mathematik*, **22**, 341–348, 1974.

[KVM78]   T. Kailath, A. Viera, M. Morf, Inverses of Toeplitz Operators, Innovations, and Orthogonal Polynomials, *SIAM Review*, **20, 1**, 106–119, 1978.

[L38]   D. H. Lehmer, Euclid's Algorithm for Integers, *The American Mathematics Monthly*, **45**, 227–233, 1938.

[L47]   N. Levinson, The Wiener RMS (Root-Mean-square) Error Criterion in the Filter Design and Prediction, *Journal of Mathematical Physics*, **25**, 261–278, 1947.

[L83]   R. Loos, Computing Rational Zeros of Integral Polynomials by p-adic Expansion, *SIAM Journal on Computing*, **12**, 286–293, 1983.

[L92]   F. T. Leighton, *Introduction to Parallel Algorithms and Architectures: Arrays·Trees·Hypercubes*, Morgan Kaufmann Publishers, San Mateo, California, 1992.

[L94]   H. Lu, Fast Solution of Confluent Vandermonde Linear Systems, *SIAM Journal on Matrix Analysis and Applications*, **15, 4**, 1277–1289, 1994.

[L95]   H. Lu, Fast Algorithms for Confluent Vandermonde Linear Systems and Generalized Trummer's Problem, *SIAM Journal on Matrix Analysis and Applications*, **16, 2**, 655–674, 1995.

[L96]   H. Lu, Solution of Vandermonde-like Systems and Confluent Vandermonde-like Systems, *SIAM Journal on Matrix Analysis and Applications*, **17, 1**, 127–138, 1996.

[L-AK84] H. Lev-Ari, T. Kailath, Lattice Filter Parametrization and Modelling of Nonstationary Processes, *IEEE Transactions on Information Theory*, **IT-30, 1**, 2–16, 1984.

[L-AKC84] H. Lev-Ari, T. Kailath, J. Cioffi, Least Squares Adaptive Lattice and Transversal Filters; a Unified Geometrical Theory, *IEEE Transactions on Information Theory*, **IT–30**, 222–236, 1984.

[LDC90]   G. Labahn, Dong Koo Choi, S. Cabay, The Inverses of Block Hankel and Block Toeplitz Matrices, *SIAM Journal on Computing*, **19**, 98–123, 1990.

[LP01]   K. Li, V. Y. Pan, Parallel Matrix Multiplication on a Linear Array with a Reconfigurable Pipelined Bus System, *IEEE Transactions on Computing*, **50, 5**, 519–525, 2001.

[LT85]   P. Lancaster, M. Tismenetsky, *The Theory of Matrices*, Academic Press, New York, 1985.

[M73]   R. Moenck, Fast Computation of GCDs, *Proceedings of 5th ACM Annual Symposium on Theory of Computing*, 142–171, ACM Press, New York, 1973.

[M74]   M. Morf, Fast Algorithms for Multivariable Systems, Ph.D. Thesis, *Department of Electrical Engineering, Stanford University*, Stanford, CA, 1974.

[M75]   J. Makhoul, Linear Prediction: A Tutorial Review, *Proceedings of IEEE*, **63, 4**, 561–580, 1975.

[M80]   M. Morf, Doubling Algorithms for Toeplitz and Related Equations, *Proceedings of IEEE International Conference on ASSP*, 954–959, IEEE Press, Piscataway, New Jersey, 1980.

[M81]   B. R. Musicus, Levinson and Fast Choleski Algorithms for Toeplitz and Almost Toeplitz Matrices, *Internal Report*, Lab. of Electronics, M.I.T., Cambridge, Massachusetts, 1981.

[M82]   L. Mirsky, *An Introduction to Linear Algebra*, Dover, New York, 1982.

[MB72]   R. Moenck, A. Borodin, Fast Modular Transform via Divison, *Proceedings of 13th Annual Symposium on Switching and Automata Theory*, 90–96, IEEE Computer Society Press, Washington, DC, 1972.

[MB79]   J. Maroulas, S. Barnett, Polynomials with Respect to a General Basis, I. Theory, *Journal of Mathematical Analysis and Applications*, **72**, 177–194, 1979.

[MC79]   R. T. Moenck, J. H. Carter, Approximate Algorithms to Derive Exact Solutions to Systems of Linear Equations, *Proceedings of EUROSAM, Lecture Notes in Computer Science*, **72**, 63–73, Springer, Berlin, 1979.

[MP96]   B. Mourrain, V. Y. Pan, Multidimensional Structured Matrices and Polynomial Systems, *Calcolo*, **33**, 389–401, 1996.

[MP98]   B. Mourrain, V. Y. Pan, Asymptotic Acceleration of Solving Polynomial Systems, *Proceedings of 27th Annual ACM Symposium on Theory of Computing (STOC'98)*, 488–496, ACM Press, New York, May 1998.

[MP00]   B. Mourrain, V. Y. Pan, Multivariate Polynomials, Duality, and Structured Matrices, *Journal of Complexity*, **16, 1**, 110–180, 2000.

[MS77]   F. J. MacWilliams, N. J. A. Sloane, *The Theory of Error-Correcting Codes*, North-Holland, New York, 1977.

[N29]   R. Nevanlinna, Über Beschränkte Analytishe Functionen, *Annales Academiae Scientiarum Fennicae*, **32, 7**, 1–75, 1929.

[N00]   M. B. Nathanson, *Elementary Methods in Number Theory*, Springer, New York, 2000.

[Oa]   V. Olshevsky, Pivoting for Structured Matrices with Applications, *Linear Algebra and Its Applications*, to appear.

[OP98]   V. Olshevsky, V. Y. Pan, A Unified Superfast Algorithm for Boundary Rational Tangential Interpolation Problem, *Proceedings of 39th Annual IEEE Symposium Foundations of Computer Science (FOCS'98)*, 192–201, IEEE Computer Society Press, Los Alamitos, California, 1998.

[OP99]   V. Olshevsky, V. Y. Pan, Polynomial and Rational Evaluation and Interpolation (with Structured Matrices), *Proceedings of 26th Annual International Colloqium on Automata, Languages, and Programming (ICALP'99)*, *Lecture Notes in Computer Science*, **1644**, 585–594, Springer, Berlin, 1999.

[OS88]   A. M. Odlyzko, A. Schönhage, Fast Algorithms for Multiple Evaluations of the Riemann Zeta Function, *Transactions of American Mathematical Society*, **309, 2**, 797–809, 1988.

[OS99]   V. Olshevsky, M. A. Shokrollahi, A Displacement Approach to Efficient Decoding of Algebraic-Geometric Codes, *Proceedings of 31st Annual Symposium on Theory of Computing (STOC'99)*, 235–244, ACM Press, New York, May 1999.

[OS99a]   V. Olshevsky, M. A. Shokrollahi, A Unified Superfast Algorithm for Confluent Tangential Interpolation Problem and for Structured Matrices, *Proceedings of Advanced Signal Processing Algorithms, Architecture and Implementation IX*, Denver, Colorado, SPIE Publications, 1999.

[OS00]   V. Olshevsky, M. A. Shokrollahi, A Displacement Approach to Efficient Decoding of Algebraic-Geometric Codes, *Proceedings of 14th International Symposium on Networks and Systems (MTNS'2000)*, University of Perpignan, Perpignan, France, 2000.

[OS00a]   V. Olshevsky, M. A. Shokrollahi, Matrix-vector Product for Confluent
          Cauchy-like Matrices with Application to Confluent Rational Interpola-
          tion, *29th Annual Symposium on Theory of Computing (STOC'2000)*, 573–
          581, ACM Press, New York, 2000.

[OS00b]   V. Olshevsky, M. A. Shokrollahi, A Superfast Algorithm for Confluent
          Rational Tangential Interpolation Problem via Matrix-Vector Multiplica-
          tion for Confluent Cauchy-like Matrices, *Proceedings of 14th International
          Symposium on Networks and Systems (MTNS'2000)*, University of Perpi-
          gnan, Perpignan, France, 2000.

[P16]   G. Pick, Über die Beschräkungen Analytisher Functionen, Welche Durch
          Vorgegebene Functionswerte Bewirkt Werden, *Math. Ann.*, **77**, 7–23, 1916.

[P66]   V. Y. Pan, On Methods of Computing the Values of Polynomials, *Us-
          pekhi Matematicheskikh Nauk*, **21**, **1(127)**, 103–134, 1966. [Transl. *Rus-
          sian Mathematical Surveys*, **21**, **1(127)**, 105–137, 1966.]

[P84]   V. Y. Pan, How to Multiply Matrices Faster, *Lecture Notes in Computer
          Science*, **179**, Springer, Berlin, 1984.

[P84a]   V. Y. Pan, How Can We Speed Up Matrix Multiplication?, *SIAM Review*,
          26, 3, 393–415, 1984.

[P85]   V. Y. Pan, Fast and Efficient Parallel Algorithms for the Exact Inversion of
          Integer Matrices, *Proceedings of 5th Annual Conference on Foundation of
          Software Technology and Theoretical Computer Science, Lectures Notes in
          Computer Science*, **206**, 504–521, Springer, Berlin, 1985.

[P87]   V. Y. Pan, Sequential and Parallel Complexity of Approximate Evaluation
          of Polynomial Zeros, *Computers & Mathematics (with Applications)*, **14,
          8**, 591–622, 1987.

[P87a]   V. Y. Pan, Complexity of Parallel Matrix Computations, *Theoretical Com-
          puter Science*, **54**, 65–85, 1987.

[P89]   V. Y. Pan, Fast and Efficient Parallel Evaluation of the Zeros of a Polyno-
          mial Having Only Real Zeros, *Computers & Mathematics (with Applica-
          tions)*, **17, 11**, 1475–1480, 1989.

[P89a]   V. Y. Pan, Fast and Efficient Parallel Inversion of Toeplitz and Block
          Toeplitz Matrices, *Operator Theory: Advances and Applications,* **40**, 359–
          389, Birkhäuser, Basel, 1989.

[P90]   V. Y. Pan, On Computations with Dense Structured Matrices, *Mathematics
          of Computation*, **55, 191**, 179–190, 1990. Proceedings version: *Proceed-
          ings of International Symposium on Symbolic and Algebraic Computation
          (ISSAC'89)*, 34–42, ACM Press, New York, 1989.

[P91]   V. Y. Pan, *Complexity of Algorithms for Linear Systems of Equations*, in *Computer Algorithms for Solving Linear Algebraic Equations (The State of the Art)*, edited by E. Spedicato, *NATO ASI Series, Series F: Computer and Systems Sciences*, **77**, 27–56, Springer, Berlin, 1991.

[P92]   V. Y. Pan, Parametrization of Newton's Iteration for Computations with Structured Matrices and Applications, *Computers & Mathematics (with Applications)*, **24, 3**, 61–75, 1992.

[P92a]   V. Y. Pan, Complexity of Computations with Matrices and Polynomials, *SIAM Review*, **34, 2**, 225–262, 1992.

[P92b]   V. Y. Pan, Parallel Solution of Toeplitz-like Linear Systems, *Journal of Complexity*, **8**, 1–21, 1992.

[P93]   V. Y. Pan, Decreasing the Displacement Rank of a Matrix, *SIAM Journal on Matrix Analysis and Applications*, **14, 1**, 118–121, 1993.

[P93a]   V. Y. Pan, Concurrent Iterative Algorithm for Toepliz-like Linear Systems, *IEEE Transactions on Parallel and Distributed Systems*, **4, 5**, 592–600, 1993.

[P94]   V. Y. Pan, Simple Multivariate Polynomial Multiplication, *Journal of Symbolic Computation*, **18**, 183–186, 1994.

[P95]   V. Y. Pan, An Algebraic Approach to Approximate Evaluation of a Polynomial on a Set of Real Points, *Advances in Computational Mathematics*, **3**, 41–58, 1995.

[P95a]   V. Y. Pan, Optimal (up to Polylog Factors) Sequential and Parallel Algorithms for Approximating Complex Polynomial Zeros, *Proceedings of 27th Annual ACM Symposium on Theory of Computing (STOC'95)*, 741–750, ACM Press, New York, 1995.

[P96]   V. Y. Pan, Parallel Computation of Polynomial GCD and Some Related Parallel Computations over Abstract Fields, *Theoretical Computer Science*, **162, 2**, 173–223, 1996.

[P96a]   V. Y. Pan, Optimal and Nearly Optimal Algorithms for Approximating Polynomial Zeros, *Computers & Mathematics (with Applications)*, **31, 12**, 97–138, 1996.

[P96b]   V. Y. Pan, Effective Parallel Computations with Toeplitz and Toeplitz-like Matrices Filled with Integers, *Proceedings of AMS-SIAM Summer Seminar: Mathematics of Numerical Analysis: Real Number Algorithms* (J. Renegar, M. Shub, and S. Smale, editors), Park City, Utah, 1995. *Lectures in Applied Mathematics*, **32**, 591–641, American Mathematical Society, Providence, Rhode Island, 1996.

[P97]   V. Y. Pan, Solving a Polynomial Equation: Some History and Recent Progress, *SIAM Review*, **39, 2**, 187–220, 1997.

[P97a]   V. Y. Pan, Fast Solution of the Key Equation for Decoding the BCH Error-Correcting Codes, *Proceedings of 29th ACM Symposium on Theory of Computing (STOC'97)*, 168–175, ACM Press, New York, 1997.

[P98]   V. Y. Pan, Approximate Polynomial GCD's, Padé Approximation, Polynomial Zeros, and Bipartite Graphs, *Proceedings of 9th Annual ACM-SIAM Symposium on Discrete Algorithms (SODA'98)*, 68–77, ACM Press, New York, and SIAM Publications, Philadelphia, 1998.

[P98a]   V. Y. Pan, Some Recent Algebraic/Numerical Algorithms, *Electronic Proceedings of IMACS/ACA'98*, http://www-troja.fjfi.cvut.cz/aca98/sessions/approximate.

[P99]   V. Y. Pan, A Unified Superfast Divide-and-Conquer Algorithm for Structured Matrices over Abstract Fields, MSRI Preprint No.1999–033, *Mathematical Sciences Research Institute*, Berkeley, California, 1999.

[P00]   V. Y. Pan, Nearly Optimal Computations with Structured Matrices, *Proceedings of 11th Annual ACM-SIAM Symposium on Discrete Algorithms (SODA'2000)*, 953–962, ACM Press, New York, and SIAM Publications, Philadephia, 2000.

[P00a]   V. Y. Pan, New Techniques for the Computation of Linear Recurrence Coefficients, *Finite Fields and Their Applications*, **6**, 43–118, 2000.

[P00b]   V. Y. Pan, Parallel Complexity of Computations with General and Toeplitz-like Matrices Filled with Integers and Extensions, *SIAM Journal on Computing*, **30**, 1080–1125, 2000.

[P00c]   V. Y. Pan, Matrix Structure, Polynomial Arithmetic, and Erasure-Resilient Encoding/Decoding, *Proceedings of ACM International Symposium on Symbolic and Algebraic Computation (ISSAC'2000)*, 266–271, ACM Press, New York, 2000.

[P01]   V. Y. Pan, A Homotopic Residual Correction Process, *Proceedings of the Second Conference on Numerical Analysis and Applications* (L. Vulkov, J. Waśnievsky, and P. Yalamov, editors), *Lecture Notes in Computer Science*, **1988**, 644–649, Springer, Berlin, 2001.

[Pa]   V. Y. Pan, Computation of Approximate Polynomial GCD's and an Extension, *Information and Computation*, to appear.

[Pb]   V. Y. Pan, Matrix Structure and Loss-Resilient Encoding/Decoding, *Computers and Mathematics (with Applications)*, to appear.

[Par80] B. N. Parlett, *The Symmetric Eigenvalue Problem*, Prentice-Hall, Englewood Cliffs, New Jersey, 1980.

[Par92]   B. N. Parlett, Reduction to Tridiagonal Form and Minimal Realizations, *SIAM Journal on Matrix Analysis and Applications*, **13**, **2**, 567–593, 1992.

[PACLS98] V. Y. Pan, M. AbuTabanjeh, Z. Chen, E. Landowne, A. Sadikou, New Transformations of Cauchy Matrices and Trummer's Problem, *Computers & Mathematics (with Applications)*, **35**, **12**, 1–5, 1998.

[PACPS98] V. Y. Pan, M. AbuTabanjeh, Z. Chen, S. Providence, A. Sadikou, Transformations of Cauchy Matrices for Trummer's Problem and a Cauchy-like Linear Solver, *Proceedings of 5th Annual International Symposium on Solving Irregularly Structured Problems in Parallel (Irregular 98)*, (A. Ferreira, J. Rolim, H. Simon, and S.-H. Teng, editors), *Lecture Notes in Computer Scence*, **1457**, 274–284, Springer, August 1998.

[PB73]   J. S. Pace, S. Barnett, Comparison of Algorithms for Calculation of GCD of Polynomials, *International Journal of Systems Science*, **4**, 211–223, 1973.

[PBRZ99] V. Y. Pan, S. Branham, R. Rosholt, A. Zheng, Newton's Iteration for Structured Matrices and Linear Systems of Equations, *SIAM Volume on Fast Reliable Algorithms for Matrices with Structure* (T. Kailath and A. H. Sayed, editors), 189–210, SIAM Publications, Philadelphia, 1999.

[PC99]   V. Y. Pan, Z. Chen, The Complexity of the Matrix Eigenproblem, *Proc. 31th Annual ACM Symposium on Theory of Computing (STOC'99)*, 507–516, ACM Press, New York, 1999.

[PCZ98]   V. Y. Pan, Z. Chen, A. Zheng, The Complexity of the Algebraic Eigenproblem, MSRI Preprint No. 1998–071, *Mathematical Sciences Research Institute*, Berkeley, California, 1998.

[PKRCa]   V. Y. Pan, M. Kunin, R. Rosholt, H. Cebecioğlu, Homotopic Residual Correction Processes, to appear.

[PLS92]   V. Y. Pan, E. Landowne, A. Sadikou, Univariate Polynomial Division with a Reminder by Means of Evaluation and Interpolation, *Information Processing Letters*, **44**, 149–153, 1992.

[PLST93] V. Y. Pan, E. Landowne, A. Sadikou, O. Tiga, A New Approach to Fast Polynomial Interpolation and Multipoint Evaluation, *Computers & Mathematics (with Applications)*, **25**, **9**, 25–30, 1993.

[PR01]   V. Y. Pan, Y. Rami, Newton's Iteration for the Inversion of Structured Matrices, *Structured Matrcies: Recent Developments in Theory and Computation* (D. A. Bini, E. Tyrtyshnikov, and P. Yalamov, editors), Nova Science Publishers, USA, 2001.

[PRW00]   V. Y. Pan, Y. Rami, X. Wang, Newton's Iteration for Structured Matrices, *Proceedings of 14th International Symposium on Mathematical Theory of Networks and Systems (MTNS'2000)*, University of Perpignan Press, Perpignan, France, June 2000.

[PRWa]   V. Y. Pan, Y. Rami, X. Wang, Structured Matrices and Newton's Iteration: Unified Approach, *Linear Algebra and Its Applications*, to appear.

[PS72]   G. Pólya, G, Szegö, *Problems and Theorems in Analysis*, Springer, Berlin, 1972.

[PS91]   V. Y. Pan, R. Schreiber, An Improved Newton Iteration for the Generalized Inverse of a Matrix, with Applications, *SIAM Journal on Scientific and Statistical Computing*, **12, 5**, 1109–1131, 1991.

[PSD70]   P. Penfield Jr., R. Spencer, S. Duinker, *Tellegen's Theorem and Electrical Networks*, MIT Press, Cambridge, Massachusetts, 1970.

[PT84]   V. Ptack, Lyapunov, Bezout and Hankel, *Linear Algebra and Its Applications*, **58**, 363–390, 1984.

[PWa]   V. Y. Pan, X. Wang, Inversion of Displacement Operators, to appear.

[PWb]   V. Y. Pan, X. Wang, Inverse Displacement Operators, to appear.

[PY99]   V. Y. Pan, Y. Yu, Certified Computation of the Sign of a Matrix Determinant, *Proceedings of 10th Annual ACM-SIAM Symposium on Discrete Algorithms (SODA'99)*, 715–724, ACM Press, New York, and SIAM Publications, Philadelphia, 1999.

[PY01]   V. Y. Pan, Y. Yu, Certified Numerical Computation of the Sign of a Matrix Determinant, *Algorithmica*, **30**, 708–724, 2001.

[PZ00]   V. Y. Pan, A. Zheng, Superfast Algorithms for Cauchy-like Matrix Computations and Extensions, *Linear Algebra and Its Applications*, **310**, 83–108, 2000.

[PZACP99] V. Y. Pan, A. Zheng, M. Abu Tabanjeh, Z. Chen, S. Providence, Superfast Computations with Singular Structured Matrices over Abstract Fields, *Proceedings of Second Workshop on Computer Algebra in Scientific Computing (CASC–99)* (V. G. Ganzha, E. W. Mayr, and E. V. Vorontsov, editors), 323–338, Springer, May 1999.

[PZHD97] V. Y. Pan, A. Zheng, X. Huang, O. Dias, Newton's Iteration for Inversion of Cauchy-like and Other Structured Matrices, *Journal of Complexity*, **13**, 108–124, 1997.

[PZHY97] V. Y. Pan, A. Zheng, X. Huang, Y. Yu, Fast Multipoint Polynomial Evaluation and Interpolation via Computations with Structured Matrices, *Annals of Numerical Mathematics*, **4**, 483–510, January 1997.

[Q94]   M. J. Quinn, *Parallel Computing: Theory and Practice*, McGraw-Hill, New York, 1994.

[R22]   J. F. Ritt, Prime and Composite Polynomials, *Transactions of American Mathematical Society*, **23**, 51–63, 1922.

[R85]   V. Rokhlin, Rapid Solution of Integral Equations of Classical Potential Theory, *Journal of Computational Physics*, **60**, 187–207, 1985.

[R88]   V. Rokhlin, A Fast Algorithm for the Discrete Laplace Transformation, *Journal of Complexity*, **4**, 12–32, 1988.

[R90]   L. Reichel, A Matrix Problem with Application to Rapid Solution of Integral Equations, *SIAM Journal on Scientific and Statistical Computing*, **11**, 263–280, 1990.

[R92]   T. J. Rivlin, *The Chebyshev Polynomials*, John Wiley, New York, 1992.

[S33]   G. Schultz, Iterative Berechnung der Reciproken Matrix, *Z. Angew. Meth. Mech.*, **13**, 57–59, 1933.

[S69]   V. Strassen, Gaussian Elimination Is Not Optimal, *Numerische Mathematik*, **13**, 354–356, 1969.

[S71]   A. Schönhage, Schnelle Berechnung von Kettenbruchentwicklungen, *Acta Informatica*, **1**, 139–144, 1971.

[S72]   M. Sieveking, An Algorithm for Division of Power Series, *Computing*, **10**, 153–156, 1972.

[S73]   V. Strassen, Die Berechnungskomplexetät von elementarysymmetrischen Funktionen und von Interpolationskoeffizienten, *Numerische Mathematik*, **20, 3**, 238–251, 1973.

[S74]   P. H. Sterbenz, *Floating-Point Computation*, Prentice-Hall, Englewood Cliffs, New Jersey, 1974.

[S77]   A. Schönhage, Schnelle Multiplikation von Polynomen über Korpern der Characteristik 2, *Acta Informatica*, **7**, 395–398, 1977.

[S80]   J. T. Schwartz, Fast Probabilistic Algorithms for Verification of Polynomial Identities, *Journal of ACM*, **27, 4**, 701–717, 1980.

[S82]   A. Schönhage, Asymptotically Fast Algorithms for the Numerical Multiplication and Division of Polynomials with Complex Coefficients, *Proceedings of EUROCAM* (J. Calmet, editor), *Lecture Notes in Computer Science*, **144**, 3–15, Springer, Berlin, 1982.

[S83]   V. Strassen, The Computational Complexity of Continued Fractions, *SIAM Journal on Computing*, **12**, 1–27, 1983.

[S85]   A. Schönhage, Quasi-GCD Computation, *Journal of Complexity*, **1**, 118–137, 1985.

[S86]   A. Schrijver, *Theory of Linear and Integer Programming*, Wiley, New York, 1986.

[S88]   A. Schönhage, Probabilistic Computation of Integer Polynomial GCD's, *Journal Algorithms*, **9**, 365–371, 1988.

[S94]   W. F. Stewart, *Introduction to the Numerical Solution of Markov Chains*, Princeton University Press, New Jersey, 1994.

[S96]   Y. Saad, *Iterative Methods for Sparse Linear Systems*, PWS Publishing Company, Boston, 1996.

[S97]   L. A. Sakhnovich, *Interpolation Theory and Its Applications*, Kluwer Academic Publishers, Dordrecht, 1997.

[S98]   D. Sarason, Nevanlinna–Pick Interpolation with Boundary Data, *Integeral Equations and Operator Theory*, **30**, 231–250, 1998.

[S99]   G. Strang, Discrete Cosine Transform, *SIAM Review*, **41**, 135–147, 1999.

[SC93] . T. W. Sederberg, G. Z. Chang, Best Linear Common Divisors for Approximate Degree Reduction, *Computer-Aided Design*, **25**, 163–168, 1993.

[SKL-AC94] A. H. Sayed, T. Kailath, H. Lev-Ari, T. Constantinescu, Recursive Solution of Rational Interpolation Problems via Fast Matrix Factorization, *Integeral Equations and Operator Theory*, **20**, 84–118, Birkhäuser, Basel, 1994.

[SS71]   A. Schönhage, V. Strassen, Schnelle Multiplikation grosse Zahlen, *Computing*, **7**, 281–292, 1971.

[SS74]   T. Söderström, G. W. Stewart, On the Numerical Properties of an Iterative Method for Computing the Moore–Penrose Generalized Inverse, *SIAM Journal on Numerical Analysis*, **11**, 61–74, 1974.

[T63]   A. L. Toom, The Complexity of a Scheme of Functional Elements Realizing the Multiplication of Integers, *Soviet Mathematics Doklady*, **3**, 714–716, 1963.

[T64]   W. F. Trench, An Algorithm for Inversion of Finite Toeplitz Matrices, *Journal of SIAM*, 12, 515–522, 1964.

[T81]   M. Tismenetsky, Bezoutians, Toeplitz and Hankel Matrices in the Spectral Theory of Matrix Polynomials, *Ph.D. Thesis, Technion*, Haifa, Israel, 1981.

[T86]   M. Trummer, An Efficient Implementation of a Conformal Mapping Method Using the Szegö Kernel, *SIAM Journal on Numerical Analysis*, 23, 853–872, 1986.

[T90]   W. F. Trench, A Note on a Toeplitz Inversion Formula, *Linear Algebra and Its Applications*, **29**, 55–61, 1990.

[T94]   E. E. Tyrtyshnikov, How Bad Are Hankel Matrices? *Numerische Mathematik*, **67**, **2**, 261–269, 1994.

[T96]   E. E. Tyrtyshnikov, A Unifying Approach to Some Old and New Theorems on Distribution and Clustering, *Linear Algebra and Its Applications*, **232**, 1–43, 1996.

[TZa]   E. E. Tyrtyshnikov, N. L. Zamarashkin, Estimates for Eigenvalues of Hankel Matrices (Ozenki Sobstvennyh Znachenii dlia Hankelevyh Matriz, in Russian), *Matematicheskii Sbornik*, **4**, 2001, to appear.

[UP83] S. Ursic, C. Patarra, Exact Solution of Systems of Linear Equations with Iterative Methods, *SIAM Journal on Algebraic and Discrete Mathematics*, **4**, 111–115, 1983.

[VHKa] M. Van Barel, G. Heinig, P. Kravanja, A Stabilized Superfast Solver for Nonsymmetric Toeplitz Systems, Report TW293, *DCS, Katolieke Universiteit Leuven*, Heverlee, Belgium, 1999.

[VK98] M. van Barel, P. Kravanja, A Stabilized Superfast Solver for Indefinite Hankel Systems, *Linear Algebra and Its Applications*, **284**, 335–355, 1998.

[VL92] C. F. Van Loan, *Computational Framework for the Fast Fourier Transform*, SIAM, Philadelphia, 1992.

[W63]   J. H. Wilkinson, *Rounding Errors in Algebraic Processes*, Prentice-Hall, Englewood Cliffs, New Jersey, 1963.

[W65]   J. H. Wilkinson, *The Algebraic Eigenvalue Problem*, Clarendon Press, Oxford, 1965.

[W86]   D. H. Wiedemann, Solving Sparse Linear Equations Over Finite Fields, *IEEE Transactions on Information Theory*, **IT–32, 1**, 54–62, 1986.

[W90]   H. K. Wimmer, On the History of the Bezoutian and the Resultant Matrix, *Linear Algebra and Its Applications*, **128**, 27–34, 1990.

[W93]   D. H. Wood, Product Rules for the Displacement of Nearly-Toeplitz Matrices, *Linear Algebra and Its Applications*, **188, 189**, 641–663, 1993.

[Wan81] P. Wang, A *p*-adic Algorithm for Univariate Partial Fractions, *Proceedings of ACM Symposium on Symbolic and Algebraic Computation*, 212–217, ACM Press, New York, 1981.

[WB94] S. B. Wicker, V. Bhargava, *Reed–Solomon Codes and Their Applications*, IEEE Press, New York, 1994.

[WGD82] P. Wang, M. Guy, J. Davenport, *p*-adic Reconstruction of Rational Numbers, *SIGSAM Bulletin*, **16**, 2–3, ACM Press, New York, 1982.

[Y76]   D. Y. Y. Yun, Algebraic Algorithms Using p-adic Constructions, *Proceedings of ACM Symposium on Symbolic and Algebraic Computation*, 248–259, ACM Press, New York, 1976.

[Z79]   R. E. Zippel, Probabilistic Algorithms for Sparse Polynomials, *Proceedings of EUROSAM'79, Lecture Notes in Computer Science*, **72**, 216–226, Springer, Berlin, 1979.

[Z90]   R. E. Zippel, Interpolating Polynomials from Their Values, *Journal of Symbolic Computation*, **9, 3**, 375–403, 1990.

[Z93]   R. E. Zippel, *Effective Polynomial Computation*, Kluwer Academic Publishes, Norwell, Massachusetts, 1993.

[ZR99]   Z. Zilic, K. Radecka, On Feasible Multivariate Polynomial Interpolation over Arbitrary Fields, *Proceedings of ACM International Symposium on Symbolic and Algebraic Computation (ISSAC'99)*, 67–74, ACM Press, New York, 1999.

Index

A_u, 37–40
adaptive compression, 195
adjoint (adjugate) matrix, 230
algebra
    Frobenius matrix, 36
    of a matrix, 104–106, 115
    of an f-circulant matrix, 34–36
    reciprocal in, 37
    τ, 106, 112
algebraic
    codes, 93–96, 111, 114, 115, 168,
        172
    computation, 17
    functions, 225, 236
    implementation of algorithms, 17
algorithm
    algebraic, 17
    Bareiss, 235
    Chinese Remainder, 82, 83, 110
    classical, 12, 58
    Cook, 214
    divide-and-conquer, 12
    Euclidean (see EA)
    Extended Euclidean (see EEA)
    fast, 12, 20
    Fast Multipole, 89
    Karatsuba, 58
    numerical, 15, 16
    numerically stable, 15
    MBA, 12, 20, 21, 156, 170–171
    Moenck–Borodin, 73–76, 109
    nearly optimal (see superfast)
    Newton's, 12, 21, 179–239
    parallel, 56
    Schönhage–Strassen's, 58
    superfast, 12, 20, 21, 54, 170–171
    superfast parallel 229, 235– 239

Toom's, 27, 36
approximate
    gcd, 45
    polynomial decomposition, 236
    polynomial division 36, 59
arithmetic parallel time 18, 143, 200,
    201, 227–229, 235–239

backward error analysis, 16
balanced CRTF (BCRFT), 158
basic subroutine for compression, 195
Berlekamp–Massey, 49
Bezoutian, 52, 57
bilinear expression, 8, 120–137, 149,
    188
binary segmentation, 28, 56, 59
binary split FFT algorithm, 26
bit-operation, 17, 230–235
block triangular factorization, 158
block Toeplitz matrices, 7

Caratheodory-Fejér problem, 98
Cauchy
    determinant, 90, 110
    factorization, 90–92, 110, 114
    inversion 90–92
    linear system, 92
    matrix, 3
    matrix computations, 88–93
    rational interpolation, 87
    type structure, 7
Cauchy-like matrix, 7–9
characteristic polynomial, 227, 236
Chebyshev, 106, 109, 113
Chinese Remainder Algorithm and
    Theorem, 82, 83, 110
circulant matrix, 33, 59, 65

closed form diagonalization, 105
co-factor polynomials, 41
companion matrix, 36
compatible operator pair, 11
complete decomposition of a
        polynomial, 225, 236
complete recursive triangular
        factorization (CRFT), 158
composition
    of polynomials, 81, 109
    power series, 81–82, 109
compress stage of displacement rank
        approach, 9, 117–141
compression
    algebraic, 139–141
    by elimination, 184–186
    by truncation, 182–184
    in Newton iteration, 193–195
    numerical, 137–139
    of displacement generator, 137
    subroutines, 182–186
computational model, 56
conditioning of a problem, 15
condition number, 15, 16, 109
confederate matrix, 105, 113
confluent
    Cauchy-like matrix, 100, 111
    matrix, 130–134
    Nevanlinna–Pick problem, 97
    Vandermonde matrix, 78, 109
continued fraction, 45, 57, 60
convolution 27, 62
Cook, 214
CRTF, 158

DCT, 107, 113
decomposition of a polynomial, 224,
        236
decompress stage of displacement rank
        approach, 9, 117–137
descending process, 157
DFT, 24
    generalized, 29
diagonal matrix, 5, 23
diagonalization, 103–108, 116, 123

discrete transforms, (see DCT, DFT,
        DST)
displacement, 4, 119
    generator, 8, 134, 136
    inversion of, 122–137
    key features of, 7–8
    of basic structured matrices, 5, 6,
        120–122
    of the inverse matrix, 10
    operator, 4, 119, 120
    orthogonal generator of, 8, 137–
        139
    rank, 5, 21
    rank approach 4, 9, 20, 21
    transformation approach 14, 21,
        144–150, 153
divide-and-conquer algorithm 12,
        155–162, 170–171
divided differences, 114
division
    integer, 180, 214
    modulo a polynomial, 40
    polynomial, 30–32
DST, 106, 113

ϵ-length of a generator, 138, 149
ϵ-rank of a matrix, 138, 149
EA, 41
EEA, 44–45
    applications, 45–50
eigendecomposition, 103, 112, 138
eigenproblem, 104, 236
eigenvalues, 103, 138
eigenvectors, 138
encoding/decoding, 93–95, 111, 114,
        115, 168, 172
erasure-resilient (see loss-resilient)
Euclidean algorithm (see EA)
evaluation-interpolation, 27, 36, 59
explicit reconstruction condition, 48,
        86
extended Euclidean algorithm (see
        EEA)

f-circulant matrix, 24, 33–36
f-potent of order n, 34, 124, 187

factorization (also see CRTF)
  of a Cauchy matrix, 90–92, 110, 114
  of a Hankel matrix, 114
  of a Vandermonde matrix, 80, 81, 91, 109
fan-in/fan-out, 74, 75
fast Fourier transform (FFT), 25, 26, 56
  matrix version, 26
  polynomial version, 25
Fast Multipole Algorithm, 89
flop, 17
Frobenius
  matrix, 36
  norm, 178, 188, 189

Gauss transform, 157
gcd
  matrix approach to, 40, 41
  of integers, 60
  of polynomials, 40, 48
general integer matrix, 230–232
general matrix, 189–193, 214, 226
generalized
  DFT, 29
  inverse, 137, 191
  Pick matrix, 97, 98
  reversion of a power series, 223, 236
  Taylor expansion, 33
generator, 8
  length, 8, 136–141
  matrices, 8
  orthogonal, 8, 137, 139, 149
generic rank profile, 156, 164
global State-Space representation, 99
Gohberg–Semencul's formula, 54, 57
greatest common divisor (see gcd)

Hadamard bound, 230, 237
Hankel
  factorization, 114
  matrix, 3
  type structure, 7
Hankel-like matrix, 7

Hartley matrix, 116
Hartley transform, 107, 113
Hensel's lifting, 231, 237
Hermit evaluation, 77
Hermit interpolation, 84, 86
Hermitian
  matrix, 138, 139
  transpose, 23
Hessenberg matrix, 104
homopotic process, 195–201, 215
  for a singular input, 201
  initialization of a, 196–197
  step sizes of a, 198–199
  tolerance values, 199
Huffman's tree, 77, 109

IEEE precision, 17
IFFT (inverse FFT), 62–65
ill conditioned matrix and problem, 16
implementation
  algebraic, 17
  numerical, 15–17
  parallel, 18
indefinite matrix, 139
integer
  division, 180, 214
  gcd, 60
  matrix, 230
  multiplication, 58
  operations with, 28, 110
  polynomial, 28
internally regular operator, 135–137
interpolation
  at roots of unity, 79
  matrix algorithm, 80, 92
  polynomial, 79, 109
  rational, 86–89, 92, 110, 111 (also see Nevanlinna–Pick)
inverse
  DFT and FFT, 24, 25, 62–65
  Fourier transform (see inverse DFT, inverse FFT)
inversion
  f-circulant matrix, 35
  formulae, 20, 55
  of a Cauchy matrix, 90–92

of a displacement operator, 122–137, 149

of a general integer matrix, 230

of a general matrix 189–193, 214, 226, 227, 236

of a Toeplitz matrix 54, 55

of a structured integer matrix, 233–236

of a triangular Toeplitz matrix, 29–30, 69–71

of a Vandermonde matrix, 80, 81, 91, 109

irregularity set, 134, 136

Kalman, 98, 111

Karatsuba algorithm, 58

Krylov

matrix, 24, 25

space, 24, 25, 227, 236

Lagrange

interpolation formula, 79

polynomial interpolation, 78

lcm, 40

leading principal submatrix, 156

least common multiple (see lcm)

length of a generator, 8, 136–141

L-generator, 8, 136–141

L-like matrix, 7

L-rank, 5, 136

L-type (of structure), 7

lifting process, 157

linear

inverse operator, 123, 124

recurrence, 49

Loewner matrix, 9, 20

loss-resilient code, 93–95, 111, 114, 115

lower triangular Toeplitz matrix, 24, 63

machine epsilon, 193

matrix

adjoint (adjugate), 230

algebra, 104

basic structured, 3

Cauchy, 3, 6

regularized, 136

Cauchy-like, 7

circulant, 33, 59, 65

companion (see Frobenius)

condition number of a, 15

confederate, 105

confluent, 130–134

confluent Vandermonde, 78, 109

diagonal, 5, 23

eigenvalues of a, 103

eigenvectors of a, 138

factorization of a, 158

f-circulant, 24, 33–36

Frobenius, 36

generalized inverse of a, 137, 191

generalized Pick, 97, 98

generator, 8

generic rank profile of a, 156, 164

Hankel, 3, 6

Hankel-like, 7

Hartley, 116

Hermitian, 138, 139

Hessenberg, 104

ill conditioned, 16

indefinite, 139

integer, 230

Krylov, 24, 25

L-like, 7

Loewner, 9, 20

lower triangular Toeplitz, 24, 63

multiplication, 149

Nehari problem, 100–101, 112, 169, 172

Nehari–Takagi problem, 101, 112

norm, 178, 214

non-negative definite (see semi-definite)

normal forms, 237

null space of a, 134

operator, 4

Pick, 96

polynomial, 104

polynomial Vandermonde, 78, 105, 109

positive definite, 138

preconditioner, 164, 171

random, 164, 171, 203
rectangular, 119
real symmetric 139
reflection 24
resultant (see Sylvester)
semi-definite 138
singular
    value decomposition of a, 137
    values of, 137
    vectors of, 137
skew-circulant, 24, 66
skew-Hankel-like, 163
spectrum of, 123
strongly non-singular, 156
structured integer, 233–239
subresultant, 51
Sylvester, 37
Toeplitz, 3, 5
Toeplitz-like 7
Toeplitz+Hankel 121, 149–151
Toeplitz+Hankel-like, 130, 135,
    149–152
transpose of a, 23
transposed Vandermonde, 80, 81
unitary, 137
unit f-circulant, 5, 24
unit lower triangular Toeplitz, 24
Vandermonde, 3, 6
Vandermonde-like, 7
well conditioned, 16
MBA algorithm, 12, 20, 21, 156
Möbius transformation, 98
model of computation, 56, 149
modified
    Cauchy rational
        interpolation, 87
    Hermite rational
        interpolation, 87
    Padé approximation, 86
    rational interpolation, 87
modular
    division, 40
    reconstruction, 82–83
    reduction, 77
    rounding-off, 223

Moenck–Borodin algorithm, 73–76,
    109
Moore–Penrose generalized inverse,
    137, 191
multilevel matrix, 112
multiplication
    Cauchy-by-vector, 88
    integer, 58
    matrix, 149
    polynomial, 27, 28, 58
    structured matrix, 35, 141–144
    Toeplitz-by-vector, 27, 66–68
    Vandermonde-by-vector, 73–76
multiplicative transformation, 144–
    147
multipoint evaluation, 73
multivariate polynomial, 112
    interpolation, 101–103, 112
    multiplication, 56, 112

nearly block diagonal operator, 11
nearly optimal algorithm (see superfast
    algorithm)
Nehari problem, 20, 100–101, 112,
    169, 172
Nehari–Takagi problem, 101, 112, 169
Nevanlinna–Pick
    algorithms, 111, 169, 172
    applications, 111
        problems, 20, 96, 98, 111
    boundary, 98, 111
    confluent, 97, 111, 115
    tangential, 96, 111
Newton's
    algebraic iteration, 219–239
    generic algorithm, 220–222
    identities, 32
    interpolation, 109, 114
    lifting, 230, 237
    numerical iteration, 177–204, 214
    polygon process, 225
Newton's iteration, 12, 21, 179–239
    complexity, 183, 185
    compression policy, 182–186,
        193–195, 214
    convergence rate, 183, 185

for a general matrix 189–193, 214, 226, 227, 236

homotopic (see homotopic process)

initialization, 190, 214

overall work of, 194, 203

scaled, 191, 214

variations, 199–201, 236

Newton-Gohberg-Semencul iteration, 186

Newton-Structured Algebraic Iteration, 228

Newton-Structured Iteration, 181–186, 214

Newton–Toeplitz Iteration, 18, 185, 186, 214

nilpotent operator matrix, 124

node polynomial, 95

non-negative definite

(see semi-definite)

non-singular operator, 123

norm of
inverse operator, 186–189
matrix, 178, 214
operator, 179
vector, 26

null space, 134

numerical
algorithms 15
experiments 202–204
generalized inverse 192
implementation 15–17
orthogonal generator 139
stability 15, 138, 171

numerically stable algorithms 15, 163

off-diagonally regular operator, 135–137

operate stage of displacement rank

approach, 9

operator
compatible pairs, 11
displacement, 4, 119, 120
internally regular, 135–137

matrix, 4
-matrix pair, 144–149
nearly block diagonal, 11
non-singular, 123
norm of a matrix, 178
off-diagonally regular, 135–137
of Stein type, 4
of Sylvester type, 4
partly regular, 134–137
regular, 125
singular, 132–137, 149

ops, 17

orthogonal
displacement generator, 12, 149
generator, 8, 137, 149
Hermitian generator, 139
L-generator, 138
real symmetric generators, 139

output set, 159

overall work, 194, 203

Padé
approximation, 46, 86
matrix approach, 52–55
table, 47

parallel
algorithm, 56, 149, 235–239
arithmetic time, 18, 143, 200, 201, 227–229, 235–239
Boolean time complexity, 56, 149, 215
computational model, 57, 149
implementation, 18, 200
matrix computation, 57

parallelization of algorithms, 30

partial fraction, 84, 85, 110

Pick matrices, 96

polynomial
composition, 81, 109
decomposition, 224, 236
division with remainder, 30–32
evaluation, 73–76, 109
gcd, 40, 48
interpolation, 79, 80, 91, 109
minimum, 229
modular representation of, 82

multiplication (also see convolution), 27, 28, 62
pseudo remainder sequence, 45
reciprocal, 30, 31, 223
reciprocal modulo a polynomial, 37–39
remainder sequence, 45
reversion modulo a power, 223, 236
root of a, 224, 236
sparse, 102
squaring, 58
Vandermonde matrix, 78, 105, 109
zeros, 222, 236
positive definite, 138
power series
    composition of, 81, 82, 109
    generalized reversion of, 233, 236
powering operator matrices, 34, 124, 187
power sums, 32–33
preconditioner matrices, 169, 171
prime, 230
primitive root of unity, 25
processor
    bounds, 143, 200, 201, 215, 227–229, 235–239
    communication, 30
    synchronization, 30
pseudo remainder sequence, 45
pseudocodes, 61–71

random
    matrix, 164, 171, 203
    prime, 232, 233, 237
randomization, 164–168, 171, 232, 233
randomized algorithms, 164–168, 171, 232, 233
randomly sampled, 164
rational
    approximation, 110 (also see Nehari and Nehari–Takagi problems)
    function reconstruction, 46, 110

function table, 47
interpolation, 86–88, 92, 110, 111 (also see Nevanlinna–Pick)
    interpolation matrix algorithms, 92
    number reconstruction, 49
reciprocal of a polynomial, 30, 31
    modulo a polynomial, 38, 39
reciprocal in algebra A_u, 38, 39
recursive triangular factorization, 158
reflection matrix, 24
regularization, 136, 162–168, 232, 233
regular operator, 125
residual correction algorithm, 180, 214
resultant, 38
    matrix (see Sylvester matrix)
reversion of a polynomial, 236
root of unity, 24
    primitive, 25

Schönhage–Strassen algorithm, 58
Schur complement, 157, 161–162
semi-definite matrix, 138
similarity transformation, 35, 103–108, 147–148
singular
    operator, 132–137, 149
    value decomposition (SVD), 137
    values 137 (suppression of), 192
    vectors, 137
skew-circulant, 24, 66
skew-Hankel-like, 163
sparse multivariate polynomial, 102
sparse multivariate polynomial interpolation, 101–103, 112
spectral approach, 187
spectrum, 123
squaring matrices and polynomials, 58
stability (see numerical stability)
State-Space, 98, 99, 111, 169
Stein type operators, 4, 119
stiff 0.7-down policy, 200
straight line algorithm, 109
strongly non-singular, 156
structure
    of Cauchy type, 7

of Hankel type, 7
of *L*-type, 7
of Toeplitz type, 7
of Vandermonde type, 7
structured matrices
applications list, 1, 19, 20
basic operations with, 9–11
classes of, 13, 18, 20
four features of, 3
random, 17
rectangular, 20
samples, 3
subresultant matrix, 51
superfast
algorithm, 12, 20, 21, 234
EA, 42
parallel algorithms, 229, 235, 238, 239
SVD (see singular value decomposition)
Sylvester
linear system, 38
matrix, 38
type operators, 4, 119
symbolic
displacement operator, 9
operator approach, 14, 21
symmetric
function algorithm, 76
matrix, 138
symmetrization, 162, 163, 196

tangential Nevanlinna–Pick
algorithms, 169, 172
problem 96–100, 111, 115 (also see Nevanlinna–Pick)
Taylor's
expansion, 33
shift of the variable, 29
Tellegen's theorem, 80, 109
Toeplitz+Hankel-like matrix, 121, 130, 135, 149–152
Toeplitz+Hankel matrix, 149–151
Toeplitz-like matrix, 7
Toeplitz matrix 3, 5
Toeplitz type structure, 7

Toom 27, 58
transfer function (realization), 98
transformation approach (see displacement transformation approach)
transpose of a matrix, 23
triangular
factorization, 158
Toeplitz matrix, 24, 63
trigonometric transforms (see DCT, DST)
Trummer's problem, 89, 110
truncation of singular values, 182–184

unification
by displacement transformation, 14
by symbolic operators, 14
unified Newton–Structured
iteration, 181
superfast algorithm, 20, 21, 155–239
unit
f-circulant matrix, 5, 24
lower triangular Toeplitz matrix, 24
roundoff, 193
unitary matrix, 137
unreduced Hessenberg matrix, 105
upper Hessenberg matrix, 104

Vandermonde
factorization, 80–81, 91, 109
inversion, 80, 81, 91, 109, 114
linear system, 78
matrix, 3, 6
multiplication by a vector, 73–76
type structure, 7
Vandermonde-like matrix, 7
variable diagonal techniques, 235, 238, 239

well conditioned matrix and problem, 16